ZEW Economic Studies

Publication Series of the Centre for European Economic Research (ZEW), Mannheim, Germany

ISBN 978-3-7908-1146-9 ISBN 978-3-642-47035-6 (eBook)
DOI 10.1007/978-3-642-47035-6

Preface

As the Kyoto conference of the parties on the United Nations Framework Convention on Climate Change once again underscored, man-made climate change has become one of the major challenges to our generation and many generations to come. Since scientific evidence on climate change can be seen as increasingly reliable, the focus of our attention has to turn more and more to the question of foreseeable damages and to possibilities to prevent and mitigate climate change. In other words, we need to analyse the economic aspects of man-made climate change and the policy options to prevent its most severe impacts.

This book reports on the findings of an international workshop on these aspects of global climate change. It was organised by the Centre for European Economic Research (ZEW) in Mannheim, Germany on March 6th and 7th 1997. In the light of the ongoing international policy-making process on climate change, we decided to publish the report after the Kyoto conference from December 1st to 10th, 1997, to include the results of the conference, which emphasise the importance of economic aspects and economic policy options when it comes to addressing the problem of man-made climate change. Thus, this book went to press in February 1998 the moment we received the official version of the Kyoto Protocol, which is reproduced in the annex.

The workshop tried to advance the international discussion between academics, economists, and policy makers in the field of the integrated assessment of man-made climate change and necessary policy responses. The results of the workshop are intended to serve as a contribution to the international scientific/economic debate and to the international as well as the German policy-making process. We hope that all interested parties find some concepts and ideas in this book helpful for their own work and decision-making.

I should like to thank all the participants of the workshop for a lively and stimulating debate and all those who contributed to the book for the efforts they made in order to produce thought-provoking manuscripts. We are specially indebted to the people at ZEW who made this workshop possible through their great organisational efforts. In particular, I wish to thank Reinhard Mechler warmly, who painstakingly edited all the manuscripts, which arrived as hard copies, files on diskettes, or as email attachments in all kinds of formats.

This workshop and the publication of this book would not have been possible without the generous financial support provided by the Centre for European Economic Research and the backing by its former director Professor Heinz König.

Mannheim, May 1998

Olav Hohmeyer

Head of the Department of
Environmental and Resource Economics,
Environmental Management
Centre for European Economic Research

Contents

Part III: Policy Options

Annex

1 Introduction: Economic Aspects of and Policy Options for Climate Protection

Olav Hohmeyer and Klaus Rennings
ZEW
Mannheim, Germany

At the time of the workshop, upon which this volume is reporting, it was not clear how the international political process on global climate change would develop. Would there be an agreed target for greenhouse gas emission reductions for the so-called Annex I countries as a result of the international negotiating process or would there just be more political rhetoric? Would economic measures like emission trading schemes get a chance or would international climate policy resort to fixed national reduction targets and command-and-control policies making it impossible to achieve international targets through an efficient environmental policy based on market-oriented instruments?

Now that the Kyoto Protocol to the United Nations Framework Convention on Climate Change has been agreed upon by the international community, we know that there is international consensus that man-made climate change is a major environmental problem to be dealt with decisively and that substantial reduction targets for the most important greenhouse gases are not just being demanded by scientists but will be established at least for the industrialised countries in the world. It is safe to say that the train has started to move in the right direction, although we know that it will have to pick up substantially more speed if we are to reach our safe destination on time. The Kyoto Protocol sets out that:

> The Parties included in Annex I shall, individually or jointly, ensure that their aggregate anthropogenic carbon dioxide equivalent emissions of greenhouse gases listed in Annex A do not exceed their assigned amounts, calculated pursuant to their quantified emission limitation and reduction commitments inscribed in Annex B in accordance with the provisions of this Article, with a view to reducing their overall emissions of such gases by at least 5 per cent below 1990 levels in the commitment period 2008 to 2012. (Kyoto Protocol, Article 3, Paragraph 1).

Considering our fears at the workshop, this can be seen as a substantial achievement for a first step, which needs to be followed by a number of further advances.

The Kyoto Protocol sets the stage for yet to be designed effective and efficient international measures to combat global climate change. The possible policy options set out in the Kyoto Protocol leave room for joint measures between Annex I countries, according to which one country may acquire emission reduction units from another Annex I country achieving more than the agreed reductions (Article 3, Paragraph 10.-12.), for early reductions to be credited against later reduction commitments for the same country (Article 3, Paragraph 13), for projects for emission reductions in non-Annex I countries financed by Annex I countries to serve as certified emission reductions to be used for the reduction purpose of the financing Annex I country (Article 12) and for emission trading among the Annex I countries included in Annex B of the Kyoto Protocol (Article 16 bis). Thus, we can now take a closer look at the economic aspects of man-made climate change and at the way effective and efficient policies have to be designed to achieve the necessary reductions in greenhouse gas emissions.

The workshop "Man-made Climate Change - Economic Aspects and Policy Options" conducted on March 6th and 7th 1997 by the Centre for European Economic Research in Mannheim addressed just these questions. The contributions to the workshop, most of which are included in this volume, gave an overview of the most important economic aspects of man-made climate change and discussed some basic policy options to achieve substantial reduction targets.

Hartmut Graßl, Director of the World Meteorological Organisation (WMO), summarises the basic scientific information on man-made climate change in his paper, setting the stage for the following discussion. At the workshop, Graßl pointed out in his opening speech that the time horizon of many economists encompasses at most a few decades. However, persistent greenhouse gases normally remain in the atmosphere for 20,000 to 50,000 years. Graßl's advice to politicians and economists: "Don't believe in economic climate change models." According to Graßl, these models will need another 10 years before they can be used to make credible predictions about the economic consequences of climate protection policies. Recent approaches of economic theory support Graßl's argument. In her contribution, Graciela Chichilnisky (Columbia University, New York) explains a new theorem of economics: the so-called "Chichilnisky criterion." This theorem assumes axiomatically that "sustainable preferences" should neither include a suppression of today's preferences nor of future preferences. Rennings and Hohmeyer (ZEW, Mannheim) try to link indicators for weak and strong sustainable development in the case of global warming as a basis for political decisions. Following the discussion of damage cost studies of global warming representing weak sustainability indicators, the "inverse scenario approach" and the "environmental space concept" are outlined as strong sustainability concepts. As a result of the comparison, the authors suggest that damage cost modules be integrated into a broader methodological framework of strong sustainability. Gonzague Pillet (ECOSYS SA, Geneva) shows several different theoretical approaches to decision-making in the field of climate protection: according to him, the economic optimisation approach can only be used with restrictions due to numerous uncertainties in climate protection policy. Procedural approaches, on the other hand focus more strongly on the process of decision-making and on making step by step progress under uncertainty.

"Sustainable Preferences" in the sense of the Chichilnisky criterion are obviously more than just a theoretical construct, as new empirical studies seem to demonstrate. This is the topic of the contributions by Ari Rabl (Ecole des Mines, Paris) and Richard Tol (Free University, Amsterdam) to the economic valuation of future damage to the climate. Rabl proposes a new method of "time-variant" discounting of the costs and benefits of climate protection, which would not lead to future generations being systematically made worse off. In other words, this means that greater efforts in climate protection are economically sensible if future benefits are valued properly. Tol explains that existing studies calculate external costs due to climate change at about two per cent of world gross product. However, in these results neither damages in developing countries nor future damages were adequately included. According to Tol, if a fair weighting of the values was to be conducted, damage costs could easily rise to 10-20% of world gross product.

But even if external costs are left aside, highly ambitious goals for climate protection can be legitimised from an economic point of view. Peter Hennicke (Wuppertal Institute) and Eberhard Jochem (ISI, Karlsruhe) show in their papers that there is a great potential for energy conservation, which in addition to protecting the climate also helps to reduce costs. However, this information has not yet been integrated sufficiently into economic macro-models. Tobias Schmidt (ZEW) shows that there may be a double (ecological *and* economical) dividend of climate protection. This thesis is still being intensively discussed. The discussion at the workshop showed a wide gap between microeconomic "bottom-up" and macroeconomic "top-down" approaches continues to exist, which should be overcome by improved co-operation in future research.

Helmuth-Michael Groscurth (ZEW) presents the results of a study on the long-term integration of renewable energies into the European energy system until the year 2050. His message is: in fifty years time the major part of the future energy demand can be covered by renewable energies, which would increase energy costs by only about 20% compared to energy provided solely by fossil or nuclear fuels. The findings of Groscurth's paper are discussed critically by N. Eyre (Eyre Energy Environment, Kirkby Lonsdale), who sees the central problem for a transition to an energy system based on renewable energies in the high degree of centralisation of our present energy supply systems, which needs to be overcome. The promotion, research, development, and increased use of new and renewable forms of energy is an important policy option to combat man-made climate change advanced explicitly in Article 2 of the Kyoto Protocol.

The potential of Joint Implementation measures and their limitations are the topic of a paper by Henning Rentz (RWE, Essen). Though there is intense discussion about these measures at international climate change conferences, globally so far only a few projects have been started. The reason for this is simply a lack of incentives. A lack of incentives is also diagnosed as a serious flaw in the construction of the policy instrument of voluntary agreements by Klaus Rennings (ZEW). The European Commission has grown sceptical of voluntary agreements as well and has advised to rather see them as tools for the implementation of environmental goals into practice in contrast to seeing them as instruments for target setting.

As an important strategy for climate protection Michael Grubb (Royal Institute of International Affairs, London) proposes CO_2-permits, which at the time of the workshop seemed a far-fetched option. In the light of the Kyoto Protocol, however, CO_2-permits seem to be one of the most important policy options.

The central question relating to a permit system is the distribution of basic emission rights, which is dealt with at length by A. Meyer (Global Commons Institute, London), who is advocating a policy of contraction (of the absolute global greenhouse gas emissions) and convergence (of the per capita emission quota assigned). This proposal will certainly gain international attention and support, at least by the non-Annex I countries.

Due to the fact that the Kyoto Protocol as well as the Framework Convention on Climate Change and their exact wording will play an important role in the development of international climate change policy, we decided to include copies of the texts as an annex in this book as additional references for the interested reader.

PART I

WHAT DO WE KNOW ABOUT CLIMATE CHANGE?

2 The Science and Impacts of Climate Change – Conclusions from the Second IPCC Assessment Report

Hartmut Grassl
World Climate Research Programme
41, Avenue Giuseppe Motta
CH 1211 Geneva 2

1 Foreword

Climate change has largely become a synonym for unsustainable behaviour of mankind, although there must be climate change without any human interference, because the strongly differing time-scales of climate system components like atmosphere, ocean, vegetation, land-ice, and distribution of continents provoke changes in mean values and statistics of deviations from these means on all time-scales longer than the slowest component's turn around time (hundreds of millions of years for continents). The problem to be discussed is therefore only the difference in the rate of change between anthropogenic and natural climate change, including the effects of interactions between both and accounting for the strong climate variability on time-scales of interest for the society.

2 Introduction

The growing concern of a potentially major man-induced climate change has stimulated the creation of a scientific assessment body in 1988. The World Meteorological Organization (WMO) and the United Nations Environment Programme (UNEP) jointly sponsor since then the Intergovernmental Panel on Climate Change (IPCC), an international, even global, high-level scientific assessment body. Its first assessment report (IPCC, 1990) submitted to the Second World Climate Conference, organised by WMO in October 1990, has stimulated the formulation of the UN Framework Convention on Climate Change (UNFCCC), which was signed by 154 countries at the United Nations Conference on Environment and Development in Rio de Janeiro in June 1992. An interim assessment of IPCC (IPCC, 1992) given to this conference underlined the radiative forcing of climate change and provided new emission scenarios of anthropogenic greenhouse gases. After the enactment of the UNFCCC in March 1994 – it had been ratified by the necessary 50 countries on December 1994 – IPCC prepared a special report (IPCC, 1995) on radiative forcing of climate to be given to the first conference of the Parties to the UNFCCC in Berlin in March/April 1995. The second major assessment report was finalised in December 1995 (IPCC, 1996a, b, c, d) and published in 1996. The following sections draw from these publications since they are, first of all, seen as the most authoritative statements available and since the author has contributed to all these reports as one out of hundreds of scientists.

3 Changed Atmospheric Composition

Since industrialisation began around the year 1750 the composition of the atmosphere has changed more strongly than during at least the last 220 000 years. All three long-lived greenhouse gases of our atmosphere have increased their concentrations: Carbon dioxide (CO_2) from 280 parts per million by volume (ppmv) to 360 ppmv, methane (CH_4) from 0.7 ppmv to 1.725 ppmv and nitrous oxide (N_2O) from 0.275 to 0.311 ppmv. Until 1994 the mean increase (averaged over 5 years) was approximately 0.5% for CO_2, 0.6% for CH_4 and 0.25% for N_2O. Since the major constituents of the atmosphere, the oxygen molecule (O_2), the nitrogen molecule (N_2) and the noble gas argon (A), which amount to 99.96 percent of the dry atmosphere, do not influence the radiation budget strongly, these above mentioned trace gas changes are of global significance. Additionally, also short-lived atmospheric constituents lice ozone (O_3), a natural greenhouse gas, nitrogen oxides ($NO + NO_2$), sulfur dioxide (SO_2) and ammonia (NH_3) have changed, i.e., increased (except for stratospheric ozone) through human activities over large areas and thus also have reached global significance in recent decades.

Whether we have caused a reaction of the most important climate gas, water vapour, is the major question of climate research. Because of the lack of reliable measurements on a global scale for longer periods we are left with the statement that the strong temperature dependence of water vapour will cause an increase at least in the lower atmosphere if the surface warms, as observed.

4 Radiation Budget Change

Any change in the concentration of a radiatively active component of the atmosphere causes a radiation budget change to which the general circulation in atmosphere and ocean as well as the ice-sheets and vegetation have to adapt. Figure 1 depicts these changes, here called radiative forcing, as they have accumulated (for the long-lived constituents) since 1850 or are forcing at present (for the short-lived constituents). The values given are a fiction in the sense that the climate system has already started to adapt to these changes, i.e., has reduced the imbalance by changed climate.

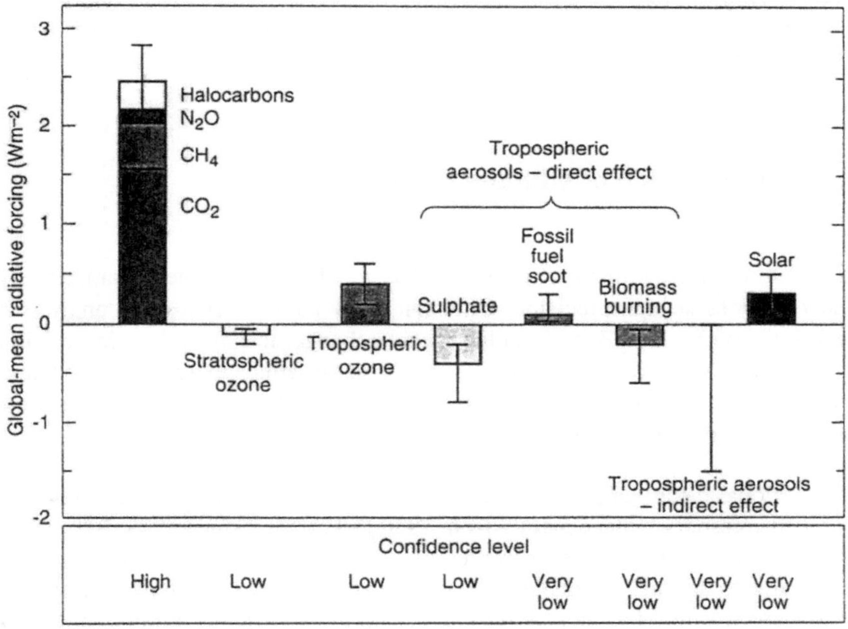

Figure 1: Estimates of the globally and annually averaged anthropogenic radiative forcing (in Wm^{-2}) due to changes in concentrations of greenhouse gases and aerosols from pre-industrial times to the present day and to natural changes in solar output from 1850 to the present day. The height of the rectangular bar indicates a mid-range estimate of the forcing whilst the error bars show an estimate of the uncertainty range, based largely on the spread of published values; our subjective confidence that the actual forcing lies within this error bar is indicated by the "confidence level".

The most reliable part of the climate change stimulators in Fig. 1 is the column for the long-lived greenhouse gas CO_2, CH_4, N_2O and CFCs (chlorofluorocarbons). + 2.5 Watts per squaremeter (Wm^{-2}) is equivalent to slightly above one percent higher solar radiation and thus not negligible at all. Please note that tropospheric ozone in the so-called photosmog is – on a global scale – as important as methane accumulation and that sulfate and biomass burning aerosols (particles suspended in air) show a negative radiative forcing since they scatter more solar radiation back to space than they hinder heat radiation to escape to space. Also the sun has, as far as we can infer from long-term sunspot records and direct measurements during the last two decades, radiated a bit more and thus has contributed to the observed warming.

Together with the latitudinally dependent radiation imbalance at the top of the atmosphere, which is the reason for the general circulation in atmosphere and ocean trying to reduce the imbalance, the stimulations by the anthropogenic radiative forcing in Fig. 1 have also changed circulation and land surface processes. This should be visible in observations.

5 Observation of Climate Change

In order to detect climate change one has to observe climate parameters over many decades at high precision. Only for a few parameters a trend analysis has been possible to date. Even if called climate change, it may still be largely long-term climate variability, if – for example – ocean-atmosphere interaction involving the deep sea with a turn-over time of hundreds of years has caused the observed trend. The following trends have been established:

- global mean temperature has risen by 0.3 to 0.6°C during this century, the increase took place mainly from 1920 to 1940 and since 1975;
- global mean sea level increased by 10 to 25 cm during this century mainly due to thermal expansion of sea water and melting of mountain glaciers;
- precipitation has increased in higher northern latitudes throughout the year and higher mid-latitudes of the Northern Hemisphere in winter during this century, while in many subtropical areas in the Northern Hemisphere precipitation decreased;
- mountain glaciers have retreated on average during this century with a recently accelerated rate (in the Alps about 50% of the glacier mass has been lost since 1850);
- most of the observed warming is due to higher night-time temperatures, i.e., the daily temperature amplitude shrank over most of the continental areas, also indicating that the frequency of very low temperatures was reduced more strongly than the frequency of very high temperatures was increased;

- temperatures in the stratosphere outside the tropics (above 10 to 15 km, depending on latitude) have fallen more strongly during recent decades than surface temperature has risen, due to both the greenhouse gas increase and the decrease of stratospheric ozone;
- Northern Hemisphere annual mean snow cover has decreased by about 10% (1.5 Mill km^2) since reliable satellite data became available in the seventies;
- stratospheric ozone content has significantly decreased in higher latitudes of both hemispheres culminating in more than a 35% loss since the sixties over Antarctica in spring. The loss is due to chlorine compounds released during the destruction of CFCs in the stratosphere.

Before we can speak more about man's influence on these trends, we should discuss climate models since they are needed for this purpose as well.

6 Projections of Future Climate

Climatologists have made tremendous progress in simulating global climate by coupling models of the three-dimensional circulation in atmosphere and ocean. If these models, integrating partial differential equations, which are non-linear, and which are based on physical laws, show climate variability near to observations during the instrumental period, they can be used for reasonable projections of future climate at given changed boundary conditions (atmospheric composition, solar output, land-use change). On time-scales to several decades these models, if run under present external forcings, approach observed climate variability to a high degree. They show for example the right probability for deviations of winter temperatures from the long-term mean for mid-latitudes, sea ice at the proper locations and the proper time and even give the observed statistics of tropical cyclones. For scenarios of observed greenhouse gas concentrations and estimates of sulfate aerosols during this century three groups (Hadley Centre, Bracknell, UK; Max-Planck-Institute for Meteorology, Hamburg, Germany; and GFDL, Princeton, USA) have hitherto found agreement between model results and observed global mean temperature trends to within decadal timescale natural variability, thus pointing to these two forcings as the most important ones.

For future scenarios of greenhouse gas emissions and aerosol loading caused by a growing world population with economic development, but under the assumption of negligible climate protection measures, these models simulate a global warming of about 2 to 3 degrees until 2100, which would constitute a warming rate not experienced since at least 10 000 years. Climate sensitivity to greenhouse gas forcing is still rather uncertain, expressed by the renewed statement that a doubling of the equivalent CO_2 content (all other gases converted to the appropriate CO_2 radiative forcing) would cause a global mean near surface air temperature rise between 1.5 and 4.5°C, if the climate system is fully adapted to the forcing. Most of this span is due to the unknown reaction of clouds to the warming. It is unknown whether they increase (thus damp) or decrease (thus enhance warming) their present cooling capacity. Such a global warming is certainly accompanied by shifts in rain belts, reduction of sea-ice and snow cover, new weather extremes for many areas, accelerated sea level rise and strongly varied river flow in most areas. A rough regionalisation is possible if the spatial scale is at least several times the mesh size of the models. All coupled models show stronger warming in the interior of continents, least warming in ocean areas with strong vertical mixing (e.g., around the Icelandic low), most show increased interannual climate variability in the tropics and all give enhanced precipitation in high northern latitudes. Sensitivity studies with coupled models for long-term thermohaline ocean circulation changes point to a reduced ocean overturning in the North Atlantic, with major implications for European climate. Global warming thus could lead to areas with some cooling.

7 Detection of the Anthropogenic Signal

If anthropogenic climate change has to be extracted from a "noisy" record, four ingredients are needed: the history of the radiative forcing, a long-term record of climate parameters, coupled ocean/atmosphere/land-models and statistical techniques to detect time-varying patterns. The essence of such studies first pursued by groups in Hamburg and San Francisco and later by others was expressed in the IPCC Second Assessment Report as follows: "The balance of evidence suggests a discernible human influence on global climate." This first detection of an anthropogenic global climate signal relies more on the geographical, seasonal and altitudinal patterns of change than on mean global warming, because neither increased solar radiation nor ocean-atmosphere interaction nor reduced volcanism can lead to the observed patterns of stratospheric cooling, stronger surface warming in higher latitudes in winter and less warming of the Northern Hemisphere. Only the joint action of ozone depletion in the lower stratosphere, sulfate aerosols over industrialised regions of the Northern hemisphere and latitudinally dependent radiative forcing by globally distributed long-lived greenhouse gases is consistent with the observed temperature changes. The main uncertainties remaining are: history of radiative forcing and representation of natural climate variability in climate models.

8 Impacts of Climate Change

Climate is largely determined by the global water cycle. Therefore, climate change will be of major importance for all our activities that depend strongly on the water cycle. These activities are: agriculture, fresh-water resource management, shipping on inland waters, hydropower generation, coastal protection, and winter sports. Since the river flow is often the small difference between two large quantities, namely precipitation minus evaporation, seemingly minor shifts in both these quantities have major consequences. An European example, the river Rhine, shall elucidate it: A group of hydrologic research institutes (including the Federal Institute of Hydrology in Koblenz) has concluded in a study for the European Union (Grabs, 1997): The Rhine will in 2050 have turned from a combined rain-snow fed river into a largely rain-fed one with major consequences for the rivers hydrograph: a higher winter and early spring peak and a lower late summer/autumn minimum discharge will make irrigation more difficult, cause more frequent and more intense winter flooding, endanger shipping in autumn after dry summers, impact on water quality. The assumptions were: Warming by 2°C, more winter rain (~ 10%–15%) and slightly less in summer (about - 10%) as taken from recent climate model runs of different groups and as varying for certain parts of the basin.

An observed impact of climate change (or long-term climate variability) is documented for the Sahel in Africa. The river Niger discharges at present on average half the water volume it discharged 30 years ago, when precipitation in its catchment was higher by about 25 percent.

A further impact of climate change will be an accelerated sea level rise amounting to roughly 0.5m ± 0.3m until the year 2100 if no climate protection measures are taken. The large uncertainty surrounding this estimate is mainly due to the lack of knowledge on the melting of the Greenland ice-sheet and the potential growth of the Antarctic ice-sheet. The consequences of such a sea level rise are especially grave for small island states and countries with major marshlands which are largely unprotected against storm surges, like Bangladesh.

Since higher temperatures at no reduction of soil moisture will cause larger crop yields climate change will also have in certain areas positive consequences as long as new weather extremes will not cause losses. New weather extremes must occur whenever a mean value of a climate parameter is changed and the distribution of deviations from the mean is either unchanged or broadened. Therefore, all impact studies have to include these considerations. Two examples may clarify what is meant: No or only a slight change in annual precipitation amount, but a redistribution within the year as given above for the river Rhine broadens the distribution and thus may turn a 10-year flood into a 5-year flood and a 20-year low water into a 10-year one. If, as observed, the daily temperature amplitude decreases nearly as much as the average shows a warming, late frost events become rarer and very high temperatures, which often damage crops, may become more frequent. However, if – as also observed – night-time minima increase more strongly than day-time maxima, the distribution is narrowed but still extreme high temperatures become more frequent. Since our infrastructures are not adapted to new extremes these events will hurt. Insurance companies warn of dramatic increases in losses due to storms and flooding in comparison to earthquakes and volcanic eruptions but a separation of the effects of changed extremes and construction in unsafe places is not yet possible.

Recently more attention has been given to the spread of vector-borne infectious diseases, like Malaria, caused by climate variability and change. This topic deserves more research efforts like many other impact areas. For instance, we need better information on climate change rates to which ecosystems can adapt, how changed agricultural practices can respond to changes in the regional water cycle, how forestry should react given a span of temperature and precipitation changes.

9 Emissions Allowed to Reach Certain Concentration Levels

The central goal of the UNFCCC is the stabilisation of greenhouse gas concentrations in the atmosphere at a level avoiding a dangerous human interference with the climate system. If one neglects for the moment the minor anthropogenic greenhouse gases, how much CO_2 are we allowed to emit in order to reach a preset stabilisation level. With global carbon cycle models of several groups all including the stimulated growth of part of the vegetation by higher CO_2 levels rather convincing but demanding answers could be given:

- At all stabilisation levels equal to or below 750 ppmv CO_2 the emissions have to fall well below present day fossil fuel related emissions of 5.5 billion tons of carbon (in the form of CO_2) during the coming century for the more stringent levels and in the 22nd century for the higher scenarios (650 and 750 ppmv).
- Different gradients of the concentration increase have only a slight influence on total accumulated emissions until stabilisation, thus allowing a rather wide range of climate policy options (also stimulating a wait-and-see attitude among certain groups).
- The influence of land-use changes (e.g., deforestation) on CO_2 concentration levels in the atmosphere is small enough (since the fossil fuel source is so dominant) to allow a clear separation of stabilisation scenarios into categories differing by 100 ppmv.
- Adopting a 450 ppmv concentration goal forces to emissions below present ones rather early in the 21st century. Only this goal would be consistent with a strategy to avoid a warming above the level in the Eemian warm period 125 000 years ago.

10 Possible Changes in the Energy Sector

Global energy growth scenarios for the coming decades have the tendency to shrink. The global energy need proposed by Greenpeace about a decade ago is now above a typical mineral oil company scenario. The main reasons are the partial or full decoupling of economic growth and energy throughput in highly developed industrialised countries, which started after the oil prize increase in 1973, and the slow economic growth of many developing countries in the seventies and eighties. A further reduction has recently been introduced by adoption of low carbon emission scenarios driven by the need for climate protection. Working Group 3 of IPCC (IPCC, 1996d) has developed such low emission scenarios until the year 2075 based on a multitude of options for an energy mix. The major assumption was: when investing to replace old systems, that have become obsolete, choose the cheapest and – if several options are available – the ecologically best suited technology. The major question then is: when will renewable, carbonfree energy sources become competitive? The answer of economists and engineers in Working Group 3 of IPCC was: Rather early in the next century if external costs are internalised progressively. As clearly visible from Fig. 2a, a low carbon emission energy supply system is feasible under the above, not very stringent assumption and already in 2025 the share of renewables is above 20% on a global scale.

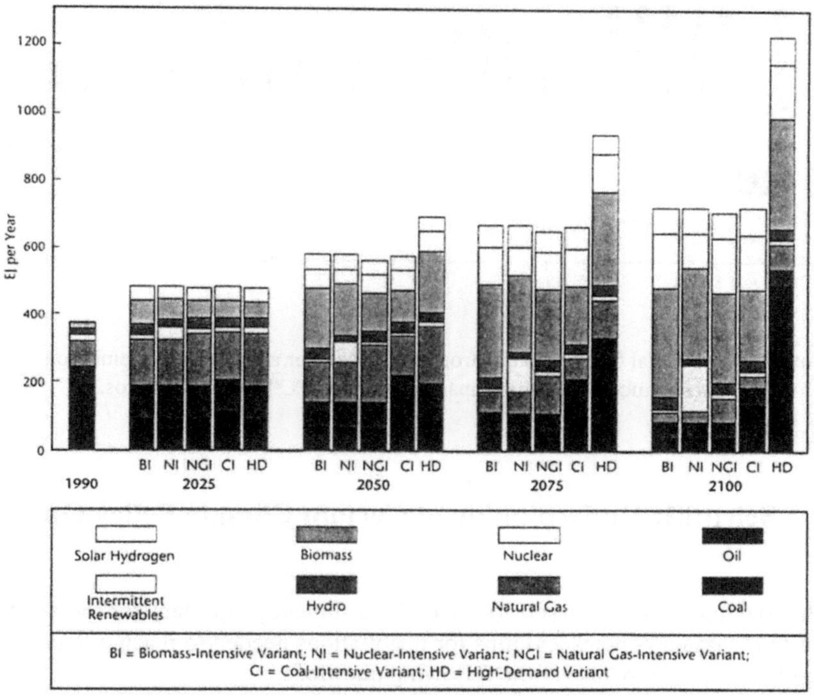

Figure 2a: Global primary energy use for alternative low emission constructions.

Compared to the 1992 scenarios of IPCC (IPCC, 1992) this is a major reduction of energy demand and a strong shift in energy sources, and the accumulated carbon emissions of the scenarios in Fig. 2b, although not driven by a given CO_2-level, are in agreement with a 450 ppmv stabilisation level.

A shift in attitude vis-à-vis the fossil fuels is needed: no longer subsidies and internalisation of externalities; this will strongly raise efficiency in fossil fuel use during the coming decades and allow the proper share for renewables.

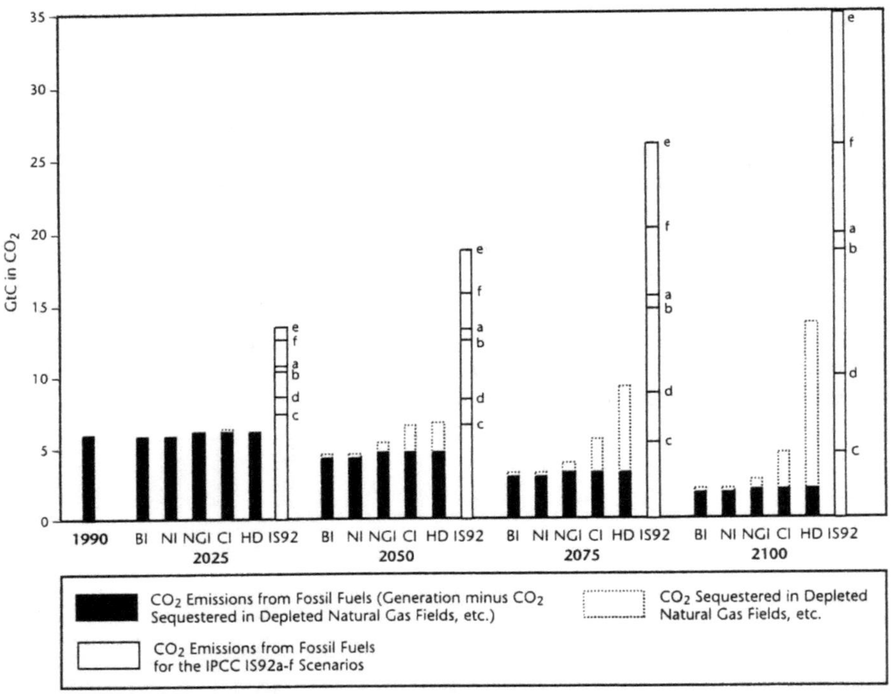

Figure 2b: Annual CO_2 emissions from fossil fuels for alternative low emission constructions, with comparison to the IPCC IS92a-f scenarios.

11 Scientific Underpinning of Certain Climate Policy Options

In this section I restrict myself to scientific arguments important for the adoption of certain climate policies and thus for negotiations needed to reach a legally binding protocol. I try to answer the following questions:

1) Which greenhouse gases should first be regulated?
2) Should joint implementation include land-use changes?
3) Which scientific uncertainties can be narrowed soon?

To 1: Emission reductions need a firm basis concerning the strength of different sources. The ranking with respect to source strength knowledge for anthropogenic greenhouse gases is: CFCs, CO_2, CH_4, N_2O. Since CFCs are already regulated the next candidate is CO_2 and, bearing in mind some restrictions, CH_4. The knowledge of sinks and their possible variation with climate change is another policy relevant information. Here the knowledge is not adequate for all gases to base policies on potential variations in the sinks as a reaction to climate change. The well-known CO_2 source fossil fuels, its dominance compared to CH_4 and N_2O favours a concentration on a CO_2-protocol which could include some fossil fuel related CH_4, because sources and partly their strengths are known to a degree allowing the control of compliance with the protocol.

To 2: Soils and forests together contain about twice the carbon in the atmosphere (1500 billion tons of carbon). Afforestation, reforestation, and changed agricultural practices thus have the potential to sequester at least for some time more anthropogenic carbon. In many regions this could, however, be largely offset by reactions of soils and forests to climate change or longer term natural climate variability. The low accuracy with which the soil carbon pool changes can be assessed, the negotiations with diverse groups outside the energy sector and the transient nature of uptake all speak against the inclusion of land-use changes into projects of joint implementation. There are numerous other ecological and economic arguments for national programmes fostering reforestation and measures against soil erosion which also confirm the exclusion of these activities from joint implementation in the energy sector.

To 3: The findings in basic climate research cannot be foreseen over decades. Nevertheless, planned satellite sensors, global ocean surveys and near future model development allow some judgement where uncertainties might be narrowed first. These areas could be: present direct radiative forcing by tropospheric aerosol particles due to a first global survey of optical depth; quantification of water vapour feedback to warming at the surface through global time series of column content and mid-tropospheric content derived from a merging of satellite and in-situ data; amplification of warming in high latitudes stemming from improved sea-ice modules in climate models; changed extremes during this century derived from a joint evaluation of long time series of national services; uptake of anthropogenic CO_2 in the global ocean retrieved from a global survey of ocean carbon content along sections of the World Ocean Circulation Experiment; CO_2 fertilisation factor for ecosystems with low management level originating from isotope studies, satellite biomass estimates and CO_2-flux measurements; mass budget of the ice-sheets in Greenland and Antarctic from present and new satellite sensors; sea level rise time series for ocean basins, hemispheres, and the globe starting in 1991 from precise satellite altimeters.

This list is not at all exhaustive and there may well be big progress soon in regionalising global climate change, if the major uncertainty related to cloud feedback is narrowed by three-dimensional observations of cloud water and ice in the atmosphere.

12 Outlook

The scientific basis for actions to protect global climate has broadened through the Second Assessment report of IPCC. The question is no longer whether we have global influence, it has turned into how we can damp anthropogenic climate change, which will have (since all societies depend on it and are interrelated) major consequences even for those not responsible for the modified atmospheric composition. The proper reaction is to start a conversion of the energy supply system into a far more efficient one and relying in growing proportions on renewable energy sources. This is best achieved by a CO_2-protocol within UNFCCC signed at the Third Conference of the Parties in Kyoto in December 1997 setting controllable targets which allow countries to find their specific route in achieving it, either by incentives or regulations.

Since the Earth System will never be fully understood, scientific advice will always be needed for more intelligent decisions. It is good to know that the Third (full) Assessment Report of IPCC is already planned for 2001.

References

Grabs, W. (1997): Impact of climate change on hydrological regimes and water resources management in Europe. Final Report of Contract EV5V–CT93–0293, International Commission for the Hydrology of the Rhine Basin (CHR).

IPCC (1990): Climate Change: The IPCC Scientific Assessment. Eds.: J.T. Houghton, G.J. Jenkins, J.S. Emphraums, Cambridge University Press.

IPCC (1992): Climate Change: The Supplementary Report to the IPCC Scientific Assessment. Eds.: J.T. Houghton, B.A. Callander, and S.K. Varney, Cambridge University Press, 200 p.

IPCC (1994): Policy Makers Summary of Radiative Forcing of Climate Change. The 1994 Report of the Scientific Assessment Working Group of IPCC, WMO, Genf.

IPCC (1995): Climate Change 1994: Radiative Forcing of Climate Change and an Evaluation of the IPCC IS92 Emission Scenario. Eds.: J.T. Houghton, L.G. Meira Filho, J. Bruce, Hoesung Lee, B.A. Callander, E. Haites, N. Harris, and K. Maskell, Cambridge University Press, 339 p.

IPCC (1996a): Climate Change 1995 – The Science of Climate Change. Summary for Policymakers. Cambridge University Press.

IPCC (1996b): Climate Change 1995 – The Science of Climate Change: Contribution of Working Group I to the Second Assessment Report of the Intergovernmental Panel on Climate Change. Eds.: J.T. Houghton, L.G. Meira Filho, B.A. Callander, N. Harris, A. Kattenberg, and K. Maskell, Cambridge University Press.

IPCC WG II (1996c): Climate Change 1995 – Impacts, Adaptations and Mitigations of Climate Change: Scientific-Technical Analyses: Contribution of Working Group II to the Second Assessment Report of the Intergovernmental Panel on Climate Change. Eds.: R.T. Watson, M.C. Zinyowera, and R.H. Moss, Cambridge University Press.

IPCC (1996d): Climate Change 1995 – Economic and Social Dimensions of Climate Change: Contribution of Working Group III to the Second Assessment Report of the Intergovernmental Panel on Climate Change. Eds: J. Bruce, Hoesung Lee, and E. Haites, Cambridge University Press.

PART II

ECONOMIC ASPECTS OF CLIMATE CHANGE

3 Decision-Making Frameworks for Addressing Climate Change

Gonzague Pillet, Ecosys, Inc., CH 1227 Geneva-Carouge
& Faculty of Economics, University of Fribourg, Switzerland

Abstract. *Decision-making Frameworks for Addressing Climate Change* is the title of Chapter 2 of the IPCC Second Assessment Report (Arrow *et al.*, 1996, pp. 53-77). That chapter discusses practical decision-making frameworks related to climate change. The implications for decision-making of the unique features of the climate change problem are first reviewed. Then the focus is placed on two interrelated approaches to decision-making: optimising quantitative models and negotiation.

In this paper, the author begins with presenting a concise account of the material of that chapter as well as of the specific implications and recommendations that have emerged as an agreed framework to address the issue (what has been said). Then he investigates "le non-dit" of the chapter — that is, what has not been said: the rather controversial and even clandestine figures that lay behind the report — in order to bring to the surface less systematically explored ways. The examination of these non-standard ways is made in the perspective of propounding further policy options.

Keywords: Decision-making, decision criteria, uncertainties, negotiation, financial instruments, insurance, sharing risks.

1 Decision Analysis Vs. Negotiation ("le dit")

In this first section, we present a concise account of decision analysis related to climate change, expose the reasons we found to evolve toward negotiating climate change decisions, and outline the specific implications for climate change decisions under the Framework Convention on Climate Change (FCCC).[1] The unique

---

[1] We will refer ad sensum to Chapter 2. At times, some short, useful sentences will appear ad verbum.

features of the climate change problem as set in Chapter 2 with respect to economic analysis include:[2]

- Scientific, socio-ecologic and socio-economic impact uncertainties which obscure, inter alia, the dynamics of climate feedback, the occurrence of weather events, how climate change will affect the relationship between human societies and the biosphere, the spread of disease, the economic and social welfare effects of climate change and abatement, or the economic valuation of resources and socio-economic interactions generally.
- Long time horizons and a substantial time lag between emissions and impacts.
- Irreversibility of effects (irreversible changes may eliminate future options).
- The global nature of the problem, which necessitates collective action by sovereign states; there is no single decision-maker: this also applies to individual states; in addition, because of uncertainties and differences in values and objectives, potential parties to a collective decision-making process will not resort to the same choice criteria.
- Social, economic, demographic, geographic, and ethical differences among the affected parties; as regards the latter difference, value judgements may diverge depending on whether society's moral authority to make decisions resorts to either an anthropocentric or an ecocentric normative view; irrespective of whether one shares such a viewpoint, this issue stresses the virtual complexity of biospheric interactions with which climate change may interfere (Parikh *et al.* 1995).

According to these concerns, cognitive and computational limitations make sense with respect to climate change. Ignoring climate change, as Caselton and Luo ironically suggested, would have simplified decision-making, but it would have been irrational to return to our previous lower level of uncertainty awareness. We thus have to accommodate the supplementary uncertainty caused by possible changes in climate and to make decisions (Caselton and Luo 1994).

1.1 Decision Analysis and Climate Change

Decision analysis relies on quantitative models for reaching the "best" choice from among a range of alternatives. It operates under a restrictive set of assumptions, among which one can designate:

- there is a single decision-maker,
- the set of decision choices is complete (and known),

---

[2] See also Chichilnisky, G., this volume, and Chichilnisky, 1994; Gassmann, 1994; Fankhauser, 1994.

- valuation of individual outcomes (options, alternatives) is consistent (in the economic theory of choice, individuals use the monetary value of final outcomes as a substitute measure of their utility) (Eeckhoudt and Gollier, 1995),
- values (hence, outcomes) are rationally ranked so that they can be compared and a choice made, based on some selection criterion (in the economic theory of choice, an individual, when facing a limited range of actions, "has in mind an ordering of all possible consequences" and chooses "that action whose consequences are preferred to those of any other" (Arrow 1951),
- uncertainty is incorporated by assigning probabilities to individual outcomes so that expected values can be computed and compared (uncertainty is in a manner of speaking consistent with probability theory), and
- risk is similarly computable.

According to the consensus expressed in the above mentioned Chapter 2, decision analysis and related quantitative models of decision-making "attempt to improve decision-making by clarifying complex decisions and making the best use of available information. Although they suffer from practical limitations, such models can offer valuable insight into many aspects of environmental decision problems like the climate change problem. Apart from their ability to generate numerical results, such models [...] can serve as an idealised reference case by which to evaluate actual decision structures and decision choices." Policy analysis can be thought, analogously, as "a set of procedures for inventing, exploring and comparing alternatives available for achieving certain [...] goals — and for inventing, exploring and comparing the alternative ends themselves — in a world limited in knowledge, in resources, and in rationality" (Rowen, 1976, as quoted by Morgan and Henrion, 1990, p. 24).

Therefore, if elements of decision analysis have considerable value in framing the decision problem and identifying its critical features, "the lack of an individual decision-maker, utility problems, and incomplete information suggest that decision analysis cannot serve as the primary basis for international climate change decision-making" (involving multiple decision-makers). Because of the large uncertainties, long time horizon, and differences between parties (objective probabilities are not available for many outcomes, and subjective probabilities would be controversial), "there may be no 'globally' optimal climate change strategy" (as no universally 'best' solution can be identified, the teleological dimension of decision analysis loses its analytical reach).

In addition, the choice of a selection criterion itself can be subjective and exogenous to the decision model. The attitude of the decision-maker toward risk is essential in affecting the valuation of decision options (outcomes) without affecting decision analysis itself. Quoting Dowlatabadi and Morgan (1992), "Prior judgement that [single] decision-makers bring to the [climate] problem, and the decision rules they employ, could be far more important in controlling the policy conclusions that they reach than are the results of scientific discoveries".

Some of the most common criteria include (after Morgan & Henrion 1990; wording is that used in Chapter 2):

- *Maximising multi-attribute utility,* which specifies a function that evaluates outcomes in terms of all their important utility attributes (these may include economic value, aesthetics, peace of mind, and other criteria affecting an individual's overall satisfaction), including uncertainties and risks. The alternative with maximum utility is selected.[3]
- *Cost-benefit,* which estimates the costs and benefits of alternatives in economic terms and chooses the one with the highest net benefit.[4]
- *Cost effectiveness,* which selects a desired performance level, perhaps on non-economic grounds, and chooses the option achieving this level at lowest cost. Cost-effectiveness analysis can aid choice between options but, unlike the cost-benefit criterion, cannot indicate whether or not any of the options are worth doing.
- *Minimax loss,* which finds the loss associated with the extreme event for each alternative, then selects the strategy that minimises the worst loss.
- *Maximin gain,* which finds the gains associated with each action and tries to maximise the minimum of these gains.
- *Minimax regret,* which chooses measures of "regret" and minimises the maximum regret. This criterion emphasises the cost of making the wrong decision.
- *Bounded cost,* which sets the maximum budget to devote to risk management activity.

To conclude, decision analysis requires that uncertainty, as defined as incomplete knowledge of the state of the world (K. J. Arrow), be quantified. At the level of an individual, this incomplete knowledge implies that he or she cannot predict with precision all the consequences of his or her actions. Of course, climate

---

3 To illustrate, decision problems under uncertainty and risk can be put together and simplified as follows. The classical ingredients of a decision problem under uncertainty are a set of feasible acts (A), a set of consequences (C), and a set of states of nature (S). The decision-maker does not know for sure which consequence will result from a chosen act (probabilities for the events are not known in advance). The role of the states of nature is to indicate the matching of feasible acts to consequences, thus resolving the uncertainty which relates acts to consequences. This makes states of natures exogenous so that probabilities can be computed. As a result, the *decision-maker basically has to assign utilities to the consequences* (utility U assigned to each consequence c in C) *and probabilities to the states* (probabilities P assigned to each state s in S). In choosing between acts, the decision-maker will elect that one with the highest expected utility—that is, the one with the highest sum of the products of probability and utility. Decision-making under risk introduces a special formulation of the decision problem to the extent that probabilities for the events are known in advance. In other words, the decision problem includes the probability distribution of an act over the consequences. Accordingly, the *decision-maker only has to determine utilities of consequences before choosing between acts.* Reducing risk entails a lower expected return while risk aversion pre-empts a higher expected return as compensation for an increase in risk. (*The New Palgrave,* pp. 70-78. *The MIT Dictionary of Modern Economics*—edited by D. W Pearce—3rd ed.)

4 Chapter 5 of the 1995 IPPC Report examines this criterion in details.

change is not a situation of total ignorance. However, objective data on the probabilities of all decision outcomes are not available, and subjective probabilities—which could conceivably be used—are dependent upon a very wide range of different views. Moreover, the figure gets complicated in as far as the climate problem involves not only a collection of countless individuals, but also sovereign states, admittedly with a set of institutional arrangements for climate change decision-making, but with many critical decision issues that have been left open for resolution in the future. Once one understands that, for these many reasons, no "globally" optimal climate change can be computed, the alternative is to rely on negotiation; that is, on some form of procedural decision-making process (in contrast to the teleological one pertaining to decision analysis), to choose responses to the problem.

1.2 Negotiating Climate Change Decisions

Since climate change related decisions entail multiple decision-makers with different objectives, and since it appears, then, that no best "globally" analytical outcome will be found, it became important, in the view of Chapter 2, "to examine negotiation and compromise as the primary basis for climate change decisions." One additional, institutional argument was the Framework Convention on Climate Change (FCCC) within which, precisely, the parties are engaged to negotiate future actions.

The withdrawal of decision analysis—beyond its significant value in framing the problem (see also below)—and the institutional step toward negotiation, gave view onto collective decisions in environmental accords, sequential decision-making under uncertainty (sequential climate decisions), as well as on further market-related instruments such as instruments for international insurance (1.3) and portfolios of climate actions (see 1.4).

1.2.1 Collective Decision-Making Within a Negotiated Framework

Internationally the Framework Convention on Climate Change establishes an initial framework within which the Parties can negotiate actions beyond the initial ones specified in the Convention. Although it requires further elaboration it established a process for linking objectives, decisions based on possible outcomes, and the decision-making process.

Elements of a collective decision-making framework

Environmental accords tend to be partial and incremental. They tend to be integrated, complex packages covering multiple, linked issues—actions, procedures, cost sharing, etc.—that are difficult to negotiate internationally. They consist of stages such as pre-negotiation, negotiation, and post-negotiation. This is due largely to scientific uncertainty concerning the problems and possible solutions, and the *a priori* need for continued learning about effects and consequences. Partial agreements are also customary because of the political uncertainty in devising fair and equitable approaches at the global level.

The Framework Convention on Climate Change shares many of these characteristics. It is a broad framework of principles that leaves many details concerning procedures and future actions to be negotiated by the Conference of the Parties (COP). The post-agreement negotiation in the Conference of the Parties can be considered the dynamics and co-operative systems, procedures, and structures institutionalised to sustain dialogue on issues that cannot, by their nature, be resolved by a single agreement.

Last, procedural/heuristic issues within a negotiated framework need to be addressed further. It results from a study by Sjöstedt (1993) that the following ones need to be dealt with by practitioners and researchers alike:

- Many new groups and individuals beyond traditional governmental officials are now important participants in the negotiation process.
- The complexity of international environmental negotiations is increasing; policy makers have linked environmental issues with other important policy concerns, such as development and trade; the implications of these linkages are often not well understood; ways must be found to develop a common understanding of these linkages.
- Bridging the differences between industrialised and developing countries will be a continuing problem, given their different interpretations of what constitutes fair and equitable solutions.
- New approaches to innovative problem solving need to be found and implemented, using the assistance of third parties or creative reasoning heuristics.
- New strategies to implement negotiated environmental agreements are needed to reduce national ratification delays and increase the likelihood of compliance (IIASA 1993).

Unlike the processes of pre-negotiation and negotiation, post-agreement negotiation has received little attention from researchers in terms of their procedural rationality. Options can be:

- Involving domestic stakeholders in negotiations from the beginning (and not at the post-agreement stage).
- Modifying the structure of the accords themselves: Clusters of simple, single-issue agreements may be more practical.

There is no social welfare function

At the world level, decisions cannot be based on a global (social) utility function (e.g., index of social welfare). There is no agreement in the literature as to how such a utility function would be constructed and applied: a social welfare function cannot exist (Arrow's impossibility theorem). In addition, there is no world sovereignty. Once it is accepted that we have no world sovereignty and no global social welfare function, the valuation and weighting of outcomes becomes a matter of negotiation among the parties.

A number of methods/tools thought to be useful to facilitating the negotiating process are found in the literature. Examples include simulation modelling, cognitive mapping and cross-impact analysis. For instance, cross-impact analysis facilitates the modelling of complex situations or stakeholder analysis uses information on the position, interest, priorities, and preferences of various stakeholders on a particular issue to facilitate analysis of the range of differences among stakeholders and the potential for coalition formation.

In general, these methods/tools are intended to help parties to reach an agreement (help diagnose the *negotiation process* itself, the procedure), not to help them negotiate the "one and best" outcome.

1.2.2 Sequentiality of Climate Decisions to Hedge Against Impact of Uncertainty and Change

Decisions can often involve a series of actions at different points in time. Future actions occur at specified or predicted dates after particular outcomes have, or have not, occurred. The FCCC is designed to allow parties to weaken or strengthen their obligations under the treaty in response to scientific information and assessment as well as relevant technical, social, and economic information.

Decision analysis can then identify the "best" *sequence* of actions given a specified decision criterion. The value of better information concerning particular outcomes can also be calculated.

Once again, however, analyses of sequential decision-making adopt the framework of a utility maximising individual decision-maker. Conceptually it is possible to define very complex decision trees featuring multiple decision points, numerous options for each decision, and numerous possible outcomes following from each choice. In practice studies that have attempted to analyse sequential decision-making *for climate change* have used very simple models typically involving two or three decision points, choices, and possible outcomes. More complex formulations quickly become intractable.

The best strategy, according to these analyses lies between the most extreme options (do nothing, aggressive action) available with the location determined by the probabilities assigned to the possible outcomes. These results assume also that none of the outcomes is irreversible (Singer, et al., 1991; Manne and Richels, 1991, 1992; Hammitt, et al., 1992; IUCC, 1993; Richels, 1994; Peck, 1994).

The strategy is to hedge against the probability of surprise such as those risks that fall below a particular threshold of probability. Such a strategy can be sequential, proceeding under an "act–then learn–then act" characterisation of the decision problem: Decisions are taken (acting) for the years to come without knowing which of the possible states of the world will occur in a sequence of decisions allowing new information (learning) to be incorporated into a complex decision process.[5]

In fact, prefatory statements in the FCCC include citations of already existing and ongoing analysis as the basis of the Convention. The actions undertaken in the Convention are thus already predicated on the basis of "learn–then act–then learn–then act".

Actually, the key point is that the sequential decision-making process—beyond some naïve utilitarian go / not-go decisions—seems most appropriate to identify *short-term strategies* to be undertaken *when facing long-term uncertainties*—because long-term decision-making is unknown and climate change-related waiting costs might prove severely high (also the cost of delay in emissions reduction by the North in terms of South's foregone opportunities could be substantial). Going to the point, decisions about how to respond to climate change should be based on the best available knowledge—now—, then the best decision strategy should be "designing actions based on existing knowledge–then learn–then act". Such a sequentiality of actions appears as the only possible framework from both a pure rationality and a procedural efficiency viewpoint when addressing collective decision-making within a negotiated framework. The flexibility of the scheme will depend upon today's decisions.

Additional reasons for sequential decision-making are (Grubb, Lave, Dowlatabadi, Hourcade, and other authors in IIASA, 1994):

- the existence of inertia (combined with risk) in technological trends,
- the existence of endogenous uncertainties,
- the existence of possible surprises, and
- the diversity of beliefs and expectations.

Climate change, knowledge related to climate change and its impacts, and developments related to mitigation and adaptation technologies are continuous processes. Hence, decisions on climate policies should be adjusted continuously in light of new information and developments and not at pre-specified intervals as in the highly stylised analyses. Recognising that options are created relative to these uncertainties—rights but no obligations to take some action in the future—would help decision-makers understand the key role of uncertainty in the timing of crucial investment decisions (Dixit and Pindyck, 1995). The option view *per se* suggests to proceed by steps.

1.2.3 The Value of Improved Information

In a sequential decision-making process better information on the nature or likelihood of potential outcomes is valuable since this allows actions to be better matched to the potential outcomes. Thus, research related to climate change and

---

[5] Hedging is not equal to "act–then learn" as learning does not necessarily take place.

its impacts and to adaptation and mitigation technologies is valuable provided that it yields results that improve future decisions. Developments of the theory consider that time conveys new information which will reduce uncertainty and thus facilitate the assignment of probabilities (in other words, in as far as probabilities have a subjective basis, they may be altered by learning).

Estimates of the value of improved climate information from these highly stylised studies ranges from a few billion to roughly 100 billion.

Under a "learn–then act" sequential decision-making, the value of new information depends on changes in the probabilities assigned to the (alternative) scenarios before and after the study. "If the probabilities of (alternative) scenarios remain equal, then the value of the study is zero; if, on the other hand, only one scenario can be selected, a study might be worth as much as 100 billion dollars" (Shlyakhter *et al.*, 1995).[6]

Under an "act–then learn–then act" sequential decision-making, there is an agreement that uncertainty *cannot be fully resolved* before a decision is taken. This does not mean that uncertainty is an obstacle for decision-making. This means that first, its resolution is at too high a cost and, second, that maybe it will not facilitate a decision—clearing up the uncertainties could make the greenhouse warming issue more difficult to resolve. However, an action now, based on current knowledge, to create options for the future should be highly valued (Dixit and Pindyck, 1995). Accordingly, uncertainty can prompt decision-makers to accelerate those measures/investments that create options. It is for instance assumed that the damage potential will remain unknown until 2010-2020. As a consequence, the appropriate emissions level would be known only well into the second decade of the twenty-first century. The actual level of investment is also influenced by the conflict between the long-term nature of the problem (and the time that would be required for resolution of the uncertainties)—and the short-term public and private decision-makers horizons (sometimes called NIMTOO phenomenon). Today decisions must thus consider the partial disclosing of the uncertainties along with the long-term effects of short-term decisions. Sequential decision-making can address this point.

Maximum accuracy, or perfect information, would eliminate the need to hedge against surprise. Accuracy of the forecast, however, will not be maximum. We therefore have to accommodate with hedging under imperfect information. In this

---

6 Values of situations before and after imperfect or perfect information is received can be computed. When there is risk neutrality, these values also represent expected utility. Suppose that expected results are as follows: (a) without information: +150; (b) imperfect information: +325; (c) perfect information: + 500. This hypothetical example "illustrates a perfectly general result: the value of perfect information (500 - 150) clearly exceeds the value of imperfect information (325 - 150)"; it also indicates that some information—albeit imperfect—is better than no information at all (Eeckhoudt and Gollier, 1995). Hourcarde and Chapuis (1994) have shown that, accounting for climate change-related surprises under a sequential decision model, a resolution of uncertainty had high value relative to no-regret potentials and technical innovation. Peck and Teisberg (1992) and Manne and Richels (1992) show similar results.

case, the research outcome will lead to a revision of our beliefs. It will help at modifying near-term decisions.

1.2.4 Application of Sequential Decision-Making

Sequentiality is critically important as regards the treatment of uncertainties. The most important result of the act-then learn-then act framing by Manne and Richels is that "we need not be overly concerned with our inability to predict the economic and technical system several decades into the future: uncertainty is important only to the extent that it confounds near-term decision-making. Today's decisions appear to be relatively insensitive to some of the more controversial longer-term uncertainties in the greenhouse debates" (Manne and Richels, 1993).

In this framework, applying the precautionary principle of the climate convention could mean: what could be the set of short-term decisions apt to increase both our learning time and its informative content so as to reconcile three long-term needs: stabilise GHGs concentrations, scientific knowledge, and appropriate trends in technical and consumption patterns to enhance the flexibility in managing transitions. In many circumstances of our daily life a rational behaviour cannot be but a sequential decision process and a "hedging" strategy which emphasises the value of new information in the future and of the "option value," namely the value of enlarging our margins of freedom for further decisions.

Literature points out four types of short-term decisions apt to improve our decision-making capabilities in the future. The first is investing in climate research simply because of the high economic value of scientific information. The second is the benefits from induced technical change (Arrow, 1962, Grubb *et al.*, 1995), which depends on short-term decisions and more generally the interest of creating options for the future. The third is related to the discussion about no-regret strategies where some kinds of quantitatively limited abatement decisions may increase our learning time in the face of controversial risks. The fourth is related to preventing risks of generating "in passing" bifurcations towards high carbon intensive development paths.

The relative insensitiveness of some short-term decisions to uncertainties about the very long-term is very "fortunate"; diverging expectations about the long run, diverging value judgements (about for example the time horizon to which developing countries should reach the development level of the richest countries), expert controversies about the relative potentials, merit, and deadlocks of large-scale use of competing techniques such as biofuels or nuclear energy could impede the formation of a consensus for action.

In this real world we are not only faced with uncertainties but with hard scientific, technical, ethical, and political controversies. In fact, as suggested very early by Allais (1953), uncertainty is not only a difficulty for decision-making, it is also a room of manoeuvre for strategic behaviours; historical lessons of other environmental disputes clearly put into evidence three main traps in this context:

- Pre-emption of the long-term by the short-term because of unchanged current behaviours.
- Dictatorship of the long-term over the short-term in the case of premature decisions which could ultimately be proved arbitrary.
- Paralysis of collective action because of endless disputes.

From a procedural viewpoint, we do not have to resolve controversies on long-term issues, to force premature agreements between very contrasted beliefs on the long run future or the too difficult disputes about burden sharing. While getting a better knowledge of impacts and enlarging the available technological spectrum, the objective of the first step of the sequence is to put us as far as possible in position to postpone technological or institutional "lock-in" and to use our negotiation time to reach wider consensus on how to approach the more difficult longer-term decisions.

1.3 Exploring Mutual Insurance Decision

Insurance[7] is another way to respond to risks and uncertainty. International insurance against climate impacts could enhance welfare. Currently, private insurance is available in some parts of the world for weather related losses such as damage due to storms and floods; but much of the world, including those most likely to be devastated by extreme impacts (as in Bangladesh) are not. Decision-making frameworks need to encompass options for better international sharing of the risks.

"Insurance as a means of sharing risks is well suited to situations where the likelihood of a damaging event for any individual party is relatively small but the potential damages are large. Since the relationship between weather-related events and climate change is not known, insurance to cover the risks of climate change per se is probably not feasible or necessary. However, insurance to cover the damages associated with weather-related events is desirable. To the extent that the frequency or severity of weather events turns out to be affected by climate change, such insurance would be a form of climate change insurance" (Arrow *et al.*, 1996).

We will consider this issue with further details in section 2.

---

[7] The word "insurance" is sometimes abused in climate debate. For instance, the insurance premium of, e.g., Manne and Richels (1992) is in fact a risk premium in the Arrow-Pratt sense (the Arrow-Pratt insurance premium describes situations where, because of risk aversion, people are willing to pay more than expected yearly damage), and the insurance scheme of AOSIS is a liability scheme.

1.4 Portfolio Analysis Analogy

Actions to limit the impacts of climate change may need to be taken for the next century or more. Numerous measures are available to limit the impacts of climate change, but no single action is clearly superior to the others. The decision-maker's problem then is to choose a portfolio of measures (Haites, 1994). The portfolio should achieve the decision-maker's goals at minimum cost taking account of the risks associated with the different measures.

The portfolio may include some relatively high cost measures to diversify the risks. If forestry measures are the lowest cost options available to a particular decision-maker, higher cost options in other sectors might be implemented as well in case the forestry measures are not as successful as expected due to disease, droughts, institutional barriers, or other factors.

Many investment decisions have implications for climate change because they "lock-in" commitments to energy use or other sources of greenhouse gas emissions possibly for several decades. The potential climate change implications of these investments do not justify delaying such decisions. However, the potential climate change implications of transportation infrastructure, land use decisions, energy-using equipment, and similar investments should be considered when the alternatives are being evaluated. The decisions should be based on the best information available at the time and properly reflect the value of flexibility.

Some of the measures in the climate change portfolio will also have relatively long lifetimes. The extent to which resources are locked in by these measures should be considered when their climate change benefits are evaluated. Although some individual measures may be inflexible, other measures can be chosen to ensure that the overall portfolio has sufficient flexibility. Thus decision-makers will be able to adjust a well chosen portfolio frequently in response to new information even though some of the measures are locked in for relatively long periods.

The key to selecting a portfolio is to understand how the measures interact over time. Mitigation measures provide more flexibility for future action, technology research can lower the cost of future action, research on climate change provides better information on the nature of the actions required.

In principle, the measures available to countries, individually or jointly, to limit climate change and its impacts include:

- Implementing low cost measures to reduce emissions of greenhouse gases; there is broad agreement that in most countries there are some measures, such as measures for increasing energy efficiency, that can reduce emissions of greenhouse gases at negative to slightly positive costs.
- Phasing out existing distortionary policies, such as some subsidies to fossil fuels, that reduce welfare and, directly or indirectly, increase greenhouse gas emissions.
- Implementing fuel switching measures from more- to less-carbon-intensive and carbon-free fuels to reduce emissions of greenhouse gases.

- Implementing measures to enhance sinks or reservoirs of greenhouse gases such as protecting and restoring forests.
- Implementing available measures and developing new techniques for reducing methane and nitrous oxide emissions including improved industrial processes, reduced emissions from landfills, agriculture, and from oil and gas extraction and transportation.
- Instituting forms of international co-operation to limit greenhouse gas emissions such as joint implementation, technology transfer, and tradable quotas.
- Planning, and implementing as appropriate, measures to adapt to the consequences of climate change.
- Undertaking research aimed at better understanding the causes and impacts of and adaptation to climate change. Economic studies suggest that such research can yield high returns by reducing the uncertainty about actions to address climate change.
- Conducting technological research aimed at enhancing energy efficiency, minimising emissions of greenhouse gases from continued use of fossil fuels, and developing commercial non-fossil energy sources. In the long-run, the cost and timing of availability of non-fossil energy technologies is one of the major determinants of the cost of addressing climate change.
- Developing institutional mechanisms, such as insurance, to share the risks of damages due to climate change.

The specific policy measures available vary from country to country; the costs and availability of measures to reduce greenhouse gas emission vary, existing distortionary policies vary, fuel switching options vary, sequestration options vary, research capabilities vary, the damages due to climatic change differ, the measures for adapting to climate change differ and international commitments vary.

Countries will select a portfolio of climate change measures that reflect, at least implicitly, a variety of objectives and constraints. A country may look for the economically optimal portfolio of climate change measures. However, considerations of the impacts of the portfolio on different economic groups, international competitiveness, international equity, and intergenerational equity are likely to come into play. These considerations will vary and have different weights from country to country.

2 Market Solutions Within a Collective Decision-Making Framework ('le non-dit')

Controversial and even clandestine figures may lie behind any report. It is the purpose of this section to account for some of those figures. Why disclosing such issues? The reason is the aim of the present workshop; that is, referring to IPCC results while trying to escape from this consensual round toward more normative views attached to decision-making processes in the perspective of propounding further policy options.

One such issue is collective decision-making for sharing risks. Another one is the use of financial markets.

2.1 Collective Decision-Making for Sharing Risks

Chapter 2 moved to collective decision-making from individual decision analysis. The reason was the existence of multiple decision-makers. In fact, one of the important potential gains from co-operating in a collective framework is that of risk sharing.

A group has much greater possibilities for coping with risk than an individual. "This is partly due to the opportunities of diversification within the group and partly due to the transfer of risk towards the least risk-adverse members (or the richest members if absolute risk aversion is decreasing)" (Eeckhoudt and Gollier, 1995). This is why organisations have always been best suited to undertake risky activities—adapting to or slowing down climate changes is a risky activity by itself due to the existence of strong uncertainties. "Without this diminution of risk aversion in an economy, thanks to the creation of risk pools, many risky projects would not have been undertaken and we would undoubtedly not have known the economic expansion that we have observed over the last two centuries" (Ibid.).

Transferring risk from one agent (country) to another or sharing risk collectively *does not change the risk but it can improve welfare*. The literature indicates that "the transfer of risk is a potential source of large improvements in economic efficiency and social welfare" rather than a zero sum game (Eeckhoudt and Gollier, 1995). Insurance shares risks and is best suited to situations where the risk of an event for any individual party is relatively small but the damages are significant. Then relatively small premiums paid by all parties are able to compensate those who suffer losses.

The classical approach to reducing the costs of risk is with insurance. But markets do not exist in which individuals or countries could insure themselves against impacts due to changes in climate or due to changes in abatement policies, and insurance on a strictly national basis would miss many of the potential benefits from co-operating in collectively sharing risks. Chichilnisky and Heal (1993) and others note several reasons why commercial insurance markets cannot adequately cover against risks associated with climate change; even to the limited extent they can, the insurance premiums would presumably fall primarily upon the potential victims and thus represent an inequitable burden. This means that people and organisations have to bear an unnecessary burden corresponding to the cost of risk-bearing. Establishing global insurance in an appropriate form could thus increase both efficiency and equity by reducing exposure to risk, and this also needs to be part of decision-making frameworks.

2.2 The Use of Financial Markets

Different perceptions of the risk involved lead in Chapter 2 to some withdrawal of the quantitative decision analysis framework. On a normative basis, however, *different perceptions of the risk involved do not preclude efficient solutions.* Differences in attitudes toward risk could merely lead to the introduction of new markets in which different risk positions are traded, with efficiency gains. Economics is about differences in preferences leading to trade. Accordingly, "betting" on particular climate outcomes could be envisaged (e.g., Pillet 1994; *see* Chichilnisky, this volume).

The characteristics of the risks associated with climate change create particular requirements for approaches to insurance. Simply on grounds of risk sharing (as distinct from equity), two classes of insurance contract need to be considered (on the same theoretical basis). One is in the nature of a mutual insurance contract, which is an agreement between parties subjects to similar risks that "those which are harmed will be compensated by others"; there are many precedents in Europe (cooperatives). In the context of climate change the parallel would be a binding agreement in which countries that are hurt by climate change more than the average (or expected) would be helped by those affected less (Chichilnisky and Heal, 1993).

The second type of "insurance contract" is required because the overall nature of the risks, as well as its specific distribution, are uncertain. The formal treatment of an appropriate insurance structure (Arrow, 1951) is complex: it requires defining "securities" for each possible outcome, which pay out only if that outcome is realised. In the climate change case the analogy would be that each country would be required to make commitments to compensate as insurance against a particular climate "outcome". To efficiently distribute the risks, however, countries would then be allowed to trade these "securities."

This (in combination with mutual insurance contracts) would not only allow a fairer distribution of the risks, but also allows for differing beliefs about the overall risks. For example, if a country genuinely believed that climate change is unlikely to have serious global impacts, it will be more prepared to see those "securities" that pay out under these conditions. These issues are discussed in detail in Heal (1993), and a formal two-country treatment is summarised in Chichilnisky and Heal (1993) and Chichilnisky (1994), who note that creating such securities would help to provide an objective test of the honesty of national positions on the risks of climate change: if a country were to argue that climate change does not involve substantial risks as part of a negotiating play to avoid onerous abatement commitments, for consistency it should be prepared to see associated high-pay off securities; and "there would be a cash cost to convincing and consistent misrepresentation of true beliefs. These cash costs could offset some of the incentive to free ride on other countries' efforts to reduce greenhouse emissions."

Whilst this Arrow-Chichilnisky-Heal analysis is primarily concerned with efficient risk-bearing, it has equity implications in terms of both outcome and process which could resolve the dilemmas. The feasibility, credibility, and equity of such securities remain to be considered and could be an important research topic.

Although insurance capable of sharing climate change risks on a global basis currently does not exist, one of the important potential gains from co-operating in a collective framework, such as the Framework Convention on Climate Change, is that of risk sharing. Insurance to cover the risks of climate change is difficult, and the international community has not yet established such sophisticated instruments. This, however, does not preclude international action to establish insurance markets sufficient for some international needs.

3 Different Decision-Making Frameworks with Different Foci

It results from this discussion that no decision-making framework is exclusively operational. Chapter 2 has legitimated to collective decision-making as if escaping from decision analysis had been imposed by the climate change problem. In fact, emerging options for dealing with that problem show that market related solutions also exist in the light of collective decision-making. In this matter, procedural decision-making comes first, due to cognitive limitations that could exclude the use of standard decision analysis. Then collective decision-making, with risk sharing, appears as the framing allowing for devising solutions which can draw back on market instruments.

Decision analysis, collective decision-making as well as procedural decision-making offer insights into decision-making with respect to climate change. Table 1 shows which difficulties these different frameworks with different foci can and cannot properly address

Table 1: Decision Frameworks, Foci, Objectives and Criteria

Framework	Primary Focus	Broad Objectives	Main decision criteria
OPTIMIZATION Individual decision making	Reaching 'the answer' (best outcome)	Expected utility maximization Valuation of outcomes	*Efficiency* Utility-based criteria Cost-benefit analysis
PROCEDURAL Decision-making process	How to reach a decision	Rules of procedure	*Procedure* Widespread acceptance
COLLECTIVE Collective decision making	Choosing weights for the member Parties Risk sharing	Valuation of outcome becomes a matter of negotiation	*Distribution* Opportunities of diversification

The Framework Convention provides for components of all three of these idealised decision-making frameworks.

Global optimisation, taking account of risk aversion, is reflected in the overall objective of stabilisation within an appropriate time frame combined with references to the need for cost-effective action[8] and the importance of respecting economic development. Chapter 5 of the IPCC report examines in detail the applicability of cost-benefit analysis and other forms of global optimisation analysis, Chapters 3, 4, and 6 discuss aspects of global valuation, and Chapter 10 discusses the development of modelling frameworks for integrated global assessment.

The need for sequential decision-making is reflected in all the provisions for review, assessment, elaboration of commitments etc.

The FCCC also refers to decision-making procedures, including principles of transparency, provision of information, participation, and sovereignty, etc.

But above all the FCCC is a framework for collective decision-making. It formalises the fundamental fact that there is no single decision-maker, so that decisions have to be reached as a collective agreement among sovereign states. This above all others is the defining context of national decision-making, but it does not of course mean that the other decision-making frameworks are irrelevant; rather, it sets the context in which countries must drawn upon the other frameworks in formulating decisions.

8 Cost-effectiveness analysis can aid the choice between options but, in opposition to cost-benefit analysis, cannot answer the question whether or not any of the options are worth doing (The MIT Dictionary of Modern Economics).

4 References

Allais, M. (1953): "Le comportement de l'homme rationnel devant le risque, critique des postulats et axiomes de l'école américaine." *Econometrica* **21**, 503-546.

Arrow, K. J (1951): "Alternative Approaches to the Theory of Choice in Risk-Taking Situations". *Econometrica* **19**: 404-437. Reprinted in Arrow, K. J (1984): *Collected Papers of Kenneth J. Arrow: 3, Individual Choice under Certainty and Uncertainty.* Cambridge, Mass.: The Belknap Press of Harvard University Press, pp. 5-41.

Arrow, K. J (1962): "The Economic Implications of Learning by Doing". *Review of Economic Studies,* **29**, 155-73. Reprinted in K. J Arrow (1985): *Collected papers of Kenneth J. Arrow, 5, Production and capital,* pp. 157-180.

Arrow, K. J, J. Parikh, and G. J Pillet [Principal Lead Authors] (1996): "Decision-making Frameworks for Addressing Climate Change". J. P Bruce, H. Lee & E. F Haites (eds), *Climate Change 1995 – Economic and Social Dimensions of Climate Change.* Cambridge University Press, Cambridge, pp. 53-77. [Published for the Intergovernmental Panel on Climate Change—IPCC.]

Caselton, B., and W. Luo (1994): "Dempster-Shafer Theory and Decision-making Under Near-ignorance." *Working paper.* University of British Columbia and British Columbia Hydro, 17 pp.

Chichilnisky, G. (1994): "Global Environmental Risks and Financial Instruments". G. J Pillet and F. Gassmann (eds), *Steps Towards a Decision-making Framework to Address Climate Change,* Paul Scherrer Institute, Würenlingen and Villigen (Switzerland), pp. 99-109.

Chichilnisky, G., and G. Heal (1992): "Financial Markets with Unknown Risks". Columbia Business School. *First Boston Working Paper Series.*

Chichilnisky, G., and G. Heal (1993): "Global Environmental Risks." *Journal of Economic Perspectives* **7** (4): 65-86.

Dixit, A. K, and R. S Pindyck (1995): "The Options Approach to Capital Investment." *Harvard Business Review* **73** (3): 105-115.

Dowlatabadi, H., and M. G Morgan (1992): "A Model Framework for Integrated Studies of the Climate Problem." *Draft manuscript.* Carnegie Mellon University, Pittsburgh, PA, Dep. of Engineering and Public Policy, 25 pp.

Eeckhoudt, L., and C. Gollier (1995): *Risk – Evaluation, Management and Sharing.* Translated by V. Lambson. Harvester Wheatsheaf, New York.

Fankhauser, S. (1994): "Global Warming Damage and the Risk of Climate Catastrophe." G. J Pillet and F. Gassmann (eds), *Steps Towards a Decision-making Framework to Address Climate Change,* Paul Scherrer Institute, Würenlingen and Villigen (Switzerland), pp. 19-29.

Gassmann, F. (1994): "Abrupt Change Scenarios". G. J Pillet and F. Gassmann (eds), *Steps Towards a Decision-making Framework to Address Climate Change,* Paul Scherrer Institute, Würenlingen and Villigen (Switzerland), pp. 3-9.

Grubb, M. , T. Chapuis and M. Ha Duong (1995): "The Economics of Changing Course." *Energy Policy,* April.

Haites, E. F. (994): "Portfolio Analysis Model of Climate Change Decision-Making." G. J Pillet and F. Gassmann (eds.), *Steps Towards a Decision-making Framework to Address Climate Change,* Paul Scherrer Institute, Würenlingen and Villigen (Switzerland), pp. 111-114.

Hammitt, J. K., R. J. Lempert, and M. E. Schlesinger (1992): "A Sequential-Decision Strategy for Abating Climate Change". *Nature* **357**: 315-18.

Hourcade, J. C., and T. Chapuis (1994): "No-Regret potentials and technical Innovation: A Viability Approach to Integrative Assessment of Climate Policies." Nakicenovic, Nordhaus, Richels and Toth (eds), *Integrative Assessment of Mitigation, Impacts and Adaptation to Climate Change.* Laxenburg: IIASA Workshop Proceedings.

IIASA (1993): "Processes of International Negotiation." *IIASA's Options,* June, pp. 4-12.

IIASA (1994): *Integrative Assessment of Mitigation, Impacts and Adaptation to Climate Change.* Nakicenovic, Nordhaus, Richels, and Toth (eds). Laxenburg: IIASA Workshop Proceedings.

IUCC (1993): "Insurance Against Climate Change? The AOSIS Proposal". *Climate Change—Fact Sheet 238.* Geneva: UNEP/WMO, 2 pp.

Manne, A. S., and R. G. Richels (1993): CO_2 Hedging Strategies—the Impact of Uncertainty Upon Emissions. Prepared for the *OECD/IEA Conference on the Economics of Climate Change.* Paris, June 14-16, 17 pp.

Manne, A. S, and R.G. Richels (1992): *Buying Greenhouse Insurance.* The MIT Press, Cambridge, Mass., 182 pp.

Morgan, M. G, and M. Henrion (1990): *Uncertainty — A Guide to Dealing with Uncertainty in Quantitative Risk and Policy Analysis.* Cambridge University Press, Cambridge, 1990.

Parikh, J., with K. J. Arrow, G. J. Pillet, and P. G. Babu, S. Kavi Kumar (1995): "Decision-making Framework to Address Climate Change." N.S.S. Narayana & Anindya Sen (eds), *Poverty, Environment and Economic Development.* Interline Publ. Priv. Ltd, Bangalore, Dehra Dun, Michigan, pp. 155-189.

Peck, S. C. (1994): "Candidate Framework for Analysis of Climate Change Decisions and Uncertainties". *Contribution to the IPCC Working Group III Writing Team II Montreux Meeting.* Montreux (Switzerland), March 3-6.

Pillet, G. J. (1994): "Betting on Climate States — The Impact of Climate Change Prospects on Decision-making under 'Strong' Uncertainty and Collective Risks". G. J. Pillet and F. Gassmann (eds.), *Steps Towards a Decision-making Framework to Address Climate Change,* Paul Scherrer Institute, Würenlingen and Villigen (Switzerland), pp. 115-137.

Richels, R. (1994): "Decision-making Under Uncertainty and the Global Climate Debate." *Contribution to the IPCC Working Group III Writing Team II Montreux Meeting.* Montreux (Switzerland), March 3-6.

Rowen, H. S. (1976): "Policy Analysis as Heuristic Aid: The Design of Means, Ends, and Institutions". L. H. Tribe, C. S. Schelling, and J. Voss (eds.), *When Values conflict: Essays on Environmental Analysis, Discourse and Decision.* Ballinger, Cambridge, Mass., pp. 137-152.

Shlyakhter, A., L. J. Valverde, Jr., and R. Wilson (1995): "Integrated Risk Analysis of Global Climate Change." *Chemosphere* **30**(8), 1585-1618.

Singer, F., R. Revelle, and C. Starr (1991): "What to do about Greenhouse Warming". *Cosmos*: 28-33.

Sjöstedt, G. (Ed.) (1993): *International Environmental Negotiation.* Sage Publications.

4 What is Sustainable Development?[1]

Graciela Chichilnisky
Columbia University
New York[2]

Abstract. The paper proposes two axioms that capture the idea of sustainable development and derives the welfare criterion that they imply. The axioms require that neither the present nor the future should play a dictatorial role in society's choices over time.

Theorem 1 shows there exist *sustainable preferences,* which satisfy these axioms. They exhibit sensitivity to the present and to the long-run future, and specify trade-offs between them. It examines other welfare criteria which are generally utilized: discounted utility, lim inf, long run averages, overtaking and catching-up criteria, Ramsey's criterion, Rawlsian rules, and the criterion of satisfaction of basic needs, and finds that none satisfies the axioms for sustainability.

Theorem 2 gives a characterization of all continuous *sustainable preferences.* Theorem 3 studies a standard dynamical system representing the growth of a renewable resource. Sustainable preferences are shown to be a natural extension of the "equal treatment criterion" for finitely many generations, in the sense that the optimal solutions for such preferences approach the "turnpike" of an equal-weight finite horizon optimization problem as the horizon increases. Theorem 3 also shows that sustainable preferences match the experimental evidence in these cases, in the sense that they imply a rate of discount that decreases and approaches zero as time goes to infinity. Theorem 4 exhibits the differences between the optimal paths according to sustainable preferences and discounted utilitarianism in an extension of the classical Hotelling problem of the optimal depletion of an exhaustible resource. Theorem 5 shows that sustainable optima can be quite different from discounted optima, no matter how small is the discount factor. Subsequent examples show the implications

---

1 This paper was prepared for an invited presentation at a seminar on Reconsideration of Values at the Stanford Institute for Theoretical Economics, organized by K. J. Arrow in July 1993. It was also an invited presentation at the Intergovernamental Panel on Climate Change (IPCC) Seminar in Montreux, Switzerland, March 1994, at a Seminar on Inconmensurability of Values at Chateaux du Baffy, Normandy, April 1994, and at the Graduate School of Business of Stanford University in May 1994.

2 UNESCO Chair in Mathematics and Economics and Director, Program on Information and Resources, Columbia University. email: gc9@columbia.edu. Research support from NSF grant No. 92-16028, the Stanford Institute for Theoretical Economics (SITE), and from the Institute for International Studies at Stanford University is gratefully acknowledged, as are the comments of Y. Baryshnikov, P. Ehrlich, S. C. Kolm, D. Kennedy, D. Kreps, L. Lauwers, D. Starrett, L. van Liedekierke, P. Milgrom, J. Roberts, R. Wilson and H. M. Wu. Special thanks are due to Kenneth Arrow, Peter Hammond, Geoffrey Heal, Mark Machina, Robert Solow, the editor and an anonymous referee.

for shadow prices. In general sustainable growth paths cannot be approximated by paths which approximate discounted optima, and that paths which maximize the present value under a standard price system may fail to reach optimal sustainable welfare levels. Sustainable preferences and discounted utilitarianism can give rise to different value systems.

1 A Voracious Use of Resources

For the first time in history, human activity has reached levels at which it could alter the planet's climate and its biological mix. Economics is the driving force. Energy is mostly produced by burning fossil fuels, and leads to carbon emissions. The emissions generated since the second world war could alter the earth's climate, although there is as yet no scientific agreement on the precise magnitude of the effects. Biologists see the loss of biodiversity during the last fifty years as one of the four or five largest incidents of destruction of life on the planet. Originating largely in the industrial countries, this voracious use of resources has been accompanied by increasing discrepancies in resource consumption and welfare between industrial and developing countries. At the 1992 United Nations Earth Summit in Rio de Janeiro, *sustainable development* emerged as one of the most urgent subjects for international policy. One hundred and fifty participating nations endorsed UN Agenda 21, proposing as part of its policy agenda sustainable development based on the satisfaction of basic needs in developing countries. This development criterion was introduced in 1976 in the *Bariloche Model* [3] and given further impetus in 1987 when the Brundtland Commission proposed that "sustainable development is development that satisfies the needs of the present without compromising the needs of the future".[4] But *what is sustainable development*?

Solow [45] pointed out recently that discussion of sustainability has been mainly an occasion for the expression of emotions and attitudes, with very little formal analysis of sustainability or of sustainable paths for a modern industrial economy. Formal analysis requires valuation, and the economic value of a resource is usually derived from its contribution to utility. This suggest that the crux of the matter is how to describe value so that it does not underestimate the futures' interests and utilities, so that the utility of the future is given an equal treatment. This can be achieved in several ways. The challenge, however, is to develop economic theory which formalizes this aim with the level of clarity and substance achieved by neoclassical growth theory, and with the practical scope of the current approach to cost benefit analysis that is based on neoclassical growth theory.

1.1 Costs Today, Benefits Tomorrow

A well-known problem is that standard cost-benefit analysis discounts the future. It is therefore biased against policies designed to provide benefits in the very long

[3] Chichilnisky [9], [10] and [32].

[4] [8], Chapter 2, para 1.

run. A sharp example is the evaluation of projects for the safe disposal of waste from a nuclear power plant. Another is policies designed for the prevention of global warming. The benefits of both may be at least fifty to a hundred years into the future. The costs, however, are here today. In these cases, the inherent asymmetry between the treatment of present and future makes it hard to justify investment decisions that large numbers of individuals and organizations clearly feel are well merited. Recent experimental evidence sheds new light on the matter.

1.2 Experimental Evidence

Several experiments have measured how people value the long run (see for example Lowenstein and Thaler [39] , Cropper et. all. [23] and the references in Lowenstein and Elster [38]). Their findings clash with the traditional discounted approach. People are shown to value the present and the future differently, but not as the standard analysis would predict. The experimental evidence indicates that the present and the future are treated more even-handedly. Typically we do discount the future, but the trade-off between today and tomorrow blurs as we move into the future. Tomorrow acquires increasing relative importance as time progresses. It is as if we viewed the future through a curved lens. The relative weight given to two subsequent periods in the future is inversely related to their distance from today. The period-to-period rate of discount is inversely related to the distance into the future. The experimental evidence shows that rate of discount between period t and period $t+1$ decreases with t. Interestingly, studies of human responses to sound summarized in the Weber-Fechner law, indicate similar responses to changes in sound intensity. The human ear responds to sound stimuli in an inverse relation to the initial stimulus.

How to explain this experimental evidence? How to make sense of our sensitivity to time and integrate it into a criterion of optimality? Several interesting alternatives to the discounted utility analysis have been proposed. None reached the clarity and consistency of the discounted utilitarian criterion used in cost-benefit analysis, nor its analytical tractability. Prominent examples are the "overtaking criterion" and Ramsey's criterion. Both, however, are seriously incomplete, failing to rank many reasonable paths. The ordering induced by the overtaking criterion cannot be represented by a real valued function, making it impractical to use. As a result, they lack the corresponding "shadow" prices to evaluate costs and benefits in an impartial fashion. These criteria therefore fail on practical grounds.

1.3 A Criterion for Sustainability

This paper proposes simple axioms which capture the concept of sustainability, and derives the welfare criterion which they imply. The criterion that emerges is complete, analytically tractable and represented by a real valued function. In optimization it leads to well-defined shadow prices which can be used for a "sustainable cost-benefit analysis". The axioms provide internal consistency and ethical clarity. They imply a more symmetric treatment of generations in the sense that neither the "present" nor the "future" should be favored over the other. They neither accept the romantic view which relishes the future without regards for the present, nor the consumerist view which ranks the present above all. The axioms lead to a complete characterization of *sustainable preferences*[5], which are sensitive to the welfare of all generations, and offer an equal opportunity to the present and to the future. Trade-offs between present and future consumption are allowed. The characterization of sustainable preferences appears in Theorem 1. Theorem 2 shows that sustainable preferences are different from all other criteria used so far in the analysis of optimal growth and of markets. Theorem 3 studies a standard dynamical system representing the growth of a renewable resource. Sustainable preferences are shown to be a natural extension of the "equal treatment criterion" for finitely many generations, in the sense that the optimal solutions for such preferences approach the "turnpike" of an equal-weight finite horizon optimization problem as the horizon increases. Theorem 3 also shows that sustainable preferences match the experimental evidence in these cases, in the sense that they imply a rate of discount that decreases and approaches zero as time goes to infinity. Theorem 4 investigates the relationship between the optimal paths according to sustainable preferences and discounted utilitarianism in an extension of the classical Hotelling problem of the optimal depletion of an exhaustible resource.

Theorem 5 shows that sustainable optima can be quite different from discounted optima, no matter how small is the discount factor. Subsequent examples show the implications for shadow prices.

2 Two Axioms for Sustainable Development

The two following axioms are non-dictatorship properties.[6] Axiom 1 requires that the present should not dictate the outcome in disregard for the future: it requires sensitivity to the welfare of generations in the distant future. Axiom 2 requires that the welfare criterion should not be dictated by the long-run future, and thus requires sensitivity to the present.[7] To offer a formal perspective a few definitions are required.

Each generation is represented by an integer g, $g = 1, 2, \ldots\infty$. An infinitely lived world obviates the need to make decisions contingent on an unknown terminal date. Generations could overlap or not. Indeed agents could be infinitely long lived and evaluate development paths for their own futures.

For ease of comparison, I adopt a formulation which is as close as possible to that of the neoclassical model. Each generation g has a preference that can be represented by a utility function u_g for consumption of n goods, some of which could be environmental goods such as water, or soil, so that consumption vectors are in R^n, and $u_g : R^n \to R$. The availability of goods in the economy is constrained in a number of ways, for example by a differential equation which represents the growth

---

5 An alternative name suggested by Robert Solow is "intertemporally equitable preferences".

6 See Arrow [1], Chichilnisky [18]. In this case we are concerned with fairness across generations, see also Solow [44], Lauwers [36], Lauwers and van Liederkierke [37].

7 Chichilnisky [12], [18].

of the stock of a renewable resource,[8] and/or the accumulation and depreciation of capital. Ignore for the moment population growth, although this issue can be incorporated at the cost of simplicity, but with little change in the results.[9] The space of all feasible consumption paths is indicated by F.

$$F = \{x : x = \{x_g\}_{g=1,2...},\ x_g \in R^n\}. \tag{1}$$

We chose a utility representation so that each generation's utility functions is bounded below and above: $u_g : R^n \to R^+$, and $\sup_{x \in R^N}(u_g(x)) < \infty$. This choice is not restrictive: it was shown by K. Arrow [3] that when ranking infinite streams of utilites as done here one should work with bounded utility representations since doing otherwise could conflict with the continuity of choice.[10] Utility across generations is assumed to be comparable. In order to eliminate some of the most obvious problems of comparability I normalize the utility functions u_g so that they are non-negative and all share a common bound, which I assume without loss of generality to be 1:

$$\sup_g (u_g(x_g))_{x_g \in R^n} \leq 1. \tag{2}$$

The space of **feasible utility streams** Ω is therefore

$$\Omega = \{\alpha : \alpha = \{\alpha_g\}_{g=1,2...}, \alpha_g = u_g(x_g)\}_{g=1,2,...} \text{and } x = \{x_g\}_{g=1,2...} \subset F\} \tag{3}$$

Each utility stream is a sequence of positive real numbers bounded by the number 1. The space of all utility streams is therefore contained in the space of all infinite bounded sequences of real numbers, denoted ℓ_∞.[11] Our welfare criterion W should rank elements of Ω, for all possible $\Omega \subset \ell_\infty$.

2.1 Sensitivity and Completeness

The welfare criterion W is **complete** if it is represented by an increasing real valued

---

8 see [15], [6].

9 Populations growth and utilitarian analysis are well known to make an explosive mix, which is however outside the scope of this paper.

10 A preference admits more than one utility representation; among these one choses a bounded representation. The need to work with bounded utility representations in models with infinitely many parameters was pointed out by K. Arrow [3], who required boundedness to solve the problem that originally gave rise to Daniel Bernouilli's famous paper on the "St. Petersburg paradox", *Utility Boundedness Theorem*, page 27 of [3]. If utilities are not bounded, one can find a utility stream for all generations with as large a welfare value as we wish, and this violates standard continuity axioms.

11 Formally: $\Omega \subset \ell_\infty$, where $\ell_\infty = \{y : y = \{y_g\}_{g=1,...} : y_g \in R^+, \sup_g \mid y_g \mid < \infty\}$. Here $\mid . \mid$ denotes the absolute value of $y \in R$, which is used to endow ℓ_∞ with a standard Banach space structure, defined by the norm $\|.\|$ in ℓ_∞

$$\|y\| = \sup_{g=1,2...} \mid y_g \mid . \tag{4}$$

The space of sequences ℓ_∞ was first used in economics by G. Debreu [24].

function defined on all bounded utility streams[12] $W : \ell_\infty \rightarrow R^+$. It is **sensitive**, or increasing, if whenever a utility stream α is obtained from another β by increasing the welfare of some generation, then W ranks α strictly higher than β.[13]

2.2 The Present

How to represent the present? Intuitively, the present is represented by all the utility streams which have no future: for any given utility stream α, its "present" is represented by all finite utility streams which are obtained by cutting α off after any number of generations. Formally:

Definition 1 *For any utility stream $\alpha \in \ell_\infty$, and any integer K, let α^K be the "$K-$cutoff" of the sequence α, the sequence whose coordinates up to and including the $K-th$ are equal to those of α, and zero after the $K-th$.*[14]

Definition 2 *The present consists of all feasible utility streams which have no future, i.e. it consists of the cutoffs of all utility streams.*

2.3 No Dictatorial Role for the Present

Definition 3 *A welfare function $W : \ell_\infty \rightarrow R$ gives a dictatorial role to the present, i.e. W is a dictatorship of the present, if W is insensitive to the utility levels of all but a finite number of generations, i.e. if W is only sensitive to the "cutoffs" of utility streams, and it disregards the utility levels of all generations from some generation on.*

Formally,

Definition 4 *The $K-th$ tail of γ is the sequence whose coordinates up to and including the $K-th$ are zero and equal to those of γ after the $K-th$ generation.*[15]

For any two $\alpha, \gamma \in \ell_\infty$, let (α^K, γ_K) be the sequence defined by summing up or "pasting together" the $K-th$ cutoff of α with the $K-th$ tail of γ. W is a dictatorship of the present if for any two utility streams[16] α, β

$$W(\alpha) > W(\beta) \Leftrightarrow$$

---

[12] The representability of the order W by a real valued function can be obtained from more primitive assumptions, such as e.g. transitivity, completeness and continuity conditions on W.

[13] Formally: if $\alpha > \beta$ then $W(\alpha) > W(\beta)$.

[14] In symbols: $\alpha^K = \{\sigma_g\}_{g=1,2...}$ such that $\sigma_g = \alpha_g$ if $g \leq K$, and $\sigma_g = 0$ if $g > K$.

[15] In symbols: $\alpha_K = \{\sigma_g\}_{g=1,2...}$ such that $\sigma_g = 0$ if $g \leq K$, and $\sigma_g = \alpha_g$ if $g > K$.

[16] Recall that all utility streams are in ℓ_∞ and they are normalized so that $\sup_{g=1,2...}(\alpha(g)) = \|\alpha\| < 1$ and $\sup_{g=1,2...}(\beta(g)) = \|\beta\| < 1$.

$$\exists\, N = N(\alpha, \beta) \text{ s.t. if } K > N,\ W(\alpha^K, \gamma_K) > W(\beta^K, \sigma_K)$$

for any utility streams[17] $\gamma, \sigma \in \ell_\infty$.

The following axiom eliminates dictatorships of the present:

- **Axiom 1: No dictatorship of the present.**

2.4 The Future

By analogy, for any given utility stream α, its "future" is represented by all infinite utility streams which are obtained as the "tail" resulting from cutting α off for any finite number of generations.

2.5 No Dictatorial Role for the Future

Definition 5 *A welfare function $W : \ell_\infty \to R$ gives a dictatorial role to the future, or equivalently W is a dictatorship of the future, if W is insensitive to the utility levels of any finite number of generations, or equivalently it is only sensitive to the utility levels of the "tails" of utility streams.*
Formally:[18]*for every two utility streams α, β*

$$W(\alpha) > W(\beta) \Leftrightarrow$$
$$\exists\, N = N(\alpha, \beta.) : \text{ if } K > N,$$
$$W(\gamma^K, \alpha_K) > W(\sigma^K, \beta_K), \forall \gamma, \sigma \in \ell_\infty.$$

The welfare criterion W is therefore only sensitive to the utilities of "tails" of streams, and in this sense the future always dictates the outcome independently of the present. The following axiom eliminates dictatorships of the future:

- **Axiom 2: No dictatorship of the future.**

Definition 6 *A sustainable preference is a complete sensitive preference satisfying Axioms 1 and 2.*

3 Existence and Characterization of Sustainable Preferences

Why is it difficult to rank infinite utility streams? Ideally one would give equal weight to every generation. For example, with a finite number N of generations,

---

[17] Recall that all utility streams are in ℓ_∞ and they are normalized so that $\|\gamma\| < 1$ and $\|\sigma\| < 1$.

[18] An equivalent definition for our pruposes would be obtained by replacing " $\exists N$" by "for any K".

each generation can be assigned weight $1/N$. But when trying to extend this criterion to infinitely many generations one encounters the problem that, in the limit, every generation is given zero weight.[19]

What is done usually to solve this problem is to attach more weight to the utility of near generations, and less weight to future ones. An example is of course the sum of discounted utilities. Discounted utilities give a bounded welfare level to every utility stream which assigns each generation the same utility. Two numbers can always be compared, so that the criterion so defined is clearly complete. However, the sum of discounted utilities is not even-handed: it disregards the long run future. I show below that it is a dictatorship of the present.

Another solution is the criterion defined by the long-run average of a utility stream, a criterion used frequently in repeated games. However, this criterion is not even-handed either: it is biased in favor of the future and against the present. It is insensitive to the welfare of any finite number of generations.

Here matters stood for some time. Asking for the two axioms together, the non-dictatorship of the present and the non-dictatorship of the future, as I do here, appears almost as if it would lead to an impossibility theorem. Not quite.

Let us reason again by analogy with the case of finite generations. To any finite number of generations one can assign weights which decline into the future, and then assign some extra weight to the last generation. This procedure, when extended naturally to infinitely many generations, is neither dictatorial for the present nor for the future. It is similar to adding to a sum of discounted utilities, the long run average of the whole utility stream. Neither part of the sum is acceptable on its own, but together they are. This is Theorem 1 below. Theorem 2 proves that under regularity conditions this procedure gives a complete characterization of all continuous sustainable preferences.

Formally, this section establishes that the sum of a dictatorship of the present plus a dictatorship of the future is in fact neither. The first part is sensitive to the present, and the second is sensitive to the future. Furthermore such a sum admits trade-offs between the welfare of the present and of the future. It is represented diagrammatically in Figure 1, which shows the trade-offs between the present's and the future's utilities. The three axes represent the utility levels of generations 1, 2, and, figuratively, ∞. The two triangular planes represent two indifference surfaces. One gives more utility to generations 1 and 2, and under a dictatorship of the present these choices would prevail; however the second surface gives more utility to the long run, so that under certain conditions the second surface is chosen over the first. Theorem 1 makes this reasoning rigorous.

The second part of Theorem 1 shows that all known criteria of optimality used until now fail to satisfy the axioms postulated here. Therefore the sustainable preferences defined here perform a role satisfied by no previously used criterion.

What is perhaps more surprising is that the sustainable welfare criteria constructed here, namely the sum of a dictatorship of the present and one of the future,

---

19 Interestingly, R. Solow has suggested making the discounting factor smaller than one and decreasing: it is possible to show that *under certain conditions* this is exactly what must be done to ensure the existence of solutions.

exhaust all the continuous utilities which satisfy my two axioms. This means that any continuous sustainable preference must be of the form just indicated. This is Theorem 2, proved in the Appendix.

3.1 The Existence of Sustainable Preferences

Theorem 1 *There exists a sustainable preference $W : \ell_\infty \rightarrow R$, i.e. a preference which is sensitive and does not assign a dictatorial role to either the present or the future:*

$$W(\alpha) = \sum_{g=1}^{\infty} \lambda_g \alpha_g + \phi(\alpha), \tag{5}$$

where $\forall g$, $\lambda_g > 0$, $\sum_{g=1}^{\infty} \lambda_g < \infty$, and where $\phi(\alpha)$ is the function $lim_{g\rightarrow\infty}(\alpha_g)$ extended to all of ℓ_∞ via Hahn-Banach theorem.[20]
The following welfare criteria are not sustainable preferences: (a) the sum of discounted utilities, for any discount factor, (b) Ramsey's criterion, (c) the overtaking criterion, (d) lim inf, (e) long run averages, (f) Rawlsian rules, and (g) basic needs.

Proof. In the Appendix.

An intuitive explanation of this result follows. The preference defined in (5) is sustainable because it is complete, its first term is sensitive to the present, in fact it increases with increases in the welfare of every generation, and its second term is sensitive to the long run future. The next step is to show why all other criteria fail the sustainability axioms proposed here.

(a) The sum of discounted utilities is a dictatorship of the present because for every $\varepsilon > 0$, there exist a generation N so that the sum of discounted utilities of all other generations beyond N is lower than ε for all utility streams since all utilities are bounded by the number 1. Now, given any two utility streams α, β, if $W(\alpha) > W(\beta)$ then $W(\alpha) > W(\beta) + \varepsilon$ for some $\varepsilon > 0$; therefore there exists a generation N beyond which the utilities achieved by any generation beyond N do not count in the criterion W. This is true for any discount factor.

(b) The Appendix establishes that the Ramsey's criterion is incomplete; this derives from the fact that the distance to Ramsey's bliss path is ill-defined for many paths.

(c) The Appendix establishes that the overtaking criterion is incomplete: see also Figure 2.[21]

---

20 The linear map "$\lim_{g\rightarrow\infty}(\alpha_g)$" is defined by using the Hahn-Banach theorem, as follows: define first the function on the closed subset of ℓ_∞ consisting of those sequences α_g which have a limit, as that limit; the function is then extended continuously to all sequences in the space ℓ_∞ by using the Hahn-Banach theorem, which ensures that such an extension exists and can be constructed while preserving the norm of the function on the closed subspace of convergent subsequences.

21 Other interesting incomplete intergenerational criteria which have otherwise points in common with sustainable preferences are found in Asheim [4], [5].

(d) and (e) Lim inf and long run averages are dictatorships of the future; furthermore the long run average is also incomplete.[22]

Both (f) and (g), Rawlsian and basic needs criteria, are insensitive because they rank equally any two paths which have the same infimum even if one assigns much higher utility to many other generations.

3.2 A Complete Classification of Sustainable Preferences

The following result characterizes sustainable preferences. Additional conditions on the welfare criterion W are now introduced: W is *continuous* when it is defined by a continuous function $W : \ell_\infty \to R$.[23] Continuity has played a useful role in social choice theory in the last ten years, in effect replacing the axiom of independence of irrelevant alternatives and allowing a complete characterization of domains in which social choice exists, Chichilnisky [12] , [13] and Chichilnisky and Heal [20] . A similar role is found here for continuity: the following theorem gives a full characterization to all sustainable criteria which are continuous.

A standard property of neoclassical analysis is that the rate of substitution between two generations—which is generally dependent on their level of consumption—is independent of their levels of utility. This is a widely used property: indeed the sum of discounted utilities, the most widely used welfare criterion, certainly satisfies it. A welfare criterion W satisfying this property is called *independent.*[24] The characterization of a sustainable criterion W in the following theorem is simplest when W is continuous and independent. In consistency with neoclassical analysis, we therefore assume independent welfare criteria.

The following theorem decomposes a sustainable criterion into the sum of two functions. The first is a discounted utility with a variable discount rate and the second is a generalization of long run averages or a lim inf, and is called a "purely finitely additive" measure. For definitions and examples of purely finitively additive measures the reader is referred to the Appendix, Example 4: this second function assigns all welfare weight to the very long run. In particular, it assigns the value zero to any sequence with has finitely many non zero terms.[25]

Theorem 2 *Let $W : \ell_\infty \to R^+$ be a continuous independent sustainable prefer-*

---

22 For example the two sequences (1,0,0,1,1,1,0,0,0,0,1,1,1,1,1,0,0,0,0,0,0,1,...) and (0,1,1,0,0,0,1,1,1,1, 0,0,0,0,0,1,1,1,1,1,1,0,...) are not comparable according to the long-run averages criterion.

23 A function W which is continuous with respect to the standard norm of the space of sequences ℓ_∞.The norm is $\|\alpha\| = \sup_{g=1,2...} \mid \alpha_g \mid$, and was defined above. Different forms give rise to different notions of continuity but in the context of equitable treatement of generations the sup norm is a natural candidate.

24 See the Appendix. This simply means that the indifference surfaces of the welfare criterion W are hypersurfaces so that it is possible to represent it by a linear function on utility streams: $W(\alpha + \beta) = W(\alpha) + W(\beta)$. Note that this does not restrict the utilities of the generations, u_g, in any way; in particular the $u_g' s$ need not be linear and so the marginal rate of substitution between consumption at different dates is non-constant and depends *inter alia* on consumption levels..

25 A finitely additive measure on the integers Z is a function μ defined on subsets of the integers, satisfying $\mu(A \cup B) = \mu(A) + \mu(B)$ when $A \cap B = \emptyset$; μ if called purely finitely additive when it assigns measure zero to any finite subset of integers. See also the Appendix for definitions and examples.

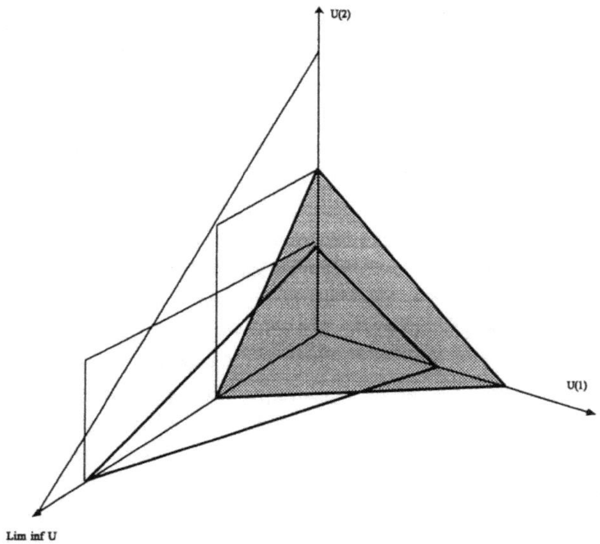

Figure 1: The 3 axes represent the utilities of finite generations (U(1) and U(2)) and the limiting utility value Lim inf U. Two level sets of the ranking are shown. One dominates over finite generations but has a lower lim inf. As the weights on finite generations are fixed, the ranking in these dimensions can be represented by the intersection of the level set restricted to the U(1)-U(2) plane with the vertical axis. The overall ranking is then shown as the sum of this ranking (the countably additive measure) with the ranking in the lim inf dimension (the purely finitely additive measure).

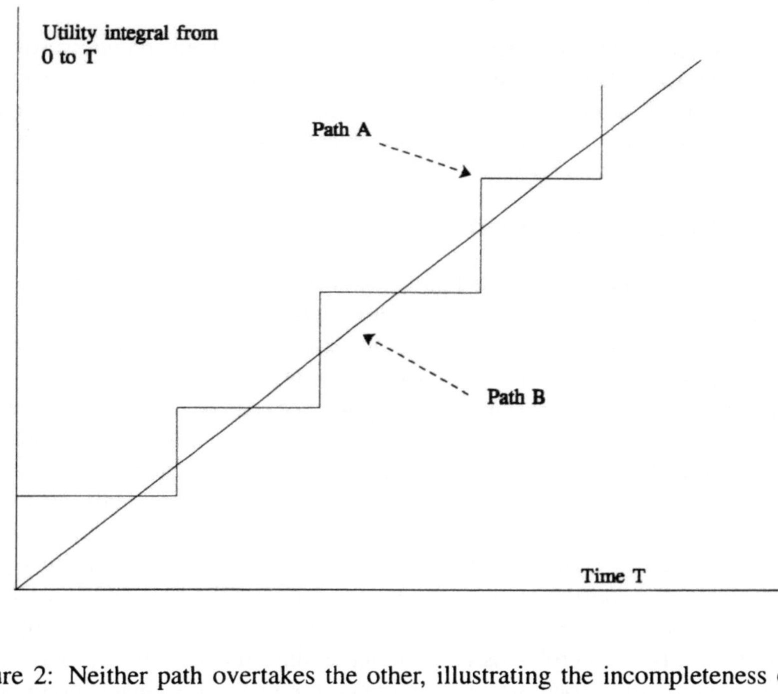

Figure 2: Neither path overtakes the other, illustrating the incompleteness of the overtaking criterion.

ence. Then W is of the form $\forall \alpha \in \ell_\infty$:

$$W(\alpha) = \sum_{g=1}^{\infty} \lambda_g \alpha_g + \phi(\alpha) \tag{6}$$

where $\forall g\ \lambda_g > 0$, $\sum_{g=1}^{\infty} \lambda_g < \infty$, *and* ϕ *is a purely finitely additive measure.*

Proof. In the Appendix.

4 Are Sustainable Preferences Reasonable?

We saw that sustainable preferences emerge from well-defined and uncontroversial axioms. But how do they fit economic intuition and empirical evidence? The following subsection will show that sustainable preferences fit well our economic intuition about finite horizon optimization. The next subsection will show that they also fit the empirical evidence rather well.

4.1 A Turnpike Theorem

Our economic intuition is grounded on finite horizons. Life on earth will certainly be of finite duration, although it is difficult to determine its final date. It is therefore important to determine whether sustainable preferences are merely an artifact of infinite horizons, or are reasonable within a finite world.

This section will show that sustainable preferences can be seen as a suitable generalization to infinite horizons of an intuitively appealing criterion for finite horizons, one which values all generations equally. This we call the "equal treatment" criterion. Indeed, for a general class of dynamic optimization problems, we will see that the limit of the optimal solution according to a sustainable preference has two interesting properties: (i) it is the "green golden rule" (see Beltratti Chichilnisky and Heal [6]), i.e., the configuration of the economy giving the maximum sustainable utility level and (ii) as the finite horizon increases, the optimal solutions of equal treatment finite horizon problems spend an increasing amount of time progressively closer to the limit of the path which is optimal according to sustainable preferences. In other words, the optimum according to sustainable preferences determines a direction in which finite horizon equal treatment optima increasingly move as the horizon increases. We refer to this property as a "turnpike" property.

To see this we formalize a typical problem of optimizing a sustainable preference over a constraint imposed by the dynamics of a renewable stock. A renewable stock s_t grows over time t according to its own biological dynamics, with growth rate $r(s_t)$; c_t of it is extracted for consumption. The utility depends on consumption and the level of the stock, as in for example Krautkraemer [34] or Heal [31] . We are assuming the stock of the resource to be an argument of the utility function, so that the resource is in the category of environmental assets such as forests, landscapes, biodiversity, etc. which provide services and value to human society via their stocks as well as via a flow of consumption. The problem is therefore:

$$\left.\begin{array}{c}\max \alpha \int_0^\infty u\left(c_t, s_t\right) e^{-\delta(t)t} dt + (1-\alpha) \lim_{t \to \infty} u\left(c_t, s_t\right) \\ \text{s.t. } \overset{\bullet}{s}_t = r\left(s_t\right) - c_t,\ s_0 \text{ given.}\end{array}\right\} \tag{7}$$

To study the asymptotic properties of a maximum for this problem it will be useful to introduce the following definition, see also Beltratti, Chichilnisky and Heal [6] :

Definition 7 *The green golden rule g^* is a stationary path $g^* = \{c^*, s^*\}$ which achieves the maximum utility level which is sustainable forever, i.e.:*

$$g^* = Max_c\, u(c, s)$$

$$\textit{subject to } c \leq R(s).$$

Equivalently:

$$g^* = \max_s u(R(s), s)$$

so that it satisfies:

$$\frac{\partial u}{\partial c}\frac{\partial R}{\partial s} = -\frac{\partial u}{\partial s}, \text{ or } \frac{\partial R}{\partial s} = \frac{\frac{\partial u}{\partial s}}{\frac{\partial u}{\partial c}}. \tag{8}$$

Definition 8 *The equal treatment problem for horizon T is:*

$$\left.\begin{array}{c}\max \int_0^T u\left(c_t, s_t\right) dt \\ s.t.\ \overset{\bullet}{s}_t = r\left(s_t\right) - c_t,\ s_0 \textit{ given.}\end{array}\right\} \tag{9}$$

Its solution is called the equal treatment optimum over T generations.

Definition 9 *The discounted utilitarian problem has the same constraint set as problem (7) but the function to be maximized is, instead, the integral of utilities discounted by a constant positive discount rate:*

$$\int_0^\infty u(c, s)e^{-\delta t}dt, \textit{ for a fixed } \delta > 0. \tag{10}$$

Theorem 3 *The optimal path for problem (7) with a sustainable preference exists if and only if the discount rate $\delta(t)$ approaches zero in the limit, in which case the optimum is the "turnpike" of finite horizon problems (9) in which each generation is treated equally. This means that as the number of generations T increases, the equal* *treatment optima for T generations approach closer to a plan which is asymptotically approached by the optima of (7) with sustainable preferences. This plan is the green golden rule. Formally:*

(1) The green golden rule g^* has a "turnpike property" for equal treatment optima, i.e., for any $\varepsilon > 0, \exists T_\varepsilon$ such that for $T \geq T_\varepsilon$, the equal treatment optimum for T spends a positive period of time within ε of g^*.

(2) The optimal solution to problem (7) exists if and only if $\delta(t) \to 0$ as $t \to \infty$, in which case it is a path $(\widehat{c}_t, \widehat{s}_t)$ which converges asymptotically to the green golden rule g^*, and so to the turnpike of the equal treatment optima.

(3) By contrast, the discounted utilitarian optimum does not have the turnpike property for equal treatment optima, for any positive (fixed) discount rate δ: the optimal path for a positive discount rate δ is uniformly bounded away from g^* and so from the turnpike of the equal treatment optima. For a zero discount rate, the discounted utility problem has no solution.

Proof. See Heal [31] . ■

Figure 3 illustrates the results. The green golden rule is the pair of consumption and stock levels at which the marginal conditions for static optimality (8) are satisfied. The solutions for the equal treatment criterion for finite time horizons T are indicated: these solutions are found by noting that the stock at T should be zero, and then finding the appropriate initial conditions by solving the Euler-Lagrange equations. As $T \to \infty$, these paths comes closer and closer to the green golden rule. This is the turnpike property.

4.2 Empirical Evidence Matches Sustainable Preferences

Sustainable preferences help explain the empirical evidence on time preferences. Recent empirical evidence on time preferences clashes with standard discounted utility maximization but is consistent with the solutions obtained from optimizing sustainable preferences in the context of a renewable resource, in the sense that an optimum exists if and only if the discount rate fall over time.

There is a growing body of empirical evidence that suggests that the discount rate which people apply to future projects depends upon, and declines with, the futurity of the project. See for example Lowenstein and Thaler [39] or Cropper et. al. [23] . Over relatively short periods up to perhaps five years, they use discount rates which are higher even than commercial rates—in the region of 15% or more. For projects extending about ten years, the implied discount rates are closer to standard rates—perhaps 10%. As the horizon extends the implied discount rates drops, to in the region of 5% for thirty to fifty years and down to of the order of 2% for one hundred years. The evidence for these statements is still tentative, and more research is needed to document fully how people trade off different degrees of futurity against

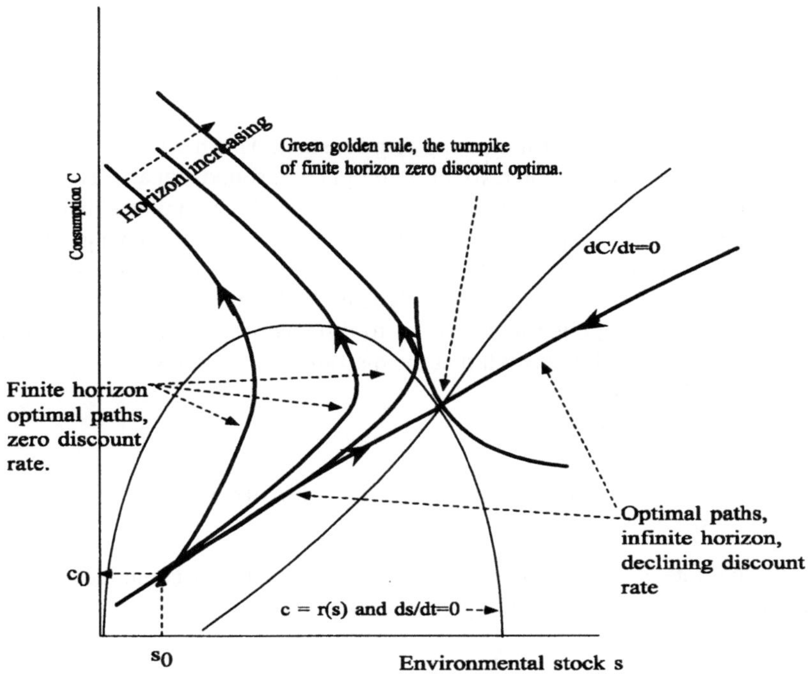

Figure 3: The utilitarian solution with a renewable resource, stock an argument of the utility function and discount rate falling to zero, asymptotes to the green golden rule. This is also the turnpike of finite horizon equal treatment (zero discount) optima.

the present. However, it is of great interest that a framework for intertemporal optimization that is sensitive to both present and future generates an implication for discounting that may rationalize a form of personal behavior that hitherto has been found irrational.

This empirically-identified behavior is consistent with results from natural sciences which find that human responses to a change in a stimulus are non-linear, and are inversely proportional to the existing level of the stimulus. For example, the human response to a change in the intensity of a sound is inversely proportional to the initial sound level: the louder the sound initially, the less we respond to a given increase. This is an example of the Weber-Fechner law, which is formalized in the statement that human response to a change in a stimulus is inversely proportional to the pre-existing stimulus. In symbols,

$$\frac{dr}{ds} = \frac{K}{s} \text{ or } r = K \log s$$

where r is a response, s a stimulus and K a constant. This has been found to apply to human responses to the intensity of both light and sound signals. The empirical results on discounting cited above suggest that something similar is happening in human responses to changes in the futurity of an event: a given change in futurity (e.g., postponement by one year) leads to a smaller response in terms of the decrease in weighting, the further the event already is in the future. In this case, the Weber-Fechner law can be applied to responses to distance in time, as well as to sound and light intensity, with the result that the discount rate is inversely proportional to distance into the future.

In our economic problem

$$Max \int_0^\infty u(c,s)\Delta(t)dt + \lim u(c,s),$$

the discount rate is $q\,(t) = \dot{\Delta}\,(t)\,/\Delta\,(t)$, where $\Delta\,(t)$ is the discount factor. One can formalize the Weber-Fechner interpretation as follows:

$$q\,(t) = \frac{1}{\Delta}\frac{d\Delta}{dt} = \frac{K}{t} \text{ or } \Delta\,(t) = e^{K \log t} = t^K$$

where K is a negative constant.

Such a discount factor can meet all of the conditions required for the existence of sustainable optima: the discount rate q goes to zero in the limit, the discount factor $\Delta\,(t)$ goes to zero and the integral $\int_1^\infty \Delta\,(t)\,dt = \int_1^\infty e^{K \log t}dt = \int_1^\infty t^K\,dt$ converges for K negative, as it always is.

A discount factor $\Delta\,(t) = e^{K \log t}$ has an interesting interpretation: the replacement of t by $\log t$ implies that we are measuring time differently: by equal proportional increments rather than by equal absolute increments. This is quite consistent with the approach taken in for example acoustics, where in response to the Weber-Fechner law sound intensity is measured in decibels which respond to the logarithm of the energy content of the sound waves. In general, non-constant discount rates can be interpreted as a non-linear transformation of the time axis.

5 Differences between Standard and Sustainable Optima

5.1 Renewable Resources

The next step in exploring the properties of sustainable preferences, is to study the difference between sustainable and discounted utilitarian optima. We do this first for the case of optimization with renewable resources, comparing the optimal path for sustainable preferences, which solves (7) above, with that which is optimal for the integral of discounted utilities, namely (10) above. In fact most of the elements needed for this comparison are in place. Figure 4 shows the optimal path for sustainable preferences, which asymptotes to the green golden rule, the point of tangency between an indifference curve of the utility function $u(c, s)$ and the growth function $r(s)$.

A stationary solution to the utilitarian case is characterized by the following conditions:

$$\left.\begin{array}{r} r(s_t) = c_t \\ \frac{u_2'(s_t)}{u_1'(c_t)} = \delta - r'(s_t) \end{array}\right\} \quad (11)$$

The first equation in (11) just tells us that a stationary solution must lie on the curve on which consumption of the resource equals its renewal rate: this is obviously a prerequisite for a stationary stock. The second gives us a relationship between the slope of an indifference curve in the $c - s$ plane and the slope of the renewal function at a stationary solution: the indifference curve cuts the renewal function from above. Such a configuration is shown in figure 5.

This involves a lower long-run stock and a higher long-run level of consumption than the sustainable optimum, shown in figure 4 and in this sense is less conservative: it also involves a lower long-run utility level. In the utilitarian solution for a constant discount rate, as the discount rate falls to zero, the stationary solution moves to the green golden rule: however, for a constant discount rate of zero, the utilitarian problem has no solution. Note that in both cases, because of the presence of the stock of the resource as an argument of the utility function, the stationary stock exceeds that giving the maximum sustainable yield. This is an obvious consequence of the valuation of the stock in its own right.

5.2 Exhaustible Resources

Consider now the rather simpler case of exhaustible resources, in essence an extension of the familiar Hotelling case. For sustainable preference we shall consider the

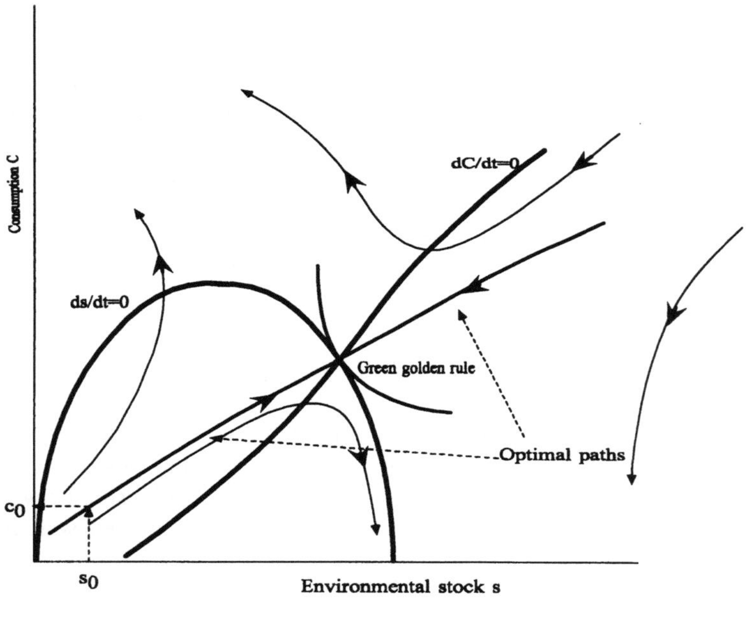

Figure 4: The green golden rule is the stationary solution of the autonomous system (7), and is a saddle point.

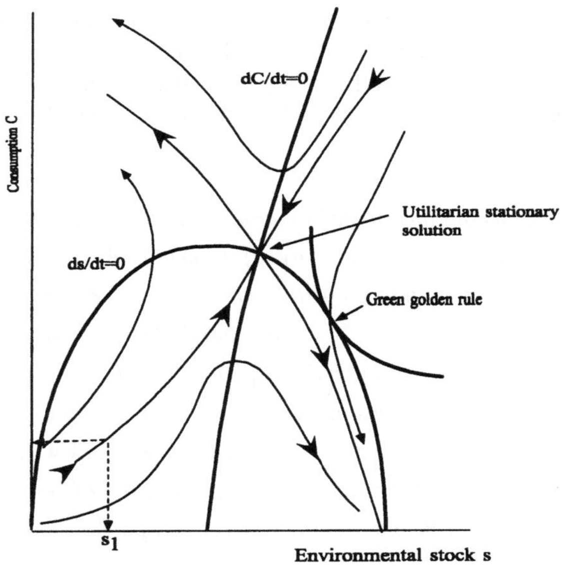

Figure 5: Dynamics of the utilitarian solution. The green golden rule - the highest sustainable utility level - is the point of tangency between an indifference curve and the growth curve.

problem

$$\left.\begin{array}{c}\max \alpha \int_0^{\infty} u\left(c_t, s_t\right) e^{-\delta t} dt+(1-\alpha) \lim_{t \rightarrow \infty} u\left(c_t, s_t\right), \ \alpha>0, \\ \text{subject to } \stackrel{\bullet}{s}_t=-c_t, \ s_t \geq 0 \forall t.\end{array}\right\} \tag{12}$$

and we shall contrast this with the equivalent discounted utilitarian formulation

$$\left.\begin{array}{c}\max \int_0^{\infty} u\left(c_t, s_t\right) e^{-\delta t} dt \\ \text{subject to } \stackrel{\bullet}{s}_t=-c_t, \ s_t \geq 0 \forall t.\end{array}\right\} \tag{13}$$

In both cases we are retaining the stock of the resource as an argument of the utility function, assuming that the resource is in the category of environmental assets such as forests, landscapes, biodiversity, etc. which provide services and value to human society via their stocks as well as via a flow of consumption. The solution to the utilitarian case is summarized in the following theorem, which is illustrated in figures 6 and 7:

Theorem 4 *Consider an optimal solution to problem (12) when the utility function is additively separable, $u(c, s) = u_1(c) + u_2(s)$. A sufficient condition for this to involve the preservation of a positive stock for ever is that the marginal utility of consumption at zero is finite, $u_1'(0) < \infty$, and that there exists a finite stock level s^*, the optimal stationary stock, such that $u_1'(0)\,\delta = u_2'(s^*)$. In this case, if the initial stock $s_0 > s^*$, then total consumption over time will equal $s_0 - s^*$: if $s_0 \leq s*$, then consumption will always be zero and the entire stock will be conserved on an optimal path. If on the other hand the marginal utility of consumption at $c = 0$ is infinite, then it will not be optimal to conserve any positive stock level indefinitely.*

The determination of the stock s^* which is conserved for ever is illustrated in figure 6. Figure 7 shows in the space of consumption and the remaining stock the phase diagram for the differential equations which define the conditions necessary for optimality in the utilitarian case (13). The initial stock s_0 is given: the optimal initial consumption is that level which leads asymptotically to $s = s^*$ and $c = 0$ along a path satisfying the necessary conditions for optimality.

When do sustainable preferences lead to more conservation, to a larger long-run stock, than the discounted utilitarian framework with the same discount rate? This will occur whenever the weight $(1 - \alpha)$ on the limiting value of utility in the maximand is strictly positive. Figure 8 compares the optimal paths of s in this problem for the alternative optimality criteria that we have examined. For the case of sustainable preferences, one can show that the same differential equations as in the utilitarian case are necessary for optimality, but that the initial and final conditions will be different from those in the utilitarian case. In particular, the final stock will be higher, the initial consumption level lower, and the initial shadow price higher (see Heal [31]).

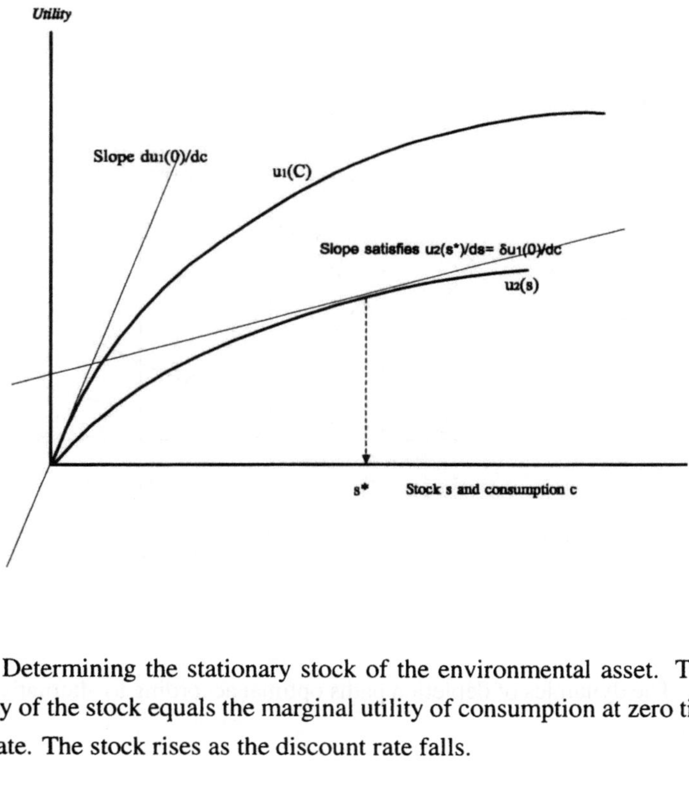

Figure 6: Determining the stationary stock of the environmental asset. The marginal utility of the stock equals the marginal utility of consumption at zero times the discount rate. The stock rises as the discount rate falls.

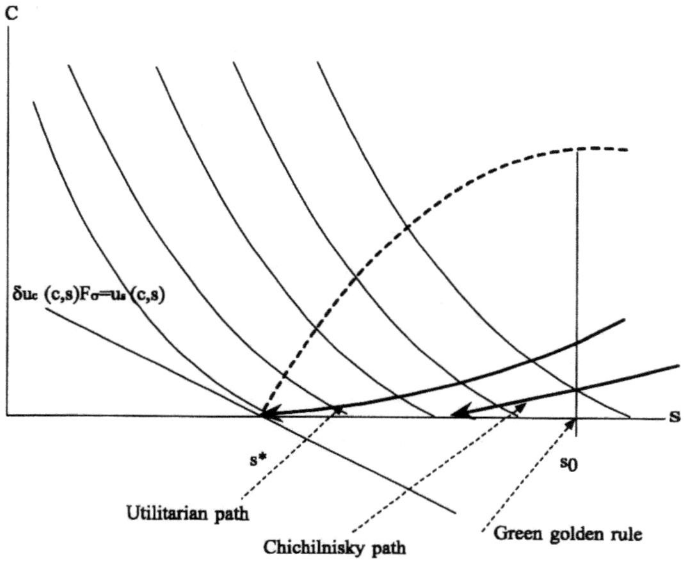

Figure 7: The dynamics of depletion paths optimal according to alternative optimality criteria.

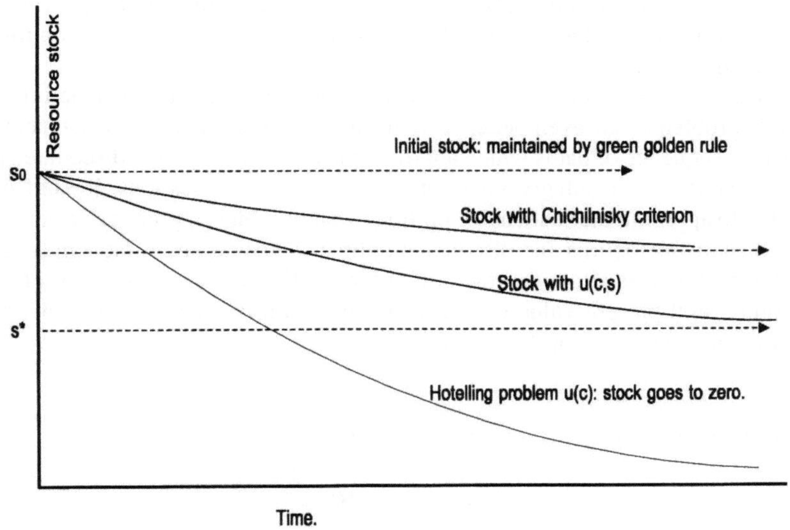

Figure 8: The time paths the environmental stock under alternative optimality concepts.

6 Sustainable Optima can be Far from Discounted Optima

I already showed that sustainable preferences are substantially different from other welfare criteria which have been used in the literature. It remains however to study how different they are in practice, for example whether the optimal solutions of problems which maximize sustainable preferences are substantially different from the optimal solutions to discounted problems.

To answer this question I shall compare problems which are defined over the same constraint set, but each of which maximizes different welfare criteria. The purpose is to explore what difference this makes in practice. An optimal problem which maximizes a sustainable preference will be called a *sustainable problem.* If the welfare criterion is a discounted sum of utilities as defined in (19), I call this a *discounted problem.* The corresponding solutions are called sustainable optima and discounted optima.[26]

Can one always approximate a sustainable optimum by paths which optimize discounted problems? Or even better: can one always approximate a sustainable optimum by a sequence of paths which approximates the solutions of a discounted problem? The following result gives a negative answer to these questions. It is not always possible to approximate sustainable optima by paths which approach discounted optima. Sustainable optima and discounted optima can be far apart. Nonetheless, sustainable preferences allow for trade-offs between the welfare of the present and of the future, and present value of profit maximization is a good approximation for maximizing the welfare of the present.[27]

Theorem 5 *Consider a sustainable problem:*

$$Max_{\{\alpha:\alpha\in\Omega\}} W(\alpha_g), \tag{14}$$

where $W(\alpha) = \sum_{g=1}^{\infty} \lambda_g \alpha_g + \Phi(\alpha)$ *is a sustainable preference and* $\Omega \subset \ell_\infty$ *is the set of all feasible utility streams. Let* α^* *be a unique sustainable optimum of (14), and* $W^* = W(\alpha^*)$. *Assume also that there exist a unique discounted optimum* β^*, *for*

[26] Formally: let F be a convex and closed subset of feasible paths in a linear space X Such as for example $X = \ell_\infty$, or $X = R^N$. A vector $\beta \in F$ is called optimal in F if it maximizes the value of a function $U : X \to R$. The vector β is called a **discounted optimum** if $U : \ell_\infty \to R$ is a discounted sum of utilities as defined in (19); β is called a **sustainable optimum** if $U : X \to R$ is a sustainable preference.

[27] This result has also implications for the support and decentralization of sustainable optima (Corollary 1 and Example 4). It may not be possible to approximate sustainable optima by paths which approximately maximize the present value of profits under any standard sequence of prices. However, level independent sustainable preferences can be used in certain cases to define shadow prices at which the optimal paths maximize value. These are "generalized" prices, but they can be given a precise mathematical formulation and a precise economic interpretation. They measure the value both of the short and of the long run, and can be called "sustainable prices".

the problem of maximizing over the same set Ω the discounted utility $\sum_{g=1}^{\infty} \lambda_g \alpha_g$.[28] *Then in general the sustainable optima α^* cannot be approximated by a sequence of feasible utility streams $\{\beta^n\}_{n=1,2...}$ which approximates the discounted optimum β^*. This is true for any sequence of discount factors $\{\lambda_g\}$.*

Proof. In the Appendix.

The intuition for this result is in the following example:

Example 1 *Consider an economy which uses trees as a necessary input to production or consumption; without this input the economy's utility is zero. The dynamics of tree reproduction requires that unless the first N periods the economy refrains from cutting more than a certain number of trees, the species becomes extinct after $K + N$ periods, in which case there is zero utility at every period from there on. The economy's feasible set of utility streams Ω is described then as follows: a minimum investment denoted $\epsilon > 0$ is required during each of the first N periods to ensure that the utility levels in all periods from the $(N + K) - th$ on is above zero. Once this threshold is reached, then all utility levels in each period after the $(N + K) - th$ exceed ϵ. Then for every discount factor, there is an N, K for which the sum of discounted utilities is maximized at a path which leads to the eventual elimination of the forest. Instead, for a sustainable preference which gives sufficient weight to the long run the optimum will keep the forest alive and yielding a minimum utility level ϵ forever. Therefore the two optima are apart by at least ϵ; any sequence of paths which approaches the discounted optimum will not approach the sustainable solution.*

7 Sustainability and Value

The following corollary and example show that the notion of value derived from sustainable preferences is rather distinctive. Paths which are optimal under sustainable preferences may not maximize present discounted value according to any standard price system. Therefore, environmental resources with a large value in the long run, may not appear valuable under a standard notion of present value profit maximization.

The following corollary explores the connection between sustainable optima and the maximization of present value.[29] A **standard price** p is a sequence of prices,

[28] This result aims at showing that the result of optimizong a sustainable preference may be quite different from that of optimizing its "discounted" part. This is why we compare the optimum of W with that of $\sum \lambda_g \alpha_g$. One could allow the case where the discount factor varies over all possible discount factors; this latter case is considered in Example 4 below.

[29] Here I consider a special case where the utilities u_g are linear; the problem can then be formulated readily without introducing any further notation. The general case can be analyzed along similar lines,

$p = (p_1, ..., p_g, ...)$ which assigns a well-defined present value to every stream, $p(\gamma) = \sum_{g=1}^{\infty} p_g \gamma_g$, for all $\gamma \in \ell_\infty$.[30] There exist continuous linear functions on ℓ_∞ which cannot be represented in this manner. An example is the long run average of a sequence α: this assigns a present value zero to every sequence with finitely many terms. If the long run average was representable as a sequence, in the limit, it would assign zero value to every sequence, which is obviously untrue.[31] The Corollary below and the following Example 4 construct two specific economic examples.

Corollary 6 *There exists a problem*[32]

$$Max_{\alpha \in \Omega} (W(\alpha)) \tag{15}$$

with the property that its sustainable optimum does not maximize present value within Ω at any standard price $p = (p_1, p_2, ...)$. Therefore for any standard price system p there are suboptimal paths which have strictly large present value than the sustainable optimum α^*, i.e. for all p, $\exists \beta^* \in \Omega$:

$$\sum_{g=1}^{\infty} p_g \alpha_g^* < \left(Sup_{\beta \in \Omega} \sum_{g=1}^{\infty} p_g . \beta_g \right).$$

This result is true for any discount factor.

Proof. This follows readily from Theorems 1 and 2. Without loss of generality, consider the case where the utility functions u_g are the identity map. In this case, the set of feasible utility streams Ω coincides with the set of feasible consumption streams. Furthermore, since the welfare function W defines an independent sustainable preference, the first term of W defines a sequence of standard prices with the desirable properties, namely $p = \{\lambda_g\}_{g=1,2...}$ well-defines the present value $\sum_g \lambda_g \alpha_g$ for each sequence of ℓ_∞.[33] Theorem 3 showed that in general the value maximizing sequence β^* is different from α^*. In particular, this implies that the sustainable optimum α^* does not generally maximize value for the standard price system p. Since the results of Theorem 3 are true for any sequence of "discount factors" $\{\lambda_g\}_{g=1,2...}$ satisfying $\forall g$, $\lambda_g > 0$ and $\sum_{g=1}^{\infty} \lambda_g < \infty$, the corollary follows. ■

at the cost of more notation. General formulations of the problem of optima and intertemporal profit maximization can be found in the literature (Debreu [24]); a simple formulation in infinite dimensional spaces that fits well our purposes is in Chichilnisky and Kalman [11].

30 In some cases, the present value coincides with intertemporal utility maximization, see Chichilnisky and Kalman [11]. Note that if p is a standard price system, represented by the sequence $(p_1, p_2, \dots)$, then by definition this sequence satisfies $\sum_{g=1}^{\infty} p_g < \infty$.

31 In fact, all purely finitely additive measures on Z define real valued functions on sequences which cannot be represented by sequences.

32 $\forall g, u_g \neq 0$.

33 It can be shown that the value $p(\alpha) = \sum_g \lambda_g \alpha_g$ can be interpreted as the intertemporal profit of the "plan" α.

7.1 A Sustainable Optimum Which Does Not Maximize Expected Value At Any Standard Price System

The results of Corollary 1 can be strengthened further by means of another example. Consider a feasible path $\beta \in \ell_\infty$ which maximizes a continuous concave utility function U within a convex set $F \subset \ell_\infty$, but such that at no standard price system p does β maximize present value.[34] For $c \in [0, \infty)$ let

$$u_t(c) = 2^t c \text{ for } c \leq 1/2^{2t}, \text{ and } u_t(c) = 1/2^t \text{ for } c > 1/2^t.$$

Now, for any sequence $c \in \ell_\infty^+$ let $U(c) = \sum_{t=1}^{\infty} u_t(c_t)$, which is well defined, continuous, concave, and increasing on ℓ_∞^+. Let $\beta \in \ell_\infty^+$ be defined by

$$\beta_t = 1/2^{2t+1}$$

and let[35]

$$F = U^\beta = \{\gamma \in \ell_\infty : U(\gamma) \geq U(\beta)\};$$

F is a closed convex subset of ℓ_∞. Now assume that $p = \{p_t\}_{t=1,2\ldots}$ is a standard supporting price system for the set U^β, $p_t \geq 0$, i.e. $p.\gamma \geq p.\beta \; \forall \gamma \in U^\beta$. By the usual marginal rate of substitution arguments,

$$p_t = p_1 2^{t-1}. \tag{16}$$

I shall show that p_1 must be zero, so that the whole sequence $\{p_t\}_{t=1,2\ldots}$ must be zero. Assume to the contrary that $p_1 \neq 0$. Define $z \in \ell_\infty^+$ by

$$z_t = 1/p_t$$

and $z^n \in \ell_\infty^+$ by

$$z_t^n = \alpha_t^* \text{ if } t \leq n, \text{ and } 0 \text{ otherwise.}$$

Then $\forall n$, $z \geq z^n$ so that

$$p(z) \geq p(z^n), \tag{17}$$

but

$$\sum_{t=1}^{\infty} p_t z_t^n = n > p(z) \text{ for some } n \text{ sufficiently large, contradicting (17).}$$

The contradiction arises from the assumption that p_1 is not zero. Therefore $p_1 = 0$ and by (16) the entire price sequence $p = \{p_t\}_{t=1,2\ldots}$ is identically zero. It is

[34] This is from Example 1 in Chichilnisky and Heal [20] p. 369, which is reproduced here for the reader's convenience. This example deals with the minimization rather than the optimization of a function over a set, but the results are of course equivalent. The example constructs a feasible set $F \subset \ell_\infty$ which is non-empty, closed and concave, and a continuous concave function $U : \ell_\infty^+ \to R$ which attains an infimum $U(\beta)$ at β in F, such that the only sequence of prices $p = \{p_n\}_{n=1,2\ldots}$ which can support β in F (i.e. such that $p(\eta) \geq p(\beta) \; \forall \eta \in F$) is identically zero.

[35] We call this set U^β in sympathy with the notation of Chichilnisky and Heal [20] .

therefore not possible to support the concave set U^β with a non-zero standard price system.[36] ■

8 Conclusions

I have defined a set of axioms which capture the idea of sustainability, and characterized the sustainable preferences that they imply. I also analyzed other criteria used in the literature, and found that they do not satisfy my axioms. Discounted utility fails to satisfy the non-dictatorship of the present. This agrees with the viewpoint of many practitioners, who have pointed out the inadequacy of discounted utility for analyzing sustainable growth.[37] Rawlsian and basic needs criteria are insensitive, since they only regard the welfare of the generation which is least well-off. The *overtaking* criterion and its relative the *catching up* criterion are incomplete as orders and cannot be represented by real valued functions. They fail to compare many reasonable alternatives. This decreases their value as aids for decision making. Ramsey's criterion has a similar drawback: it is defined as the integral of the distance to a "bliss" utility level, but this integral can be ill-defined.[38]

The sustainable preferences proposed here and characterized above circumvent all of these problems. From the practical point of view, they satisfy two desirable criteria: they fit our intuition of finite horizon problems, because in important examples they have a turnpike property with respect to equal treatment finite horizon problems. In addition, they fit rather well empirical observations that indicate that people's perceptions of the future imply lower discount rates as time progresses. Important classes of dynamic problems have a solution according to sustainable preferences only if the implies discount rates are decreasing through time.

I also showed that sustainable preferences give rise to optimal solutions which are different from those obtained by discounted optimization criteria. A path which is optimal under a sustainable preference may not be approximated by paths which approximate discounted optima.

The notion of value derived from sustainable preferences is distinctive. Paths which are optimal under sustainable preferences may not maximize value according to any standard price system. Therefore, environmental resources with a large value in the long run, may not appear valuable under a standard notion of profit maximization.

These results may help to disentangle the apparent contradictions in values which were discussed in the beginning of this paper. We noted that governments and international organizations appear seriously concerned about global environmental prob-

36 Further examples of phenomena related to the results in Theorem 2 and Corollary 1 can be found in Dutta [26]

37 E.g. Dasgupta and Heal [25] , Broome [7] , Cline [21] .

38 Hammond (see [28]) has defined agreeable paths as those which are approximately optimal for any sufficiently long horizon, in the sense that the welfare losses inflicted by considering only finite horizons go to zero as the length of the finite horizon goes to infinity. The criterion is not designed as a complete order but rather as a way of identifying acceptable paths. A similar issue arises with the overtaking criterion, which is ill-defined in many cases.

lems which lie so far into the future that with current discounted utility measures do no lead to substantial economic loss. The axioms for sustainable preferences proposed here may help resolve this contradiction. Discounted profit maximization and sustainability lead to different value systems. Some trade-offs are possible, but the two values are not the same. The empirical evidence we have today is more in favor of sustainable preferences than discounted utility.

Solow has proposed that sustainability should allow intergenerational trade-offs, but no generation should be favored over any other. This standard is met by sustainable preferences when applied to the "present" and to "future" generations. The long run does matter and so does the short run. Indeed, independent sustainable preferences can define *shadow prices* for sustainable optima, which can be used for project evaluation and for the characterization of optimal solutions. Several of the aims of this paper have therefore been reached, and several of the questions that we posed have been answered. But perhaps the results open up at least as many new questions.

It remains to understand the concern for the long run future which is observed in practice, and which appears formalized in the axioms proposed here and their implied preferences. Nobody alive today, not even their heirs, has a stake on the welfare of 50 generations into the future. Yet many humans care about the long run future of the planet, and the results of this paper indicate that axioms which formalize this concern are not altogether unacceptable. One may then ask: whose welfare do sustainable preferences represent?

Perhaps an answer for this riddle may be found in a wider understanding of humankind as an organism who seeks its overall welfare over time. Such proposals have been advanced in the concepts of a "selfish gene", or, more practically, in Eastern religions such as Buddhism which view the unity of humankind as a natural phenomenon. If such unity existed, humankind would make up an unusual organism, one whose parts are widely distributed in space and time and who is lacking a nervous system on which the consciousness of its existence can be based. Perhaps the advances in information technology described at the beginning of this article, with their global communications and processing reach, are a glimmer of the emergence of a nervous system from which a global consciousness for humankind could emerge.

9 Appendix

9.1 Continuity

In practical terms the continuity of W is the requirement that there should exist a sufficient statistic for inferring the welfare criterion from actual data. This is an expression of the condition that it should be possible to approximate as closely as desired the welfare criterion W by sampling over large enough finite samples of utility streams. Continuity of a sustainable criterion function $W : \ell_\infty \rightarrow R$ is not needed in Theorem 1; it is used solely for the characterization in Theorem 2. Continuity is defined in terms of the standard topology of ℓ_∞ : the norm defined by $\|\alpha\| = \sup_{g=1,2...} \mid \alpha(g) \mid$.

9.2 Independence

The welfare criterion $W : \ell_\infty \to R$ will be said to give independent trade-offs between generations, and called **independent**, when the marginal rate of substitution between the utilities of two generations g_1 and g_2 depends only on the identities of the generations, i.e. on the numbers g_1 and g_2, and not on the utility levels of the two generations. Independence of the welfare criterion is not needed in Theorem 1. It is used solely in the characterization of Theorem 2, to allows us to obtain a simple representation of all sustainable preferences. Formally: let ℓ_∞^* be the space of all continuous real valued linear functions on ℓ_∞.

Definition 10 *The welfare criterion $W : \ell_\infty \to R$ is independent if $\forall \alpha, \beta \in \ell_\infty$,*

$$W(\alpha) = W(\beta) \Leftrightarrow$$
$$\exists \lambda \in \ell_\infty^*,\ \lambda = \lambda(W),\ \textit{such that}\ \lambda(\alpha) = \lambda(\beta)$$

This property has a simple geometric interpretation, which is perhaps easier to visualize in finite dimensions. For example: consider an economy with n goods and 2 periods. Let $\alpha = (\alpha_1, \alpha_2), \beta = (\beta_1, \beta_2) \in R^2$ denote two feasible utility streams. Then α and β are equivalent according to the welfare criterion $W : R^2 \to R$, i.e. $W(\alpha) = W(\beta)$, if and only if there exists a number $\mu = \mu(W)$, $\mu > 0$, such that

$$\frac{\alpha_2 - \beta_2}{\alpha_1 - \beta_1} = \mu \tag{18}$$

The geometric interpretation of (18) is that the indifference surfaces of W are affine linear subspaces of R^2. Level independence implies that the indifference surfaces of the welfare function W are affine hyperplanes in ℓ_∞. In particular, W can be represented by a linear function on utility streams, i.e. $W(\alpha + \beta) = W(\alpha) + W(\beta)$. Examples of welfare criteria which satisfy this axiom are all time-separable discounted utility functions, any linear real valued non-negative function on ℓ_∞, and the welfare criteria in Theorem 2. As already mentioned, this axiom is used to provide a tight representation of sustainable preferences, but is not strictly necessary for the main results.

Definition 11 *A continuous independent sustainable preference is a complete, sensitive preference satisfying Axioms 1and 2 and which is continuous and independent.*

9.3 Definition of Previous Welfare Criteria

To facilitate comparison, this section defines some of the more widely used welfare criteria.

A function $W : \ell_\infty \to R$ is called a **discounted sum of utilities** if it is of the form:

$$W(\alpha) = \sum_{g=1}^{\infty} \lambda_g \alpha_g, \ \forall \alpha \in \ell_\infty, \tag{19}$$

where $\forall g, \lambda_g \geq 0$ and $\sum_{g=1}^{\infty} \lambda_g < \infty$; λ is called the discount factor. **Ramsey's welfare criterion** [42] ranks a utility stream $\alpha = \{\alpha_g\}_{g=1,2...} \in \ell_\infty$ above another $\beta = \{\beta_g\}_{g=1,2...} \in \ell_\infty$ if the utility stream α is "closer" to the bliss path, namely to the sequence $\varsigma = \{1, 1, ..., 1, ...\}$, than is the sequence β. Formally:

$$\sum_{g=1}^{\infty} (1 - \alpha_g) \leq \sum_{g=1}^{\infty} (1 - \beta_g). \tag{20}$$

A Rawlsian rule [43] ranks two utility streams according to which has a higher infimum value of utility for all generations.[39] This is a natural extension of the criterion proposed initially by Rawls [43] . Formally: a utility stream α is preferred to another β if

$$\inf\{\alpha_g\}_{g=1,2...} > \inf\{\beta_g\}_{g=1,2...} \tag{21}$$

The criterion of **satisfaction of basic needs** introduced in Chichilnisky [9] ranks a utility stream α over another β if the time required to meet basic needs is shorter in α than in β. Formally:

$$T(\alpha) \leq T(\beta), \tag{22}$$

where $T(\alpha) = \min\ \{t : \alpha_g \geq b\ \forall g \geq t\}$, for a given b which represents basic needs. The **overtaking criterion** [48] ranks a utility stream α over another β if α eventually leads to a permanently higher level of aggregate utility than does β. Formally: α is preferred to β if $\exists N$:

$$\forall M > N, \ \sum_{g=1}^{M} \alpha_g \geq \sum_{g=1}^{M} \beta_g. \tag{23}$$

The **long-run average** criterion can be defined in our context as follows: a utility stream α is preferred to another β if in average terms, the long run aggregate utility[40] achieved by α is larger that achieved by β. Formally: $\exists N, K > 0$:

$$\frac{1}{T}\left(\sum_{g=M}^{T+M} \alpha_g\right) \geq \frac{1}{T}\left(\sum_{g=M}^{T+M} \beta_g\right), \ \forall T > N \text{ and } M > K. \tag{24}$$

39 Related rawlsian rules are discussed in Asheim [4] .

40 This is only one of the possible definitions of long run averages. For other related definitions with similar properties see Dutta [26] .

9.4 Countable and Finitely Additive Measures

Definition 12 *Let $(S, \sum)$ denote the field of all subsets of a set S with the operations of unions and intersections of sets. A real valued, bounded additive set function on $(S, \sum)$ is one which assigns a real value to each element of $(S, \sum)$, and assigns the sum of the values to the union of two disjoint sets .*

Definition 13 *A real valued bounded additive set function is called countably additive if it assigns the countable sum of the values to a countable union of disjoint sets.*

Example 2 *Probability measures on the real numbers, R, or on the integers Z, are typical examples of such countably additive functions. Any sequence of positive real numbers $\{\lambda_g\}_{g=1,2...}$ such that $\sum_{g=1}^{\infty} \lambda_g < \infty$ defines a countably additive measure μ on the integers Z, by the rule*

$$\mu(A) = \sum_{g \in A} \lambda_g, \ \forall A \subset Z.$$

Definition 14 *A real valued bounded additive set function φ on $(S, \sum)$ is called purely finitely additive (see Yosida and Hewitt* [47] *) if whenever a countably additive function υ satisfies:*

$$\forall A \in (S, \textstyle\sum), \ \upsilon(A) \leq \varphi(A), \ \textit{then}$$
$$\upsilon(A) = 0 \ \forall A \in (S, \textstyle\sum).$$

This means that the only countably additive measure which is absolutely continuous with respect to a purely finitely additive measure, is the measure which is identically zero.

Example 3 *Any real valued linear function $V : \ell_\infty \to R$ defines a bounded additive function $\widehat{V}$ on the field $(Z, \sum)$ of subsets of the integers Z as follows:*

$$\forall A \subset Z, \ \widehat{V}(A) = V(\alpha^A) \tag{25}$$

where α^A is the "characteristic function" of the set A, namely the sequence defined by

$$\alpha^A = \{\alpha_g^A\}_{g=1,2...} \ \textit{such that}$$
$$\alpha_g^A = 1 \ \textit{if} \ g \in A \ \textit{and} \ \alpha_g^A = 0 \ \textit{otherwise}. \tag{26}$$

Example 4 *Typical purely finitely additive set functions on the field of all subsets of the integers, $(Z, \sum)$, are the lim inf function on ℓ_∞, defined for each $\alpha \in \ell_\infty$ by*

$$\liminf(\alpha) = \lim \inf_{g=1,2\ldots} \{\alpha_g\}. \tag{27}$$

Recall that the lim inf *of a sequence is the infimum of the set of points of accumulation of the sequence. The "long run averages" function is another example: it is defined for each $\alpha \in \ell_\infty$ by*

$$\lim_{K,N\to\infty} \left(\frac{1}{K} \sum_{g=N}^{K+N} \alpha_g \right) \tag{28}$$

It is worth noting that a purely finitely additive set function ϕ on the field of subsets of the integers $(Z, \sum)$ cannot be represented by a sequence of real numbers in the sense that there exists no sequence of positive real numbers, $\lambda = \{\lambda_n\}$ which defines ϕ, i.e. there is for no λ such that

$$\forall A \subset Z, \phi(A) = \sum_{n\in A} \lambda_n.$$

For example the $\liminf : \ell_\infty \to R$, defines a purely finitely additive set function on the integers which is not representable by a sequence of real numbers.

9.5 Proof of Theorem 1

Proof. To establish the existence of a sustainable preference $W : \ell_\infty \to R$, it suffices to exhibit a function $W : \ell_\infty \to R$ satisfying the two axioms. For any $\alpha \in \ell_\infty$ consider

$$W(\alpha) = \theta \left[\sum_{g=1}^{\infty} \delta^g \alpha_g\right] + (1-\theta)\left[\liminf \{\alpha_g\}_{g=1,2\ldots}\right],$$
$$\text{with } 0 < \delta < 1, \text{ and } 0 < \theta < 1.$$

W satisfies the axioms because it is a well defined, non-negative, increasing function on ℓ_∞; it is not a dictatorship of the present (Axiom 1) because its second term makes it sensitive to changes in the "tails" of sequences; it is not a dictatorship of the future (Axiom 2) because its first term makes it sensitive to changes in "cutoffs" of sequences.

The next task is to show that the following welfare criteria do not define sustainable preferences: (a) Ramsey's criterion, (b) the overtaking criterion, (c) the sum of discounted utilities, (d) lim inf, and (e) long run averages (f) Rawlsian criteria and (g) basic needs.

The Ramsey's criterion defined in (20) fails because it is not a well defined real valued function on all of ℓ_∞, and cannot therefore define a complete order on ℓ_∞. To see this it suffices to consider any sequence $\alpha \in \ell_\infty$ for which the sum in (20) does not converge. For example, let

$$\alpha = \{\alpha_g\}_{g=1,2...} \text{ where } \forall g,\ \alpha_g = (g-1)/g.$$

Then $\alpha_g \rightarrow 1$ so that the sequence approaches the "bliss" consumption path $\beta = (1, 1, ..., 1, ...)$. The ranking of α is obtained by the sum of the distance between α and the bliss path β. Since $\lim_{N\rightarrow\infty} \sum_{g=1}^{N}(1-\alpha_g) = \lim_{N\rightarrow\infty} \sum_{g=1}^{N} 1/g$ does not converge, Ramsey's welfare criterion does not define a sustainable preference.

The overtaking criterion defined in (23) is not a well defined function of ℓ_∞, since it cannot rank those pair of utility streams $\alpha, \beta \in \ell_\infty$ in which neither α overtakes β, nor β overtakes α. Figure 2 exhibits a typical pair of utility streams which the overtaking criterion fails to rank.

The long run averages criterion defined in (24) and the lim inf criterion defined in (27) fail on the grounds that neither satisfies Axiom 2; both are dictatorships of the future.

Finally any discounted utility criterion of the form

$$W(\alpha) = \sum_{g=1}^{\infty} \alpha_g \lambda_g, \text{ where } \forall g,\ \lambda_g > 0, \text{ and } \sum_{g=1}^{\infty} \lambda_g < \infty$$

is a dictatorship of the present, and therefore fails to satisfy Axiom 1. This is because

$$\forall \gamma \in \ell_\infty \text{ s.t. } \sup_{g=1,2...} (\gamma_g) \leq 1, \text{ and } \forall \epsilon > 0,$$

$$\exists N > 0,\ N = N(\epsilon) : \sum_{g=N}^{\infty} \gamma_g < \epsilon, \tag{29}$$

and therefore, since

$$W(\alpha) > W(\beta) \Rightarrow \exists \epsilon > 0 :\ W(\alpha) - W(\beta) > 3\epsilon,$$

then by (29)

$$\exists N > 0 \text{ such that } \forall \sigma, \gamma \in \Omega,\ W(\alpha^K, \sigma_K) > W(\alpha^K, \gamma_K),\ \forall K > N.$$

The function W thus satisfies the first part of the definition of a dictatorship of the present, i.e.

$$\begin{gathered} W(\alpha) > W(\beta) \Rightarrow \\ \exists\ N,\ N = N(\alpha, \beta.) :\ \forall \gamma, \sigma \in \ell_\infty \text{ with } \|\gamma\| \leq 1 \text{ and } \|\sigma\| \leq 1, \\ W(\alpha^K, \gamma_K) > W(\beta^K, \sigma_K), \forall K > N. \end{gathered}$$

The reciprocal part of the definition of dictatorship of the present is immediately satisfied, since if $\forall \sigma, \gamma \in \ell_\infty$ such that $\|\alpha\| \leq 1, \|\beta\| \leq 1$, $W(\alpha^K, \sigma_K) > W(\alpha^K, \gamma_K)$, and obviously this implies $W(\alpha) > W(\beta)$. Therefore W is a dictatorship of the present and violates Axiom 1.

Finally the Rawlsian welfare criterion and the criterion of satisfaction of basic needs do not define independent sustainable preferences: the Rawlsian criterion defined in (21) fails because it is not sensitive to the welfare of many generations: only to that of the less favored generation. Basic needs has the same drawback. ■

9.6 Proof of Theorem 2

Proof. Consider a continuous independent sustainable preference. It must satisfy Axioms 1 and 2. There exists a utility representation for the welfare criterion $W : \ell_\infty \to R$, defining a non-negative, continuous linear functional on ℓ_∞. As seen above in Example 3, (25) and (26), such a function defines a non-negative, bounded, additive set function denoted $\widehat{W}$ on the field of subsets of the integers Z, $(Z, \sum)$.

Now we utilize the representation theorem of Yosida and Hewitt ([47] [46]), which establishes that every non-negative, bounded, additive set function on $(S, \sum)$, the field of subsets $\sum$ of a set S, can be decomposed into the sum of a non-negative measure μ_1 and a purely finitely additive, non-negative set function μ_2 on $(S, \sum)$. It follows from this theorem that $\widehat{W}$ can be represented as the sum of a countably additive measure μ_1, and a purely finitely additive measure on the integers Z. It is immediate to verify that this is the representation in (5). To complete the characterization of an independent sustainable preference it suffices now to show that neither λ nor ϕ are identically zero in (5). This follows from Axioms 1 and 2: we saw above that discounted utility is a dictatorship of the present, so that if $\phi \equiv 0$, then W would be a dictatorship of the present, contradicting Axiom 1. If on the other hand $\lambda \equiv 0$, then W would be a dictatorship of the future because all purely finitely additive measures are, by definition, dictatorships of the future, contradicting Axiom 2. Therefore neither λ nor ϕ can be identically zero.

Consider now a preference which is sustainable but not necessarily independent. Then by Axioms 1 and 2 it defines a continuous function $W : \ell_\infty \to R$, which need not be linear. However, for each $\alpha \in \ell_\infty$, it is possible to find a continuous linear function $V^\alpha : \ell_\infty \to R$, such that $V(\alpha) = W(\alpha)$ for which the previous proof applies. This completes the proof of the theorem. ■

9.7 Proof of Theorem 5

The statement of Theorem 5 is:

Consider a sustainable optimum growth problem

$$Max_{\alpha \in \Omega}\, W(\alpha_g), \tag{30}$$

$$\text{where } \alpha_g = \{u_g(x_g)\}_{g=1,2...} \in \Omega \subset \ell_\infty.$$

where Ω is the set of all feasible utility streams and W is an independent sustainable preference. By Theorem 2 W must be of the form:

$$W(\alpha) = \sum_{g=1}^{\infty} \lambda_g \alpha_g + \phi(\alpha),\ \forall \alpha \in \ell_\infty, \tag{31}$$

where $g, \lambda_g > 0, \sum_{g=1}^{\infty} \lambda_g < \infty$, and $\phi \neq 0$ is a purely finitely additive independent measure on Z. Assume that there exists a unique solution to problem (30), denoted α^* and called a sustainable optimum, with welfare value $W^* = W(\alpha^*)$. Assume also that there exist a unique solution, denoted β^* and called a discounted optimum, for the problem of maximizing over the same set Ω the discounted utility

$$Max_{\alpha \in \Omega} (U(\alpha)) \quad \text{where } U(\alpha) = \theta \left(\sum_{g=1}^{\infty} \lambda_g \alpha_g, \right) \tag{32}$$

which is the first term defining the preference W in (31). Then in general the sustainable optima α^* cannot be approximated by a sequence of feasible utility streams $\{\beta^n\}_{n=1,2\ldots}$ which approximates the discounted optimum β^*, i.e. for all such sequences

$$\lim_{n \to \infty} \left(\theta \sum_{g=1}^{\infty} \lambda_g \beta^n \right) \neq Max_{\gamma \in \Omega} \left(\theta \sum_{g=1}^{\infty} \lambda_g \gamma \right).$$

This is true for any sequence of "discount factors" $\{\lambda_g\}_{g=1,2\ldots}$ satisfying $\forall g, \lambda_g > 0$ and $\sum_{g=1}^{\infty} \lambda_g < \infty$.

Proof. I define a family of optimal growth problems, each with a welfare function of the form (31), and each having a feasible set $\Omega \subset \ell_\infty$, all satisfying the conditions of the Theorem. For each problem in this family, the optimum α^* cannot be approximated by a sequence which approximates the optima $\beta^*$of discounted utility functions of the form (32). This is true for any discount factors $\lambda : Z \to R$ which satisfy $\forall g, \lambda_g > 0, \sum_{g=1}^{\infty} \lambda_g < \infty$.

The set of feasible utility streams Ω is as follows:

$$\Omega = \{\alpha \in \ell_\infty^+ : \alpha = \{\alpha_g\}_{g=1,2\ldots}, Sup_g(\alpha_g) \leq 1 \text{ and}$$

$$\exists \epsilon > 0, \text{ and integers } N \text{ and } K \text{ such that}$$
$$\text{if } \alpha_g \geq \epsilon \, \forall g \leq N \text{ then } \alpha_g = 0 \, \forall g > K + N,$$
$$\text{while if } \alpha_g < \epsilon \, \forall g \leq N, \text{ then } \alpha_g \geq \epsilon, \, \forall g > K + N.$$

Each set of parameters ϵ, N and K define a different feasible set of utility streams Ω.

If the welfare function W is a discounted utility of the form (32), then there exists ϵ, N and K such that the discounted optimum $\beta^* = \{\beta_g^*\}_{g=1,2\ldots} \in \Omega$ satisfies

$$\beta_g^* = 1 \text{ for } g \leq N + K, \text{ and } \beta_g^* = 0 \text{ for } g > K + N.$$

The sustainable optima α^* is quite different when in the definition of W, (31), the purely finitely level independent measure ϕ has most of the "weight", i.e. when $\theta \sim 0$. Indeed, for $\theta \sim 0$, the sustainable optimum α^* satisfies

$$\alpha_g^* \geq \epsilon \text{ for } g > K + N.$$

Since both $\alpha^*$and β^* are unique, and

$$\|\alpha^* - \beta^*\| \geq \epsilon > 0,$$

it is clear that a sequence $\{\beta^n\}$ which approaches β^* cannot approach also α^*. This completes the proof of the theorem. ■

References

[1] Arrow, K. J. (1953) *Social Choice and Individual Values*, Cowles Foundation Monographs, Wiley, New York

[2] Arrow, K. J. and M. Kurz (1970) *Public Investment, the Rate of Return and Optimal Fiscal Policy*, Johns Hopkins Press, Baltimore and London.

[3] Arrow, K. J. *Aspects of the Theory of Risk-Bearing*, Yrjö Jahnsson Lectures, Yrjo Jahonssonin Säätio, Helsinki, 1964.

[4] Asheim, G. (1988) "Rawlsian Intergenerational Justice as a Markov-Perfect Equilibrium in a Resource Economy". *Review of Economic Studies*, LV, 469-484.

[5] Asheim, G. (1991) "Unjust Intergenerational Allocations" *Journal of Economic Theory*, Vol. 54, No. 2, August 1991, 350-371.

[6] Beltratti, A., G. Chichilnisky and G. M. Heal (1995) "The Green Golden Rule: Valuing the Long Run" *Economic Letters,* Vol. 48, p. 433-439.

[7] Broome, J. (1992) *Counting the Cost of Global Warming*, London, White Horse Press.

[8] Brundtland, G. H. (1987) The U. N. World Commission on Environment and Development: *Our Common Future*, Oxford: Oxford University Press.

[9] Chichilnisky, G. (1977) "Economic Development and Efficiency Criteria in the Satisfaction of Basic Needs" *Applied Mathematical Modeling*, Vol. 1, No 6, September 1977, p. 290-297.

[10] Chichilnisky, G. (1977) "Development Patterns and the International Order" *Journal of International Affairs,* Vol. 31, No. 2, Fall-Winter 1977, p. 275-304.

[11] Chichilnisky, G. and P. J. Kalman (1980) "Application of Functional Analysis to Models of Efficient Allocation of Economic Resources" *J. Optimization Theory and Applications*, 30, 19-32.

[12] Chichilnisky, G. (1982) "Social Aggregation Rules and Continuity" *Quarterly Journal of Economics*, May 1982, p. 337-352.

[13] Chichilnisky, G. (1993) "On Strategic Control" *Quarterly Journal of Economics*, February 1993, p. 285-290.

[14] Chichilnisky, G. (1993) "The Cone Condition, Properness and Extremely Desirable Commodities" *Economic Theory*, 3, 177-182.

[15] Chichilnisky, G. (1993) "North-South Trade and the Dynamics of Renewable Resources" Paper presented to the conference *The Environment after Rio,* Courmayeur, Italy, 10-12 February 1993, in *Structural Change and Economic Dynamics*, December 1993.

[16] Chichilnisky, G. (1991) "North-South Trade and the Global Environment" Technical Report No. 31, Stanford Institute of Theoretical Economics, 1992, *American Economic Review*, September 1994.

[17] Chichilnisky, G. (1993) "The Abatement of Carbon Emissions in Industrial and Developing Countries", Paper presented at *The International Conference on the Economics of Climate Change*, OECD/IEA, Paris June 14-16th. 1993, to appear in *OECD: The Economics of Climate Change* (ed. T. Jones).

[18] Chichilnisky, G. (1994) "Social Diversity, Arbitrage and Gains from Trade: A Unified Perspective on Resource Allocation" *American Economic Review*, May 1994, p. 427-34.
[19] Chichilnisky, G. (1996) "An Axiomatic Approach to Sustainable Development" *Social Choice and Welfare*, Vol 13, No. 2, 231-257.
[20] Chichilnisky, G. and G. Heal (1993) "Competitive Equilibrium in Sobolev Spaces without Bounds on Short Sales" *Journal of Economic Theory*, 59, p. 364-384.
[21] Cline, W. R. (1992) *The Economics of Global Warming*, Washington D. C., Institute for International Economics.
[22] Coppel, J. (1993) "Implementing a Global Environmental Policy: Some Selected Issues", Economic Division, OECD, paper presented at The International Conference on the Economics of Climate Change, OECD/IEA Paris14-16, June 1993, to appear in *OECD: The Economics of Climate Change* (ed. T. Jones).
[23] Cropper, M. L. S. K. Aydede and P. Portney (1994) 11Preferences for Life Saving Programs: How the Public Discounts Time and Age" *Journal of Risk and Uncertainty*, 8, 243-265, 1994.
[24] Debreu, G. (1954) "Valuation Equilibrium and Pareto Optimum" *Proceedings of the National Academy of Sciences*, vol. 40., July 1954, p. 588-592.
[25] Dasgupta, P. and G. Heal (1979) *Economic Theory and Exhaustible Resources*, Cambridge University Press.
[26] Dutta, P. (1991) "What do Discounted Optima Converge to? *Journal of Economic Theory*, 55, 64-94.
[27] Gale, D.(1967). "On Optimal Development in a Multisector Economy" *Review of Economic Studies*, 1-18.
[28] Hammond, P. J.(1993) "Is there anything new in the concept of sustainable development" Paper presented to the conference *The Environment after Rio*, Courmayeur, Italy, 10-12 February 1993.
[29] Heal, G. (1985) "Depletion and Discounting: a Classical Issue in the Economics of Exhaustible Resources" *American Mathematical Society, Proceedings of Symposia in Applied Mathematics*, Vol. 32, 33-43.
[30] Heal, G. M. (1993) "Valuing the Very Long Run: Discounting and the Environment" Payne Webber Working Paper Series, Columbia Business School, New York.
[31] Heal, G. M. (1995) *Lectures on Sustainability*, book manuscript, also Working Paper University of Oslo.
[32] Herrera, A., H. Scolnik, G. Chichilnisky et. al. (1976) *Catastrophe or New Society: A Latin American World Model* (The Bariloche Model), International Development Research Center, Ottawa, Canada.
[33] Hicks, J. (1946) *Value and Capital*, The Clarendon Press, Oxford University Press (Second edition).
[34] Krautkramer, Jeffrey. "Optimal Growth, Resource Amenities and the Preservation of Natural Environments" *Review of Economic Studies*, 1985, 153-170.
[35] Koopmans, T. (1963). "On the Concept of Optimal Economic Growth" Cowles Foundation Discussion Paper No. 163 and 163A, December 1963.
[36] Lauwers, L. (1993) "Infinite Chichilnisky Rules" Discussion Paper, Katholik Universitaet Leuven, March, to appear *Economic Letters*.

[37] Lauwers L. and L. van Liederkierke "Monotonic Chichilnisky Rules with Infinite Populations" Discussion Paper, Katholik Universitaet, Leuven, to appear in *Social Choice and Welfare.*

[38] Lowenstein George and Jon Elster (editors). *Choice Over Time* Russell Sage Foundation, New York, 1992.

[39] Lowenstein, G. and R. Thaler, (1989) "Intertemporal Choice" *Journal of Economic Perspectives*, 3, 181-193, 1989.

[40] McFadden, D. (1967). "The Evaluation of Development Programmes", *Review of Economic Studies*, 25-50.

[41] Mirrleess, J and N. Stern (1973) *Theory of Economic Growth*, McMillan for the International Economic Association.

[42] Ramsey, F. (1928) "A Mathematical Theory of Saving" *Economic Journal* 1, 38 543-559.

[43] Rawls, J. (1971) *A Theory of Justice*, Bellknap Press, Cambridge MA.

[44] Solow, R. (1974) "Intergenerational Equity and Exhaustible Resources" *Review of Economic Studies,* Symposium on the Economics of Exhaustible Resources, 29-45.

[45] Solow, R. (1992) "An Almost Practical Step Toward Sustainability" Lecture on the Occasion of the Fortieth Anniversary of Resources for the Future, *Resources for the Future*, Washington D. C. October 1992.

[46] Yosida,K. (1974) *Functional Analysis*, Springer Verlag, Berlin, Heidelberg, New York, fourth edition.

[47] Yosida, K. and E. Hewitt (1952) "Finitely level independent Measures" *Transactions of the American Mathematical Society*, 72, 46-66.

[48] von Weizsaecker (1967) "Lemmas for a Theory of Approximate Optimal Growth" *Review of Economic Studies*, 143-151.

5 Linking Weak and Strong Sustainability Indicators: The Case of Global Warming

Klaus Rennings, Olav Hohmeyer
Zentrum für Europäische Wirtschaftsforschung (ZEW)
P.O. Box 10 34 43, D-68034 Mannheim, Germany

Acknowledgements:
Research was supported by a grant of the Commission of the European Communities within the ExternE Project of the DG XII. We are grateful to our project team members Tom Downing, Nick Eyre, Richard Tol, Giles Atkinson and our colleagues Helmuth-Michael Groscurth, Wolfgang Bräuer, and Isabel Kühn for helpful comments and suggestions.

Abstract. The aim of this paper is to describe and discuss the weak and strong sustainability approach of assessing climate change and to show reasonable applications, weaknesses, possible improvements and linkages of both approaches. Main features of "weak" and "strong" sustainability approaches are characterised. Damage cost studies of global warming representing weak sustainability indicators are discussed. Further, the examples of the "inverse scenario" approach of the German Advisory Council on Global Change (WBGU) and the environmental space concept of the Dutch Advisory Council for Research on Nature and Environment (RMNO) are described and discussed for illustrating advantages and weaknesses of strong sustainability indicators. Finally, the integration of damage cost modules into a broader methodological framework of strong sustainability is recommended.

Keywords: weak sustainability, strong sustainability, environmental space, inverse scenario, external costs, climate change, global warming, damage costs.

Non-technical Summary

The Framework Convention on Climate Change (FCCC) has been one of the results of the global summit on environment and development in Rio 1992. Within the negotiation process following the Rio-conference, the need for a protocol in which the parties commit themselves to reduction targets for the different greenhouse gases (or for many developing countries: targets to limit increases) has become evident. The role of scientific research in this process is helping decision makers to derive reasonable reduction targets. However, the appropriate decision rule depends on the underlying paradigm of sustainability which can follow an ecological or an economic approach.

- From an ecological perspective, the ("strong") sustainability rule requires that the total sum of greenhouse gas emissions should not exceed the assimilative capacity of the atmosphere and that, at least, irreversible and catastrophic effects on the global ecosystem should be avoided.
- From a welfare-theoretical perspective, a ("weak") sustainability approach is based on the principle that social welfare should be maximised and the total costs of climate change (abatement, adaptation, and damage costs) should be minimised.

In order to bring ecological and economic requirements together, it is necessary to enhance integrated scientific assessments of climate change. During the past years, several approaches for integrating ecological aspects into economic theory have been conducted by models of integrated assessment of climate change. However, these models can still be divided into two groups: economic approaches based on cost-benefit analysis, and ecological approaches based on environmental targets.

Against this background, this paper explains the weak and strong sustainability approach of climate protection and shows reasonable applications, weaknesses, possible improvements, and linkages of both approaches. In a first step, main features of "weak" and "strong" sustainability approaches towards climate stability are characterised. Then damage cost studies of global warming are discussed which represent indicators of the weak sustainability approach. Further, the examples of the "inverse scenario" approach of the German Advisory Council on Global Change (WBGU) and the environmental space concept of the Dutch Advisory Council for Research on Nature and Environment (RMNO) are described and discussed for illustrating operational indicators of strong sustainability.

Finally, the integration of damage cost modules into a broader methodological framework of strong sustainability is recommended. Economic impact assessment should be improved and included into integrated models for assessing climate change policy. Weak and strong sustainability indicators can be used complementarily in the assessment of climate change. Both can be understood as parts of broader approaches of integrated assessment models.

1 Introduction

The philosophy of a sustainable management of natural resources has been derived from the theory of environmental and resource economics. Originally the concept was developed for renewable resources, especially forests, for which maximum sustainable yields have been calculated. Hartwick (1978) has widened the sustainability principle to applications for exhaustible resources. The concept of a circular economy, as developed by Pearce and Turner (1990, p. 35), has established a broader concept which includes the function of the natural environment as a sink for emissions and waste.[1]

With regard to the assimilative capacity of the environment, ecosystem health is threatened by the greenhouse effect. The Conference of the Parties to the Framework Convention on Climate Change in Berlin 1995 has shown once more the need for a protocol in which the parties commit themselves to reduction targets for the different greenhouse gases (or for many developing countries: targets to limit increases). With regard to Article 3.1 of the Convention, targets for individual states should be defined "in accordance with their common but differentiated responsibilities and respective capabilities".

The role of scientific research in this process is helping decision makers to derive reasonable reduction targets. However, the appropriate decision rule depends on the underlying interpretation of the sustainability paradigm which can follow an ecological or an economic approach.

- From an ecological perspective, the ("strong") sustainability rule requires that the total sum of greenhouse gas emissions should not exceed the assimilative capacity of the atmosphere and that, at least, irreversible and catastrophic effects on the global ecosystem should be avoided.
- From a welfare theoretic perspective, a ("weak") sustainability approach is based on the principle that social welfare should be maximised and the total

---

[1] Beyond that some additional issues are discussed within the sustainability paradigm. For example, Daly emphasizes the role of eco-efficiency as a basic management rule: "Improving end use efficiency of resources is desirable regardless whether the resource is renewable or non-renewable" (Daly, 1990, p. 5). Additionally, the German Council of Environmental Advisors (SRU, 1994, p. 48) stresses that the aspect of protecting human health is neglected by many proponents of sustainability, and introduces health as a further important goal.

costs of climate change (abatement, adaptation, and damage costs) should be minimised.

In order to bring ecological and economic requirements together, it is necessary to enhance integrated scientific assessments of climate change. During the past years, several approaches for integrating ecological aspects into economic theory have been conducted by models of integrated assessment of climate change. However, these models can still be divided into two groups: economic approaches based on cost-benefit analysis, and ecological approaches based on environmental targets (Weyant et al. 1996).[2]

Against this background, the aim of this paper is to explain the weak and strong sustainability approach of climate protection and to show reasonable applications, weaknesses, possible improvements and linkages of both approaches. In a first step, main features of "weak" and "strong" sustainability approaches towards climate stability will be characterised. Then damage cost studies of global warming will be discussed which represent indicators of the weak sustainability approach.[3] Furthermore, the examples of the "inverse scenario" approach of the German Advisory Council on Global Change (WBGU) and the environmental space concept of the Dutch Advisory Council for Research on Nature and Environment (RMNO) will be described and discussed for illustrating operational indicators of strong sustainability. Finally, the integration of damage cost modules into a broader methodological framework of strong sustainability is recommended.

---

2 Some contributions towards a more ecological-oriented target-setting have been made by advocates of ecological economics, e.g., general contributions to integrated ecological and economic models like the one of Common and Perrings, 1992, and special contributions to the consideration of sustainability aspects of climate change into economic analysis, like the ones of Spash, 1994, and Hohmeyer, 1996.

3 Since a detailled description of the contents and results of existing studies has already been done in the IPCC Second Assessment Report (Pearce et al., 1996), this paper focuses on the critical issues of the studies and the further development of their methodological framework.

2 Weak and Strong Sustainability Approaches for Assessing Climate Change

2.1 Definition of Weak and Strong Sustainability

In several contributions, damage cost calculations of climate change like that of Nordhaus (1991), Cline (1991) and Fankhauser (1995) were criticised especially from an ecological perspective. It had been argued that mere neoclassical optimisation concepts tend to ignore the ecological, ethical, and social dimension of the greenhouse effect, especially issues of an equitable distribution and a sustainable use of non-substitutable, essential functions of ecosystems.[4]

The ecological argument addresses the use of damage cost values for computing optimal levels of emission abatement neglecting the special function of the atmosphere as a sink for greenhouse gases. This function is absolutely scarce and essential for the global ecosystem. It is feared that, by putting certain monetary values on this essential natural function, politicians may be encouraged to "sell" it in exchange for goods being possibly of higher value in a short time horizon (e.g., income).

An early proposal for considering sustainability constraints in cost-benefit analysis has been made by Barbier, Markandya, and Pearce (1990, pp. 1260 - 1261). They formulate a sustainability criterion requesting that the sum of damages done by a certain amount of projects should be zero. If E_i is the damage done by the i-th project, the criterion is

$$\sum_i E_i \leq 0 \tag{1}$$

The idea of the criterion is that any environmental damage should "be compensated by projects specifically designed to improve the environment" (Markandya and Pearce, 1991, p. 150). In terms of welfare economics, the compensation criterion is shifted from hypothetical to actual compensation.

However, the sustainability criterion of Barbier, Markandya, and Pearce is "weak" because it allows for unconstrained elasticities of substitution between different types of natural capital. For example, a further depletion of the ozone layer can be compensated by projects supporting the protection of panda bears. Such a weak sustainability criterion should be supplemented by "strong" sustainability criteria which stress more the limitations of substitutability. In this sense, "strong sustainability regards natural capital as providing some functions that are

4 Most of the critical arguments pointing out the limits of traditional cost-benefit-analysis can be found in the IPCC Second Assessment Report (IPCC 1995, WG III).

not substitutable by man-made capital. These functions, labeled 'critical natural capital', are stressed by defining sustainability as leaving the future generations a stock of natural capital not smaller than the one enjoyed by the present generation" (Cabeza Gutés 1996, p. 147).[5]

2.2 Elements of a Theoretical Foundation of Weak and Strong Sustainability

2.2.1 Sustainable Preferences and Adjusted Discount Rates

Strong sustainability, as defined above, is defined in physical terms. This definition is hardly compatible with neoclassical economics having pushed physical factors into the background. What counts in neoclassical welfare theory are subjective perceptions and preferences of people. This preferences of individuals give a certain value to man-made or natural capital. Following this logic, climate stability is a limiting factor of human development if and only if some individuals have an aversion against observed climate risks. As long as individuals do not care about climate change, climate protection does not produce any benefit for them and has, therefore, no economic value. In other words: strong sustainable development can only be translated into neoclassical economics by introducing individual preferences for the long-term protection of life-support functions of ecosystems. Thus, the protection of critical natural capital can be achieved if revealed preferences for intact ecosystems exist.

Such a translation of sustainable development into terms of welfare economics has been suggested by Chichilnisky (1996a). Chichilnisky introduces "sustainable preferences" which are defined by two axioms to rule out "dictatorial" solutions. In her approach, neither the present nor future generations should be dictatorial. The axioms are:

- no dictatorship of the present (no finite set of generations should be dictatorial), and
- no dictatorship of the future ("the very long run" should not be dictatorial).

The axioms can easily be related to the strong and weak sustainability paradigms. While the weak sustainability approach of discounted utilitarianism can be characterised as a dictatorship of the present, the strong sustainability approach

[5] Similar definitions can be found in Pearce and Atkinson (1993) and Pearce, Hamilton, and Atkinson (1996), pp. 85 - 87. It should be added that critical elements of natural capital can neither be subsituted by man-made capital nor by other elements of natural capital, as the example of the ozone layer and the panda bear may illustrate.

can be regarded as dictatorship of the future.[6] The non-dictatorial solution are the so called "sustainable preferences" or the "Chichilnisky criterion." Sustainable preferences are sensitive to the welfare of all generations. This means that:

- the conventional way of discounting future preferences may lead to unsustainable development paths and catastrophic outcomes for future generations, so special weight has to be given to future generations.
- On the other hand, zero discounting may discriminate against the present generation. If consumption in all periods would be weighted equally on an endless time scale, weights would sum to infinity.

Chichilnisky (1996b, p. 2) sees empirical evidence that sustainable preferences exist already within the present generation. If such empirical evidence can be found, the conventional way of measuring and discounting people's values for the future have to be adjusted. Such adjustments will be described more detailed in chapter 3 of this paper.

2.2.2 Uncertainty and Multi Criteria Analysis

Even if sustainable preferences are assumed, the question is whether the costs and benefits of climate change are quantifiable. In a complex and uncertain situation where irreversible damages can occur, Faucheux and Froger (1994) argue that the conventional Bayesian approach of assuming known risk probabilities is not appropriate. According to Faucheux and Froger, the assumption of bounded rationality[7] is more adequate to the problem. Bounded rationality means limited ability of the human mind to collect, remember, and evaluate informations. Assuming uncertainty and bounded rationality supports the strong sustainability paradigm, since the optimisation of outcomes can be judged as over-ambitious in situations where even valid estimates of rough future trends are hard to find. Faucheux and Froger refer to Simon (1972, p. 410) who explains the consequences of assuming bounded rationality as follows: "The decision question has been switched to the question of how much of the actor's resources should be allocated to search". Intitiating such a process of search is called procedural rationality. The methodological consequence is that satisfactory choices may become more relevant than optimal choices, and safe minimum standards may function as rules of thumb during the process of search. It is nevertheless necessary to

6 In the terminology of Beltratti, Chichilnisky, and Heal (1994), the strong sustainability criterion is called the „green golden rule". They define it as the configuration of the economy which gives the highest indefinitely maintainable level of long-run utilities. In principle, the green golden rule requires a zero functional consumption of exhaustible resources.

7 Bounded rationality is an established assumption among new institutional economists, see, e.g., Simon (1972) and Rennings (1992), p. 16.

classify and evaluate alternative outcomes in a weighting scheme. Faucheux and Froger (1994, p. 62) suggest the multicriteria analysis as an analytical tool for decision making based on procedural rationality.

2.3 Scale Issues and Ecological Carrying Capacity

Daly (1992) has adressed the categories of weak and strong sustainability by separating the policy goals of sustainable scale, just distribution, and efficient allocation. Scale refers to the ecological carrying capacity requiring that economic activities should not jeopardise the stability of ecosystems. According to Daly, processes which are relevant to the level of entropy, as, e.g., resource use or flows of matter-energy, should be restricted according to a sustainable scale. Scale has to be measured in absolute physical units:" A good scale is one that is at least sustainable, that does not erode environmental carrying capacity over time. In other words, future environmental carrying capacity should not be discounted in present value calculations. An optimal scale is at least sustainable, but beyond that it is a scale at which we have not yet sacrificed ecosystem services that are at present worth more at the margin than the production benefits derived from further growth in the scale of resource use" (Daly 1992, pp. 186 - 187). Hence, economic growth should be adjusted to the absolute carrying capacity of ecological systems. This is seen as a prerequisite for dealing with questions of distribution and allocation of natural resources.

Using Daly's categories, the valuation of external costs seems to be useful to gather indicators of efficient allocation. However, correcting market failure by estimating the "right" social costs is only the third and last step within the sequential process of addressing policy issues of sustainability, equity and efficiency. Obviously, additional ecological and social indicators are needed for the first two steps. Pursuing this, pressures endangering the long-term stability of ecosystems have to be identified and transformed into critical thresholds. Such thresholds function as "safe minimum standards" (Hampicke, 1993, p. 149; Bishop, 1978, pp. 10 - 18) for essential parts of ecosystems. Daly has used the metaphor of "plimsoll lines" to describe this function of scale limits (Daly, 1992, p. 192).[8]

---

[8] While the concept of Daly is strictly based on the law of entropy, the relevance of entropy processes to explain interactions in open economic and ecological systems seems to be vague and is still disputed (Rennings 1994, pp. 106 - 110; Binswanger 1993, pp. 220 - 229). Alternative approaches are more oriented on ecological criteria like the "resilience and stability of ecosystems" (Common and Perrings, 1992, pp. 15 - 21).

2.4 Indicators of Weak and Strong Sustainability

The weak sustainability approach is represented by external cost estimates of climate change being closely related to the economic rule of maximising welfare. Thus, external costs can be interpreted as indicators of weak sustainability. They indicate the amount of money that has to be spent for the compensation of the estimated welfare losses. The question discussed in this paper is whether these indicators are useful and which role external costs may play in a broader concept of strong sustainability.

In comparison with that, indicators of strong sustainability should offer information about critical elements of the natural capital. Firstly, they should inform about changes in quantity and quality of essential natural resources and functions. Secondly, they should reflect how far the actual use of natural resources is away from a sustainable scale.

Although the process of specifying and quantifying critical elements and thresholds evokes several problems, some progress has been made in developing physical sustainability indicators during the last years (Billharz/Moldan 1995). Rennings and Wiggering (1997) have focused on the assimilative capacity concerning acidification and eutrophication. While critical thresholds can be observed and measured for these problems, the issue becomes more difficult for linear or uncertain risks where no safe levels are obvious. With regard to climate change, the question has to be answered if and how acceptable levels of greenhouse gas emissions can be quantified.

3 Assessment of Weak Sustainability Indicators of Climate Change

3.1 Handling of Global Warming in External Cost of Energy Studies

Due to methodological and empirical problems, the major valuation studies estimating external costs in the energy sector refused to integrate damage costs of climate change into their results. Two different options have been used alternatively:

- The first alternative is the calculation of abatement costs (for specificied CO_2-reduction targets) instead of damage costs. Most advocates of an ecological paradigm of sustainable development prefer the use of abatement costs because they are normally related to CO_2-reduction targets leading to sustainable future emission paths. The abatement cost option has been chosen, e.g., by studies from de Boer/Bosch (1995), Bernow et al. (1996), and Ott (1996).

- Other research teams, being more obliged to a neoclassical paradigm of external costs, decided to renounce the use of damage cost values until more comprehensive studies and methodologies are available. Amongst these groups are the research teams of the valuation studies of the European Commission (ExternE) and of the U.S. Department of Energy (DOE-Study). As Lee (1996, p. 16), one of the authors of the DOE-Study, states: "The earlier studies include estimates of damages from climate change; the more recent studies do not include them in their summary tabulations." And in a footnote he remarks that "this conclusion does not say that damages from climate change are zero, but that precise estimates of these damages do not have a sound scientific basis because of great uncertainty".

3.2 Problems of Valuing Global Warming Damages

The report of phase II of the ExternE project (EC 1994, pp. 159 - 162) does not recommend the use of any monetary value for global warming, but describes the state of the art concerning the valuation of damages. The main results have been that:

- greenhouse gas emissions from each fuel cycle are known accurately,
- the impacts of global warming are complex, scenario dependent, very uncertain, long term and potentially very large,
- the regional variation of climatic change is poorly understood,
- the most comprehensive impact assessments (IPCC) are largely qualitative,
- the results are very sensitive to scenarios considering secondary effects, especially starvation in developing countries,
- serious ethical questions are touched which go beyond mere allocation questions of welfare theory, and
- there is no consensus about these fundamental ethical questions.

Similar conclusions have been drawn by the Intergovernmental Panel of Climate Change (IPCC) in it's Second Assessment Report being finished at the end of 1995.[9] The report cites the range of estimates of marginal damage at 5 - 125 $ per ton of carbon emitted now (Pearce et al., 1996, p. 218). The Working Group III of the IPCC has given special attention to the assessment of cost-benefit

[9] The structure of the IPCC includes three Working Groups: Working Group I (WGI) assessed the science of climate change, Working Group II (WGII) focused on the analysis of impacts and response strategies, and Working Group III (WG III) studied the socio-economic implications of impacts, adaptation, and mitigation and prepared future emissions scenarios (Arris 1996, p. 1). Each Working Group prepared a final report and a summary for policymakers (SPM) (IPCC 1995, WG I - III). The summaries are supplemented by a synthesis report covering the issues of all the three Working Groups (IPCC 1995).

analysis and the incorporation of intra- and intergenerational equity aspects. It has identified some key problems being not adequately addressed by applying traditional cost-benefit-analysis to climate change (IPCC 1995, WG III, pp. 7 - 16; Arrow/Parikh/Pillet et al. 1996, p. 59):

- large uncertainties,
- long time horizons,
- global, regional, and intergenerational nature of the problem,
- wide variations of the cost estimates of potential physical damages due to climate change,
- wide variations of the cost estimates of mitigation options,
- low confidence in monetary estimates for important consequences (especially non-market impacts),
- possible catastrophes with very small probabilities, and
- issues of intragenerational equity (especially lower values for statistical lives of people in developing countries than those in developed countries).

Besides these weaknesses, some additional methodological problems are still unsolved. While normally marginal impacts of single power plants are calculated in recent damage cost studies, marginal impacts of one power plant on the global climate seem to be insignificant (Plambeck/Hope, 1996, p. 784). Also Hohmeyer (1996) points out that valid cost-benefit optimisation is impossible because future marginal costs are impossible to derive for long-term climate change. For that reason, average values have to be used. Thus, the "bottom up"-approach and the estimation of marginal, site-specific effects has to be modified for the global warming issue.

The issue of serious ethical questions refers to normative assumptions of external cost studies which are not transparent for the user of the results. One assumption is that economic welfare is measured by people's willingness to pay or willingness to accept compensation. Thus, the welfare of rich people and nations has a greater weight in the results than the welfare of poor ones. Especially the common way of valuing human lives in developing countries lower than those in developed countries is highly disputed. Another implicit, but central judgement concerns the possibility of compensating future individuals for climate damages. Such assumptions have to be made transparent, and a representative range of assumptions should be used in the form of an ethical sensitivity analysis.

3.3 Improvements of Damage Cost Valuation

3.3.1 Intragenerational Equity

Responding to the IPCC criticism, Fankhauser and Tol (1995) and Tol (1996b) have derived a research agenda for the economic assessment of climate change impacts including:

- improved damage estimates for less developed countries;
- improved estimates for non-market losses, especially morbidity and ecosystem effects;
- assessment of the importance of variability and extreme events;
- models of the process of adaptation and the dynamics of vulnerability;
- formal uncertainty assessments and analyses of the outcomes;
- improved comparison and aggregation of estimates between countries;
- improved comparison and aggregation of estimates between generations;
- ensuring consistency between economic and non-economic impact assessment.

Following this research agenda, first progress can be observed, especially concerning the handling of intra- and intertemporal equity questions.[10] Intragenerational equity questions have been addressed by contributions from Fankhauser, Tol and Pearce (1996), and Azar and Sterner (1996). Both use an approach of equity weighting: on the basis of the existing estimates of global warming damages, willingness to pay values are adjusted in the aggregation process. While aggregating estimates for single countries or world regions to a global value, the damages are weighted by the inverse of income. Damages of rich countries are weighted down and damages of poor countries are weighted up by adjusting these damages to the average annual per capita world income. The reason for the adjustment is "decreasing marginal utility of money and for the same reason we can argue that a given (say one dollar) cost which affects a poor person (in a poor country) should be valued as a higher welfare cost than an equivalent cost affecting an average OECD citizen" (Azar and Sterner, 1996, p. 178). Thus, equity weighting leads to the result that damages and deaths in developed countries do not count more than in developing countries. Due to the fact that the annual world income is constant, it has to be used as a budget restriction.

Beyond issues on intragenerational distribution, improvements concerning intergenerational equity have been made by several authors and will be described more detailed in the next section.

---

[10] Additionally, some efforts have been made towards a more dynamic modelling of climate change damages which will not be discussed in this paper. See for details Tol (1996a; 1996c).

3.3.2 Intergenerational Equity

The results of monetary values of climate change damages depend substantially upon the choice of the discount rate. The higher the discount rate, the lower the present value of future damages. Thus, discounting is often criticised because it produces incentives to shift environmental risks from the present to the future. However, the relationship between discount rate and climate change is very ambiguous. Lowering the discount rate induces an increasing level of economic activity and investment. This would probably lead to further emissions of greenhouse gases (CEC/US Joint Study, 1993, S. 2-19). The relationship between the discount rate and environmental deterioration is known as the "conservationist's dilemma", since both, high and low discount rates, can favour environmental conservation (Norgaard/Howarth, 1991, p. 90).

Commonly, a range of discount rates is used in cost-benefit-analyses. Following Markandya, discount rates of 0, 3, and 10% represent an adequate range of parameters for the European Union (CEC/US Joint study, 1993, p. 2-22). 3% is taken as a rate for social time preference, 0% and 10% as extreme parameters for sensitivity analysis.[11] With regard to climate change, none of the three rates is satisfying: while rates of 3 to 10% lead to nearly zero costs for long term damages, a rate of 0% may evoke infinite costs.

Time-variant discount rates

The rate of 3% can be derived from the concept of social time preference (STP), a measure of the decline of social welfare or utility of consumption over time (Markandya/Pearce, 1991, p. 142). The social time preference depends on the rate of pure individual time preference (ITP) or impatience, on the growth rate of real consumption per capita (W), and on the elasticity of the marginal utility of consumption (U). The equation is:

$$STP = ITP + W \times U \qquad (2)$$

An important argument against the STP concept is that ecological "limits to (economic) growth" will set biophysical constraints on W in the long run. When choosing the rate of W, such constraints should be taken into consideration. For the EU, Markandya recommends a rate for W of around 1 or 2% as a low sustainable rate (CEC/US Joint study, 1993, p. 2-20).

It is argued from an environmental perspective that ITP should be refused in social investment decisions. This position takes the perspective of society as a whole and criticises impatience for being irrational. For a society - contrary to the

[11] Relative high (market) discount rates of 6 or more percent normally represent the concept of opportunity costs of capital.

individual view - it seems to be unreasonable to privilege present preferences above future preferences. However, a collective view conflicts with methodological individualism being a fundamental element of welfare economics.

Rabl (1993) argues that a discount rate for intergenerational effects should be defined by taking the perspective of future generations. From Rabl's point of view, market interest rates can only be taken to the extent that a market exists. Following Rabl, the longest time horizon of market transactions is 30 to 40 years. Thus, there is no inconsistency in lowering the interest rate for damages beyond that time horizon. In consequence, Rabl recommends to split STP in an

STP = ITP + W x U for short-term effects (< 30 to 40 years) (3)

and an

STP = W x U for long-term effects (> 30 to 40 years) (4)

At first glance, the splitting concept and the time horizon for market transactions chosen by Rabl seem to be very arbitrary. At second glance, a special treatment of long-term effects seems to be reasonable, because otherwise damages occuring a hundred or more years in the future will be totally ignored in monetary valuation studies. Understood as a first rule of thumb, Rabl's concept of time-variant discount rates helps to improve the treatment of long term effects in external cost studies.

Following the idea of the Chichilnisky criterion (see section 2.2.1), Heal (1996, pp. 6 - 7) has introduced the concept of logarithmic discounting with a similar consequence, namely a decreasing discount rate in time. In his approach, the discount rate is inversely proportional to distance into the future. This formalisation has been derived from the Weber-Fechner law stating that "human response to a change in a stimulus is inversely proportional to the pre-existing stimulus" (Heal, 1996, p. 6). Heal argues that a decreasing discount rate is a natural phenomenon: "postponement by one year from the next year to the year after, is clearly quite a different phenomenon from postponement from fifty to fifty one years hence. The former represents a major change: the latter a small one." Compared with the approach of Rabl, the concept of logarithmic discounting is more elaborated and sophisticated. The main advantage is that discontinous damage functions can be avoided.

It can be summarised that the introduction of time-variant discount rates is reasonable with regard to long-term environmental damages. To express it in the words of Sterner and Azar (1996, p. 174), "a constant discount rate should only be seen as a special case of the more general case where the discount rate is allowed to vary".

Markandya's estimate for ITP is around 1 or 2%. Added to W, the result is a STP of 2 to 4% (CEC/US Joint study, 1993, p. 2-20). Rabl's estimates of W are quite similar to Markandya's values (Rabl, 1993, p. 2 - 4).

Zero discount rate

A more radical position is to set the discount rate equal to zero (Pearce, 1993, pp. 57 - 61). A zero discount rate follows the rule that consumption at one point of time does not count more than welfare at another point of time. However, it is feared that a zero discount rate would imply infinite social costs and total current sacrifice (Pearce, 1993, p. 58).

Nevertheless, there is some reason to use zero discount rates for certain natural resources (W_{NR}) whose market value are expected to rise proportionally to the Gross Global Product (GGP). One important example is the demand for safety, expressed in terms of an statistical value of life (VSL). It can be assumed that the growth rate of the WTP for reducing health risks will be at least as high as the growth rate of GGP. All things considered, 0% seems to be an appropriate discount rate only if

$$W = 0 \tag{5}$$

or if

$$W > 0, \text{ and } W_{NR} \approx W \tag{6}$$

A prerequisite for (6) is that W_{NR} is not already included in calculations of the underlying cost and benefit streams. Studies expressing the cost of global warming as a percentage of GGP (Mayerhofer, 1994, pp. 2 - 5) already include W_{NR}, whereas WTP for reducing health risks can be assumed to be at least proportional to GGP but is commonly calculated with a constant statistical value of life (Rabl, 1994, pp. 1 - 2).

Discounting, equity, and distribution

The main controversial issue among the different discounting concepts is the question of compensation among generations. While the use of market interest rates assumes compensation from one generation to another for losses of natural capital, the use of lower discount rates assumes more or less that environmental protection is the only way to make these transfers (Arrow/Cline/Mäler et al. 1996, p. 133). Following this argument, intergenerational fairness can be characterised as a matter of distribution across generations. It seems to be reasonable to separate these issues of distribution from issues of efficiency. According to Daly, "the policy instrument for bringing about a more just distribution is transfers - taxes and welfare payments" (Daly, 1992, p. 186). Norgaard and Howarth argue in the same direction by pointing out that "if we are concerned about the distribution of welfare across generations, then we should transfer wealth, not engage in inefficient investments. Transfer mechanisms might include setting aside natural resources, and protecting environments, educating the young, and developing technologies for the sustainable management of renewable resources. Some of these might be viewed as worthwhile investments on the part of this generation, but if their intent is to function as transfers, then they should not be evaluated as investments. The benefits from transfers, in short, should not be discounted" (Norgaard/Howarth 1991, p. 98).

From this point of view, the discount rate should only function as a mechanism of efficient allocation of resources. Distributional aspects are separated from allocation, although they are not independent. It is plausible that transfers to future generations change relative prices. As Norgaard and Howarth remark: "With different distributions and efficient allocations, new prices arise. One can no more speak of 'the' rate of interest when societies are giving major consideration to the sustainability of development than one can speak of 'the' price of timber when deciding whether to conserve forests. Redistributions change equilibrium prices" (Norgaard/Howarth, 1991, p. 97).

3.4 Conclusions

It can be summarised that the approach of deriving weak sustainability indicators of global warming by estimating damage costs requires strong normative choices about inter- and intragenerational fairness and the handling of uncertainty. This normative choices are made in most cases implicitly, i.e., they are hidden under a veil of aggregation and discounting rules. These problems have been especially emphasised by the Second IPCC Assessement Report. However, important responses to the IPCC criticism have now been made. While the long term dynamic effects of global warming and the resulting social and economic impacts are still not well understood, at least some important contributions have been made with regard to an improved handling of intra- and intergenerational equity issues. As far as allocation is concerned, an appropriate range of discount rates should integrate

- 0% as a rate for long term effects which are expected to rise with GDP,
- 1% as a rate for STP ignoring ITP,
- 3% as a rate for STP including ITP and
- higher discount rates representing market interest rates (concept of opportunity costs of capital).

The concept of time-variant discount rates seems to be consistent with the principles of welfare theory. While 3% can be used as a standard discount rate, lower rates can be applied for the long-term global warming effects.

It is obvious that equity weighting and time-variant discounting will have a substantial influence on the fact which amount of investments for stabilising the global temparature can be justified by mere economic reasons. In the IPCC report with a cited range of 5 - 125 $ marginal per ton of carbon, the lower bound of the range is derived from the Nordhaus study. Using mainly the Nordhaus parameters and a model considering the retention of carbon in the atmosphere, Azar and Sterner (1996, p. 182) introduce time-variant discount rates and equity weighting as described above. Doing this, they calculate marginal damages in the range of 260 - 590 $ per ton of carbon. This is roughly 50 to 100 times higher than the Nordhaus value.

Nevertheless, large uncertainties concerning future climate scenarios and damage paths remain. A reasonable solution may be to link monetary indicators with more ecologically oriented approaches which will be presented in the next chapter.

4 Assessment of Strong Sustainability Indicators of Climate Change

4.1 Indicators of Environmental Space

Important early contributions to the discussion of acceptable levels of greenhouse gas emissions have been made by the Dutch Advisory Council for Research on Nature and Environment (RMNO) (Weterings/Opschoor 1994) and by the German Enquete Commission "Preventive Measures to Protect the Earth's Atmosphere." The Enquete-Commission has derived specific national reduction targets and general targets for developed and developing countries (German Bundestag 1991, Vol. 1; pp. 70-75) from the recommendations of the World Conference on Atmospheric Change in Toronto 1988. The recommendation of the conference had been (German Bundestag, 1991,Vol 2, 796 -840):

- to reduce global emissions of CO_2 and other trace gases by over 50% by the year 2050 and
- to reduce global CO_2 emissions by about 20% by the year 2005, relative to 1988 emission levels.

Furthermore, the recommendations of the RMNO and the Enquete Commission are based on a study of Krause, Bach, and Koomey (1990) estimating tolerable CO_2-emissions. The authors calculate a tolerable relative deviation of 0.1°C per decade (data on the ability of trees to migrate suggest this as a maximum rate of temperature-rise) and an absolute warming limit of 2.0-2.5°C above pre-industrial level for the next 100 years (which would lead to a maximum acceptable sea level rise of 1 meter in the forthcoming centuries). Within this temperature change the most important ecological functions are supposed to be sustained. These thresholds have been transferred into critical concentrations and critical emission paths.

In a report to the RMNO, Weterings and Opschoor have used the concept of environmental space to share the global budget for CO_2-emissions among nations. They describe the concept of environmental space as follows:

> Environmental utilisation space (or: environmental space) is a concept which reflects that at any given point in time, there are limits to the amount of environmental pressures that the earth's ecosystem can handle without irreversible damage to these systems or to the life support processes that they enable. This suggests to search for the threshold levels beyond which actual environmental systems might become damaged in the sense indicated above, and to regard this set of deductively determined critical values as the operational boundaries of the environmental space (Weterings/Opschoor 1994, p. 3).

Five different criteria have been used for the distribution of the carbon-budget over regions and nations:

- GNP,
- land area,
- current energy consumption (status quo criterion),
- current population (equity current criterion; equal emission per capita),
- current and future population (equity cumulative criterion; equal emission per capita).

Following each of the five criteria, the global carbon-budget was distributed among nations and regions. The different sustainability indicators (according to different distribution criteria) were compared to the actual and forecasted performance of the OECD countries (see table 1). The report comes to the conclusion that "the OECD does not meet the various sustainability criteria currently and is not forecasted to do so in the forthcoming decades. Nor any of the individual member-states does. Even if we forget about a more equal distribution in respect of the developing countries, the OECD emission exceeds sustainability (status quo) by more than a factor 2. From the equity perspective the OECD performance is unsustainable by a factor of 7 to 10" (van der Loo 1993, p. 65).

Obviously, the definition of strong sustainability standards requires some normative choices. Compared with "weak" optimisation concepts, the advantage is that these choices are made explicitly and are not hidden under a veil of aggregation and discounting rules.

Table 1: Sustainability Criteria for OECD-carbon release (Assumption: Sustainable world carbon budget 300 GtC as estimated by Krause et al. 1990; OECD current annual release 2.8 GtC)

Criterion	OECD Budget 1985- 2100 (% global budget)	(GtC)	OECD Annual average (GtC)	% current emission
GNP	63 %	189	1.64	57 %
Land area	24 %	72	0.63	22 %
Status Quo	47 %	140	1.17	42 %
Equity current	16 %	48	0.42	15 %
Equity cumulative	11 %	33	0.29	10 %

Source: van der Loo (1993), p 65.

4.2 The "Inverse Scenario" or "Tolerable Window Approach"

Another example of making such choices explicitly and describe them transparently is the "inverse scenario" of the German Advisory Council on Global Change (WBGU). Based on the scenario asumptions, the WBGU draws the conclusion that the acceptable absolute positive deviation from the present mean temperature on earth is 1.3 °C and a temperature change of 0.2 °C per decade is the tolerable upper limit.

The new "scenario for the derivation of global CO_2-reduction targets and implementation strategies" was published on the occasion of the Climate Conference in Berlin (WBGU, 1995a, pp. 111-128). The scenario specifies firstly tolerable stresses for humans and nature, "and then, by proceeding backwards, the long-term global reduction target is derived which would ensure that these maximum stress levels are complied with" (WBGU, 1995b, pp. 3 - 4). Thus, the "backwards mode" of the scenario follows a strong sustainability approach being based on acceptable impacts or minimum standards of climate stability. Standard scenarios are carried out in a "forwards mode" estimating future, possibly non-sustainable emission and damage paths.[12]

The "backwards scenario" or "inverse scenario" contains six steps (see figure 1):

1) In step one, a range of tolerable stresses caused by climate change is defined. Identifying tolerable impacts and damages, the "inverse scenario" starts explicitly with a *normative judgement*.
2) In step two, temperature changes are derived which assure that tolerable stresses are not exceeded.
3) In step three and four, admissible concentrations and

[12] For example, the impact pathway methodolody of the ExternE-project belongs obviously to these „forwards mode" scenarios (EC 1995, pp. 7 - 30).

4) emissions of greenhouse gases (here: only CO_2) are quantified by using models of climate dynamics and the carbon cycle.
5) In step five, the total emission reduction has to be broken down to individual states or groups of countries.
6) In a final step, a mix of efficient instruments for mitigating climate change has to be derived.

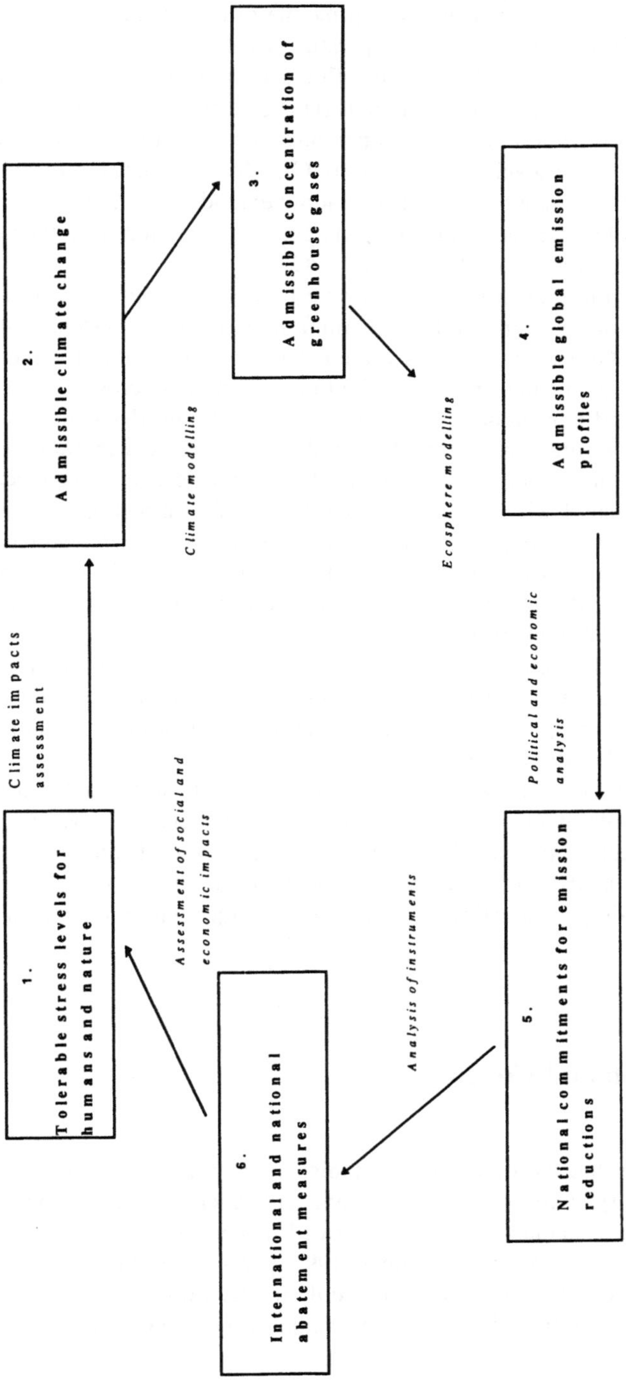

Figure 1: The Council's "Inverse Scenario"

The basic normative principles of the council are the preservation of Creation and the prevention of excessive costs. The principle of preservation of Creation is formalised in the form of a tolerable "temperature window" (WBGU 1995b, p. 7) being derived from the natural temperature fluctuation during the geological period having shaped our present environment (late Quaternary period). The minimum and maximum values of this temperature window are the last ice age (10.4°C) and the last interglacial period (16.01°C). With an extension of this temperature range by 0.5°C at either end, the window extends from 9.9°C to 16.6°C. Using these thresholds, the acceptable absolute positive deviation from the present mean temperature on earth (15.3°C) is only 1.3°C.

The principle of the prevention of excessive costs is defined very crudely in losses of GGP. On the assumption that a disruption of economic systems will take place if losses of GGP exceed 5%, this value is taken as a threshold for economic impacts. The possible unequal spatial distribution of damages across nations (e.g., for island states) and non-monetary burdens are not yet considered in this minimum-standard. Most monetary estimates of doubling CO_2-concentrations until 2100 (mean temperature increase of roughly 0.2 °C per decade) have calculated GGP-losses of around 1-2%. Considering that these calculations did not include several damage categories (e.g., extreme events) and may have underestimated the total costs, the WBGU sees "good reason to assume that with a temperature change of 0.2 °C per decade the upper limit for adaptation costs of 5% of GGP would be reached" (WBGU 1995b, p. 8).

The thresholds of the temperature window have been formulated as minimum standards for political reasons because the results should not be assessed too pessimistically. With the help of these operational criteria, a two-dimensional climate window is defined that should not be exceeded.

It is important to mention that the missing link of "backwards" and "forwards" scenarios is the assessment of social and economic impacts. This assessment is located here between step 6 and 1. Pursuing a closed circle of integrated assessment, an economic analysis of different abatement and adaptation strategies would be desirable, including a valuation of remaining damages with monetary or non-monetary values.

5 Discussion and Conclusions

It has been shown that the outcomes of studies estimating economic and social impacts in monetary units are very uncertain. However, as the alternative approaches of strong sustainability have illustrated, uncertainties do not disappear when norms are used instead. Schellnhuber (1995, p. 58), one of the developers of the "inverse scenario", states that norms can only induce maximums or minimums (e.g., a safe minimum standard) and not optimums. If the identification of optimal emission paths among minimum and maximum standards would be pursued, the strong sustainability approach should be supplemented by an economic impact assessment of damages, adaptation, and abatement strategies. This would close the circle of integrated assessment of climate change.

Given perfect information about damage paths and present as well as future preferences, impact assessment would be able to replace the normative judgements in step 1 of the "inverse scenario." However, in the light of the discussion about decision-making under uncertainty, it becomes clear that a complete substitution of normative judgement by cost-benefit-analysis or integrated assessment models will hardly be possible. Damage cost valuation techniques themselves contain central normative assumptions.

On the other hand, even normative target approaches often depend on monetary values for defining tolerable stress levels. The "invers scenario" documents this close link between acceptable emission paths and economic damages. Thus, it becomes evident that further information about the global distribution of costs and benefits of climate change is desirable for the political negotiation process. For example, what is to do when global average damages do not exceed 5% of GGP, but reach 100% of the national income for certain island states and coastal zones? And how to handle high disparities of damages between economic sectors, social groups, or species? Which damages can be compensated, which cannot?

Many of these questions can only be answered by following a broader approach of strong sustainability including damage figures as far as valid estimates are available. Within the negotiation process, tolerable stress levels depend on specific burdens and economic costs of world regions and losses of certain economic sectors or societal groups. Thus, the relevance of imperatives like preservation of creation and prevention of excessive costs for political negotiations can be improved by more disaggregated sectoral and regional information about climate change impacts.

So, it seems reasonable to improve economic impact assessment and to include it into integrated models for assessing climate change policy (WBGU1995b, p. 7). Weak and strong sustainability indicators can be used complementarily in the assessment of climate change. Both can be understood as parts of broader approaches of integrated assessment models.

The critical IPCC review of social cost studies seems to have an influence on research and has enhanced methodological progress, especially the handling of intra- and intergenerational equity issues. Further progress may lead towards more dynamic models and a multidimensional valuation of impacts. It should be noted that damage cost valuation is only one aspect in a modern interpretation of cost benefit analysis as it has been described in the IPCC report (Munasinghe et al., 1996, p. 170). According to the IPCC report, cost benefit analysis encompasses a family of decision-analysis techniques like multicriteria analysis or decision analysis. Using such a broad interpretation, social and economic impacts do not necessarily have to be described in monetary units. A main disadvantage of cost-benefit analysis is that in complex decision situations relevant multiple criteria (e.g., efficiency, equity, uniqueness of resources or health and safety) are mixed and reduced to one single criterion. Multicriteria analysis may be a better way to show trade-offs between these different policy goals (Munasinghe et al., 1996, p. 168).

It can be summarised that monetary indicators within a weak sustainability approach dominate the economic literature because cost benefit analysis has commonly been interpreted in a narrow sense. Broadly speaking, the narrow approach can be charaterised by assuming perfect markets (e.g., rationality and flexibility) with the exception that the existence of external costs is admitted. Global environmental problems and especially global warming require a broader approach considering inter alia uncertainty, bounded rationality, equity and scale effects. Obvious deficits of economic approaches with regard to these issues have promoted the search of more adequate alternatives. Ecological economics, one of the relevant scientific streams for searching new ideas, is driven by the underlying paradigm of strong sustainability. It is concluded here that, even within a strong sustainability paradigm, estimates of costs and benefits of climate protection remain to be a valuable and important tool for decision making. The prerequisite for an appropriate use of cost-benefit analysis is a move towards a more frequent use of multiple, monetary, and non-monetary valuation schemes.

References

Arris, Lelani (1996): The IPCC Second Assessment Report: A Review. In: Environment Watch Western Europe (EWWE) - Special Edition March 1996, Cutter Information Corp.

Arrow, K. J., J. Parikh, G. Pillet et al. (1996): Decision-Making Frameworks for Adressing Climate Change. In: J. J. Bruce, H. Lee, E. F. Haites: Climate Change 1995 - Economic and Social Dimensions of Climate Change. Cambridge University Press, pp. 53 -78.

Arrow, K. J., W. R. Cline, K.-G. Mäler et al. (1996): Intertemporal Equity, Discounting and Economic Efficiency. In: J. J. Bruce, H. Lee, E. F. Haites: Climate Change 1995 - Economic and Social Dimensions of Climate Change. Cambridge University Press, pp. 125 - 144.

Azar, Christian, Thoams Sterner (1996): Discounting and Distributional Considerations in the Context of Global Warming. In: Ecological Economics 19 (1996), pp. 169 - 184.

Barbier, Edward B., Anil Markandya, David W. Pearce (1990): Environmental sustainability and cost-benefit-analysis. In: Environment and Planning A, Vol. 22, pp. 1259 - 1266.

Beltratti, A., G. Chichilnisky, G. M. Heal (1994): The Green Golden Rule. Working paper, Columbia Business School 1994

Bernow, Stephen, Bruce Biewald & Paul Raskin (1994): From social costing to sustainable development: beyond the economic paradigm. In: Hohmeyer, Olav, Richard L. Ottinger (Eds.) (1994): Social costs of energy - present status and future trends. Springer, Berlin, Heidelberg, New York, pp. 373 - 404.

Bernow, Stephen, Bruce Biewald, William Dougherty, David White (1996): Counting the Costs: Scientific Uncertainty and Valuation Perspective in EXMOD. In: Olav Hohmeyer, Richard Ottinger, Klaus Rennings (Eds.). Social Costs and Sustainability. Springer, Berlin, Heidelberg, New York, pp. 200 - 231.

Billharz, Suzanne, Bedrich Moldan (eds.) (1995): Scientific Workshop on Indicators of Sustainable Development. Workshop report. Scientific Committee on Problems of the Environment (SCOPE). Wuppertal, Germany, November 15 - 17.

Binswanger, Mathias (1993): From microscopic to macroscopic theories: entropic aspects of ecological and economic processes. In: Ecological Economics, 8/1993, pp. 209 - 234

Bishop, Richard C. (1978): Endangered species and uncertainty: the economics of a safe minimum standard. In: American Journal of Agricultural Economics, vol. 60, pp. 10 - 18.

Cazeba Gutés, Maite (1996): The Concept of Weak Sustainability. In: Ecological Economics 17/1996, pp. 147 - 156.

CEC/US Joint Study on Fuel Cycle Costs (1993): Assessment of the external costs of the coal fuel cycle: Draft Position Paper, prepared for DG XII of the Commission of the European Community.

Chichilnisky, G. (1996a): An Axiomatic Approach to Sustainable Development. In: Social Choice and Welfare, 13/1996, pp. 231 - 257.

Chichilnisky, G. (1996b): What is Sustainable Development? Columbia University, New York, Discussion Paper, July 1993, revised November 1996.

Cline, William R. (1991): The Economics of Global Warming. Institute for International Economics. Washington.

Common, Mick, Charles Perrings (1992): Towards an Ecological Economics of Sustainability. In: Ecological Economics, 6/1992, pp. 7 - 34.

Daly, Hermann (1990): Towards Some Operational Principles of Sustainable Development. Ecological Economics, 2/1990, pp. 1 - 6.

Daly, Hermann E. (1992): Allocation, Distribution and Scale: Towards an Economics that is Efficient, Just and Sustainable. In: Ecological Economics, 6/1992, pp. 185 -193.

De Boer, Bart, Peter Bosch (1995): The Greenhouse Effect: an Example of the Prevention Cost Approach. Paper Prepared for the Second Meeting of the London Group on national Accounts and the Environment. Washington, March 15 - 17, 1995. Statistics Netherlands, Environment Statistics. Voorburg, the Netherlands.

EC - European Commission, DG XII (Ed.) (1994): Externalities of Fuel Cycles - ExternE Project. Summary Report. Bruxelles.

EC - European Commission, DG XII (Ed.) (1995): ExternE Externalities of Energy, Vol. 2 - Methodology. Brussels, Luxembourg.

Fankhauser, Samuel (1995): Valuing Climate Change - The Economics of the Greenhouse. Earthscan, London.

Fankhauser, Samuel, Richard S. J. Tol (1995): The Social Costs of Climate Change: The IPCC Second Assessment Report and Beyond. Instituut voor Milieuvraagstukken (IVM), Frije Universiteit Amsterdam. Discussion paper W - 95/34.

Fankhauser, Samuel, Richard S. J. Tol, David W. Pearce (1996): The Aggregation of Climate Change Damages: A Welfare Theoretic Approach. Unpublished Paper, Free University Amsterdam, Institute for Environmental Studies, the Netherlands.

Faucheux, Sylvie, Géraldine Froger (1994): Decision-making Under Environmental Uncertainty. In: G. Pillet, F. Grassmann: Steps towards a Decision Making Framework to Adress Climate Change. Paul Scherrer Institut, Bericht Nr. 94 - 10, Würenlingen and Villigen, Switzerland, pp. 51 - 65.

German Bundestag (Ed.) (1991, Vol. 1): Protecting the Earth - a Status Report with Recommendations for a New Energy Policy. Bonn.

German Bundestag (Ed.) (1991, Vol. 2): Protecting the Earth - a Status Report with Recommendations for a New Energy Policy. Bonn.

Hampicke (1993): Möglichkeiten und Grenzen der monetären Bewertung der Natur. In: H. Schnabl (Hrsg.): Ökointegrative Gesamtrechnung, Berlin, New York, pp. 135 - 155.

Hartwick, John M. (1978): Intergenerational Equity and the Investing of Rents from Exhaustible Resources. In: The American Economic Review, Vol. 67, pp. 972 - 974.

Heal, Geoffrey (1996): Sustainable Cost-Benefit Analysis. Draft Working Paper, unpublished.

Hohmeyer, Olav (1996): Social Costs of Climate Change. Strong Sustainability and Social Costs. In: Richard Ottinger, Olav Hohmeyer, Klaus Rennings (Eds.). Social Costs and Sustainability. Springer, Berlin, Heidelberg, New York, pp. 61 - 83.

IPCC (1995): Second Assessment Synthesis of Scientific-Technical Information Relevant to Interpreting Article 2 of the UN Framework Convention on Climate Change. IPCC Secretariat, Geneva.

IPCC Working Group I (1995): Summary for Policymakers of the Contribution of Working Group I to the IPCC Second Assessment Report. IPCC Secretariat, Geneva.

IPCC Working Group II (1995): Summary for Policymakers: Impacts, Adaptation and Mitigation Options. IPCC Secretariat, Geneva.

IPCC Working Group III (1995): Summary for Policymakers: Economic and Social Dimensions of Climate Change. IPCC Secretariat, Geneva.

Krause, F., W. Bach, J. Koomey (1990): Energy Policy in the Greenhouse, Vol. 1.: From Warming Fate to Warming Limit: Benchmarks for a Global Climate Convention. London.

Lee, Russell (1996): Externalities Studies: Why are the numbers different? In: Olav Hohmeyer, Richard Ottinger, Klaus Rennings (Eds.). Social Costs and Sustainability. Springer, Berlin, Heidelberg, New York, pp. 13 - 28.

Markandya, Anil, David W. Pearce (1991): Development, the environment, and the social rate of discount. In: The World Bank Research Observer, Vol. 6, No. 2, pp. 137 - 152.

Munasinghe, M., P. Meier, M. Hoel, S. W. Hong, A. Aaheim (1996): Applicability of Techniques of Cost-Benefit Analysis to Climate Change. In: J. J. Bruce, H. Lee, E. F. Haites: Climate Change 1995 - Economic and Social Dimensions of Climate Change. Cambridge University Press, pp. 145 - 177.

Mayerhofer, Petra (1994): Climate change in the framework of external costs of energy systems. Paper presented on the A & WMA international speciality conference "global climate change: science, policy and mitigation strategies", Phoenix, Arizona, U.S.A., April 5-8, 1994.

Nordhaus, William D. (1991): To Slow or not to Slow? The Economics of the Greenhouse Effect. In: The Economic Journal, Vol. 101, pp. 920 - 937.

Norgaard, Richard B., Richard B. Howarth (1991): Sustainability and discounting the future. In: Costanza, Robert (Ed.): Ecological Economics: the science and management of sustainability. New York, pp. 87 - 101.

Ott, Walter (1996): External Costs and External Price Addings in the Swiss Energy Sector. In: Richard Ottinger, Olav Hohmeyer, Klaus Rennings (Eds.). Social Costs and Sustainability. Springer, Berlin, Heidelberg, New York, pp. 176 - 183.

Pearce, David W. (1993): Economic values and the natural world. Earthscan, London.

Pearce, David W., Giles D. Atkinson (1993): Capital Theory and the Measurement of of Sustainable Development: an Indicator of 'Weak' Sustainability. In: Ecological Economics, 8/1993, pp. 103 - 108.

Pearce, David W., W. R. Cline, A. N. Achanta, S. Fankhauser, R. K. Pachauri, R. S. J. Tol, P. Vellinga (1996): The Social Costs of Climate Change: Greenhouse Damage and the Benefits of Control. In: J. J. Bruce, H. Lee, E. F. Haites: Climate Change 1995 - Economic and Social Dimensions of Climate Change. Cambridge University Press, pp. 179 - 224.

Pearce, David W., Kirk Hamilton, Giles Atkinson (1996): Measuring sustainable development: progress on indicators. In: Environmental and Development Economics 1 (1996): 85 - 101.

Pearce, David W., Robert K. Turner (1990): Economics of Natural Resources and the Environment. New York et al.

Plambeck, Erica L., Chris Hope (1996): PAGE 95 - An updated valuation of the impacts of global warming. In: Energy Policy, Vol. 24, No. 9, pp. 783 - 793.

Rabl, Ari (1993): Discounting of long term costs: what would future generations prefer us to do? Discussion Paper, Ecole des Mines, Paris. Now published together with Rabl (1994) in: Ecological Economics 17 (1996), pp. 137 - 145.

Rabl, Ari (1994): Discounting and intergenerational costs: why 0 may be the appropriate effective rate. Discussion Paper, Ecole des Mines, Paris. Now published with Rabl (1993) in: Ecological Economics 17 (1996), pp. 137 - 145.

Rennings, Klaus (1992): Zur Relevanz der Transaktionskostentheorie für die Verkehrswirtschaft. In: Klaus Rennings, Matthias Fonger, Henning Meyer: Make or Buy - Transaktionskostentheorie als Entscheidungshilfe für die Verkehrswirtschaft, S. 9 - 50.

Rennings, Klaus (1994): Indikatoren für eine dauerhaft-umweltgerechte Entwicklung. Metzler-Poeschel, Stuttgart.

Rennings, Klaus, Hubert Wiggering (1997): Steps towards Indicators of Sustainable Development: Linking Economic and Ecological Concepts. In: Ecological Economics, 20, pp 25 - 36.

Schellnhuber, Hans-Joachim (1995): Die internationale Klimawirkungsforschung auf ihrem langem Marsch zur Integrierten Modellierung. In: Peter Hennicke (Ed.): Klimaschutz- Die Bedeutung von Kosten-Nutzen-Analysen. Birkhäuser, Berlin, Basel, Boston, pp. 52 - 82.

Simon,H. A. (1972): "Theories of Bounded Rationality". In: C. B. Radner, R. Radner (Eds.) : Decision and Organization. Amsterdam, North Holland Publishing Company. Cited in Faucheaux/Froger (1994).

Spash, Clive L. (1994): Double CO_2 and Beyond: Benefits, Cost and Compensation. In: Ecological Economics, 10/1994, pp. 27 - 36.

SRU - Rat von Sachverständigen für Umweltfragen (Council of Environmental Advisors): Umweltgutachten 1994 - für eine dauerhaft-umweltgerechte Entwicklung. Metzler-Poeschel, Stuttgart.

Tol, Richard S. J. (1996a): The Damage Costs of Climate Change: Towards a Dynamic Representation. In. Ecological Economics 19 (1996), pp. 67 - 90.

Tol, Richard S. J. (1996b): The Damage Costs of Climate Change: Towards an Assessment Model, And a new Set of Estimates. Draft Paper, Instituut voor Milieuvraagstukken (IVM), Frije Universiteit Amsterdam.

Tol Richard S. J. (1996c): The Climate Framework for Uncertainty, Negotiation and Distribution (FUND). Discussion Paper 96/02, Institute for Environmental Studies, Free University Amsterdam, the Netherlands.

van der Loo, F. A. (1993): From Sustainability to Indicator: The Climate Change Case. In: Weterings, R., J. B. Opschoor (1994): Towards Environmental Performance Indicators Based on the Notion of Environmental Space. Report to the Advisory Council for Research on Nature and Environment, Rijswijk, the Netherlands, pp. 53 -68.

WBGU - Wissenschaftlicher Beirat der Bundesregierung Globale Umweltveränderungen (German Advisory Council on Global Change) (1995a): Jahresgutachten 1995 - Welt im Wandel: Wege zur Lösung globaler Umweltprobleme. Springer, Berlin, Heidelberg, New York.

WBGU - Wissenschaftlicher Beirat der Bundesregierung Globale Umweltveränderungen (German Advisory Council on Global Change) (1995b): Scenario for the derivation of global CO_2 reduction targets and implementation strategies. Secretariat of the WBGU, Bremerhaven.

Weterings, Rob, Johannes B. Opschoor (1994): Towards Environmental Performance Indicators based on the Notion of Environmental Space. Report to the Advisory Council for Research on Nature and Environment , the Netherlands, Rijswijk.

Weyant, J. et al. (1996): Integrated Assessment of Climate Change: An Overview and Comparison of Approaches and Results. In: J. Bruce, H. Lee, E. F. Haites: Climate Change 1995 - Economic and Social Dimensions of Climate Change. Cambridge University Press, pp. 367 - 396.

6 Discounting of Long Term Costs: What Would Future Generations Prefer Us to Do?[1]

Ari Rabl
Ecole des Mines, 60 boul. St.-Michel, 75272 Paris

Abstract. Decisions with intergenerational consequences may leave future generations worse off (i.e., not be Pareto-optimal) if costs and benefits are not discounted correctly. We point out that this happens if the conventional social discount rate is applied to intergenerational costs: with that rate the justification of intergenerational costs (benefits of a project are invested and the proceeds can cover future costs) leads to an inconsistency over the long term (eventually the supposed benefits of any project become larger than the entire GNP). The conventional approach commits an error by treating the time preference component of the discount rate as if it represented creation of wealth while it really involves only redistribution. The time horizon of the redistribution is limited by the duration of market transactions, especially loans. Thus we recommend a two-step discounting procedure, with the conventional social discount rate for the short term (about 30 years) and the growth rate of the economy for the long term. Data for GNP per capita growth in the US, the UK, France, Germany, Japan, and the Soviet Union since the industrial revolution, combined with data for net economic welfare per capita suggest an intergenerational discount rate in the range of 1 to 2%. For practical application the rate at which costs and benefits will grow over time is just as important as the discount rate; only the difference between these two rates matters. For many environmental costs these rates may be close to each other during the next few generations; this implies that some of the uncertainty in the prediction of rates cancels and one can treat the problem as if the effective rate were close to zero. The consequence for costs such as global warming or nuclear power can be very significant: the valuation can increase by more than an order of magnitude compared to values based on conventional discounting.

Keywords: Discounting; economic growth; environmental costs; intergenerational equity; social discount rate; cost-benefit analysis.

---

[1] Article first published in Ecological Economics Vol.17, pp. 137-145 (1996).

1 Introduction

The most troubling and controversial environmental impacts are those that can affect future generations.[2] The question of discounting is crucial for any decisions in this matter because the balance of costs and benefits can be totally altered by the choice of the discount rate. The problem is epitomised by the case of nuclear power where the legacy of radioactive waste may burden our descendants over millions of years. But even our use of fossil fuels may, through the greenhouse effect, impose severe costs on several generations.

The approach of conventional economic practice is to discount all costs and benefits at the social discount rate, with typical values in the range of 3 to 8% (Lind, 1982; Howe, 1990; Portney, 1990; Quirk and Terasawa, 1991).[3] This is highly controversial because future costs are reduced so much that many people reject the entire analysis, feeling that it fails to take the interests of future generations into account (Costanza and Wainger, 1991). As a typical expression of this criticism of intergenerational discounting we cite the statement "[...] use of a social discount rate greater than zero leads ultimately to a disenfranchising of future generations" in the *Highlights: Synthesis of the Issues* chapter of the Expert Workshop on Life-Cycle Analysis of Energy Systems (OECD, 1992).

While the literature on this subject is long and inconclusive (see, e.g., Ramsey, 1928; Markandya and Pearce, 1988), we believe we can add a fresh perspective by asking what future generations would prefer us to do. Our paper is motivated by the fact that the decision process through which we will affect future generations involves an analysis of the benefits and costs of proposed projects (including uncertainties) as perceived by the decision maker.[4] The situation is analogous to that of parents who must make a decision that will affect their children before the latter can express their own preferences. It is difficult to imagine a better approach than trying to the best of one's ability to estimate the costs and benefits imposed on the descendants. Despite the obvious difficulties of trying to predict future preferences, the underlying concept of costs and benefits is universal, as

---

2 For example, according to a recent opinion poll (Eurobarometer 1993, p.34) the possibility of irreversible consequences for future generations is the most widely stated reason for considering an environmental impact serious.

3 All rates in this paper are real (i.e., above inflation) and stated as %/year.

4 For example, the choice of energy paths will be influenced by the results of a major international project to evaluate the external costs of fuel cycles, the joint Accounting Framework/ExternE Program of the CEC and the US DOE (ORNL/RFF, 1994; EC, 1995; see also an analogous study by Rowe et al., 1995), in which the author is participating (Curtiss at al., 1995). One of the findings of this program is that intergenerational effects for nuclear and for fossil fuels can dominate the external costs, depending on the choice of the discount rate. To deal with this issue, the ground rule was adopted of carrying out the valuation for three discount rates: 0%, 3%, and 10%. The present paper shows that the range can and should be narrowed considerably.

demonstrated by its success in explaining even the behaviour of animals (Alcock, 1989).

Of course, wealth does not equal welfare. But nobody seems to know how to measure well being or utility of future generations; even for the present generation there is no consensus, as shown by the difficulties of defining a net economic welfare (see Section 4 below). In any case, all attempts at doing so take as starting point the GNP and apply corrections to it. Even if cost-benefit analysis is not ideal, we have no better alternative. Many cost-benefit analyses deserve to be criticised, for example, when they fail to account for externalities; however, the criticism concerns the practice not the principle.

Thus we focus pragmatically on the needs of decision makers and we remain within the framework of cost-benefit analysis, rather than trying to solve the discount rate problem in its generality, or trying to define criteria of sustainability (for a recent review see Krautkraemer et al., 1993) and intergenerational equity (see, e.g., Solow, 1986; Howarth and Norgaard, 1992 and 1993). We develop a two-step discounting procedure that is compatible with the current practice of discounting within our generation, yet respects the interests of future generations.[5]

We begin, in Section 2, by pointing out an inconsistency in the traditional justification of discounting (benefits of a project are invested and the proceeds cover future costs) when the conventional social discount rate is applied to intergenerational costs (eventually the supposed benefits of any single project become larger than the entire GNP). The conventional argument commits an error by treating the time preference component of the discount rate as if it represented creation of wealth while it really involves only redistribution. The time horizon of the redistribution is limited by the duration of market transactions, especially loans. Only the growth component of the discount rate is relevant for a cost-benefit analysis from the point of view of future generations.

Therefore we propose a discounting procedure that uses the conventional rate up to a time horizon t_{short} (about 30 years) and a reduced rate beyond t_{short}. As shown in Section 3, a cost-benefit analysis from the perspective of future generations implies that the preferred rate beyond t_{short} is the growth rate r_{gro} of the economy. To find guidance for the choice of a numerical value we examine, in Section 4, data for GNP growth of the US, the UK, Germany, France, Japan, and the Soviet Union since the industrial revolution. Assuming that population growth cannot continue indefinitely, we focus on the growth of GNP per capita. Together with data for net economic welfare, available for the US, this leads us to recommend rates in the range of 1 to 2%.

To highlight the consequences we point out, in Section 5, that the discount rate r_{dis} is only one of a pair of elements that effect the analysis equally. Before discounting one must predict the rate r_{esc} at which future costs and benefits will

[5] Our analysis concerns only public investments, since there are essentially no private investment decisions (other than intra-family) where intergenerational costs are taken into account.

grow. Formally, the result looks as if one had simply discounted at an effective rate $r_{eff} = r_{dis} - r_{esc}$. Many environmental costs can be expected to grow at a rate between 0 and r_{gro}. If, as is likely for the next few generations, r_{esc} is fairly close to the latter, some uncertainty of the predictions cancels. Thus in some cases our recommendation is in effect not far from the approach advocated by many environmentalists: one can treat certain long term costs with an effective rate r_{eff} which is close to zero.

Apparently our proposal is not new, even though we have not found any specific references. There have been papers proposing a reduced discount rate for intergenerational effects, for instance Burton (1993) who calls attention to "the potential for an intertemporal externality" but does not show how the intergenerational rate is to be determined. We believe that the two-step procedure of our paper has not gained favour until now because the error (of counting redistribution as new wealth) of the conventional argument had not been recognised explicitly.

2 Critique of Conventional Discounting

Suppose an investment project entails a single benefit B (net of investment) at time t = 0 and a single cost C at time t = N. Discounted at rate r, the net present value is:

$$P = B - \exp(-rN)\, C \tag{1}$$

The project is beneficial if and only if P > 0. It is instructive to look at the future value F of this project at time t = N:

$$F = \exp(rN)\, P = \exp(rN)\, B - C. \tag{2}$$

This equation highlights the traditional justification for the discounting of long term costs: the benefit B can be invested at rate r and the resulting future value is available to pay for the future cost C. We accept this justification for the principle of discounting.

However, we disagree with the practice of using for r the conventional social discount rate when intergenerational effects are involved. Roughly speaking the discount rate r arises from two distinct phenomena: the preference for present over future consumption, and the growth of the economy. This is often expressed in the classic equation (see, e.g., Nordhaus, 1994):

$$r = r_{pref} + \eta\, r_{gro} \tag{3}$$

where
r_{pref} = pure rate of time preference, r_{gro} = growth rate of real income, and:

$$\eta = - \frac{u''(c)\, c}{u'(c)} \tag{4}$$

is the elasticity of marginal utility of income, where u(c) is the utility of the consumption c. For the social discount rate it is reasonable to take values of these parameters in the vicinity of 3% for r_{pref}, 1 for η (which corresponds to a logarithmic form of u(c)), and 2% for r_{gro} (= rate of GNP growth), resulting in a discount rate r of 5%.

In models of economic growth the parameters in Eq.3 are generally not independent. For example, r_{pref} influences savings and investments which in turn influences r_{gro}. In the present paper the relation between time preference and r_{gro} is of no concern for several reasons. First of all, we are looking at a project from the perspective of the future. Looking back at any period, one can simply take the values of r_{pref}, η, and r_{gro} experienced during this period as given, regardless of how they may be related or whether they have been calculated by an economic model or extracted from historical data. Secondly, if our two-step discount procedure is implemented, the effect on growth is negligible because only intergenerational costs are affected and these account only for a small part of the economy (and the effect would arise only from differences in total investment due to intergenerational costs).

Now, looking at F of Eq. 2 from the perspective of the future, we see that there is an inconsistency if any rate r greater than the GNP growth rate r_{gro} is used for a sufficiently long time: eventually the annual benefit r exp(rN) B becomes larger than the entire GNP - clearly an absurdity. If one expresses B as a fraction of GNP, the time N_x at which the interest begins to exceed the GNP is readily found by solving:

$$\exp(r_{gro}\, N_x) = r \exp(rN_x)\, B. \tag{5}$$

To show that this has practical relevance, let us insert numbers that correspond to the nuclear power program in France - with some simplifications to render the argument more transparent. In particular let us assume steady state conditions where the plants are replaced at a constant rate, each lasting 30 years. The annual investment cost I for this replacement amounts to approximately I = 0.33% of GNP, based on the following data:

1) GNP = US\$ $1.2 * 10^{12}$ in 1990 (USDOC, 1992),
2) nuclear capacity = $60 * 10^9$ W, and

3) investment cost US$ 2/W (Ministère de l'Industrie, 1993) spread over 30 years.

Since we are focusing on intergenerational costs, let us interpret B in Eq.1 as the net present value of this investment calculated without intergenerational costs. The exact value of B depends on details of cost components that are not of interest here. For the present purpose it is sufficient to assume a value for the ratio of B/I (the benefit-cost ratio excluding intergenerational costs). We assume a value of B/I = 3.3 as an approximate value implied by cost data for nuclear power in France (Ministère de l'Industrie, 1993; EdF, 1994). Thus B = 1.1% of GNP, and Eq. 5 yields $N_x = 250$ years. This result is not very sensitive to B/I and in any case the precise numbers do not matter: the point of this exercise has merely been to demonstrate that the justification of conventional discounting breaks down well within the time span over which it is supposed to be valid.

At what point does this breakdown occur? Before N_x, but not within the present generation: the benefits can indeed be invested to bear interest at rates comparable to the social discount rate r because there exist financial instruments such as bonds that offer such returns over durations up to 30 years. This interest comes from two sources, reflecting the two components of the discount rate: there is the money people pay to be able to consume now rather than in the future, and there is the gain from economic growth. Only the latter represents creation of wealth; the former merely redistributes it. Assuming constant rates, the gain from growth continues forever. But the money from redistribution is a limited pool, essentially paid by each generation for its own consumption preferences. It would be an error to consider it as income forever.

The temporal boundaries of this redistribution are fuzzy because generations overlap and different loans have different durations. Any particular loan of duration N years redistributes money over a period of N years, and the redistributive contribution of this loan to the social discount rate stops with the end of the loan. Thus the time preference component of the discount rate cannot be counted for the compensation of damages beyond the duration of the loans that are active at the time when a project is chosen. There is a large market for loans up to 30 years (e.g., residential mortgages in the US); in addition there are funds for retirement, which run over longer periods, in some cases up to about 70 years between pay in and pay out. The longer the time span between pay in and pay out, the smaller the number of transactions.

Within the present generation the conventional rate is justified by the existence of a market of borrowing and lending activities which expresses the preferences between current and future consumption. There is no intergenerational market, and only the growth component of the discount rate is relevant for a cost-benefit analysis from the point of view of future generations. Without a market where intergenerational preferences can be expressed, the role of r_{pref} and u(c) in Eq.3 becomes irrelevant beyond t_{short} (since, as explained above, a cost-benefit analysis is the best we can do). Thus the discount rate can be one of the parameters in a cost-benefit analysis that change with time.

3 Cost-Benefit Criterion of Future Generations

Note that the absence of a market does not rule out the possibility of compensation from the present to the future. Therefore the basis of the cost-benefit approach is valid even for intergenerational effects. In particular, one can make the ethical choice that the decisions of the present generation be Pareto-optimal for future generations, i.e., no future generation should be made worse off.[6]

These considerations lead us to propose a two-step discounting procedure, using the conventional rate up to a time t_{short} (for which 30 years appears reasonable), followed by a reduced rate for intergenerational effects. Such a change in the discount rate may appear to conflict with the expectation that one obtain future loans at real interest rates comparable to current loans and the fact that the sequence of two loans for the same amount and the same interest rate, one over the first N_1 years, the other over the following N_2 years, is mathematically equivalent to a single loan over $N_1 + N_2$ years. However, this mathematical equivalence does not correspond to economic reality because there is no market of intergenerational loans.

Let us designate by P_{short} the net present value of the project, calculated from the selfish perspective of the generation that makes the decision. Thus P_{short} is the value at the time of the decision; it is calculated with the short term discount rate r_{short} (= r of Eq.3) and excludes any costs beyond t_{short}. Of course, the project is cost effective within this perspective if P_{short} is positive; in other words the short term decision criterion is:

cost-benefit criterion of present generation:

yes if $P_{short} > 0$. (6)

If the project is realised it can be considered to increase the GNP by P_{short} at the time of the decision (temporal details of its contribution to the GNP do not matter when seen from the distant future). Henceforth one can assume P_{short} to grow at the average growth rate r_{gro} of the GNP. This growth determines the future benefits that will be available to compensate intergenerational costs.

In principle the growth rate could be different for different projects, different places and different times, involving different mixes of labour, capital and natural resources. For example, a biomass plantation and a petroleum refinery may stimu-

---

6 The distribution of costs and benefits between individuals within a given generation is the responsibility of each generation.

late different growth patterns. In the present paper we will not try to disaggregate such effects, and we will only determine a single average growth rate. This seems appropriate since during the period up to t_{short} much of the growth effects from a project will have diffused throughout the economy.

Consider this project in the years beyond t_{short}, supposing the cost in year t to be C(t) (the argument holds equally if there are benefits beyond t_{short}). In the year t the amount P_{short} will have grown to $P_{short} \exp(r_{gro}\, t)$. The net benefit in the year t is obtained by subtracting from this value the costs incurred after t_{short}. These costs must likewise be corrected because their effect also grows at the rate r_{gro}. The net benefit in the year t is therefore:

$$B(t) = P_{short} \exp(r_{gro}\, t) - \int_{t_{short}}^{t} dt'\, C(t') \exp(r_{gro}\, (t-t')) \tag{7}$$

where the exponential in the integral reflects the fact that the effect of the cost C(t') grows at rate r_{gro} from t' to t. A generation in year t prefers us to choose the project if and only B(t) is positive. This is the case if and only if the quantity

$$P_{long} = [P_{short} - \int_{t_{short}}^{t} dt'\, C(t') \exp(-r_{gro}\, t')\,] \tag{8}$$

is positive. Therefore we obtain the criterion of the future generations as:

cost-benefit criterion of future generations:
yes if $P_{long} > 0$. (9)

This has a simple interpretation. The right hand side of Eq. 8 is formally the same as the present value of a project that brings an initial benefit P_{short}, followed by a series of costs beyond t_{short} that are discounted at the rate r_{gro}. Thus we obtain the cost-benefit criterion of future generations if we discount the costs beyond t_{short} at the rate r_{gro}. A project satisfies intergenerational Pareto-optimality if and only if Eq. 9 holds for each future generation. So Eq. 9 needs to be evaluated separately for each generation (although this is trivial if the stream of costs and benefits is uniform).

4 Data for r_{gro}

To apply this one needs to predict the growth rates for the future. Economic growth is a complex phenomenon. There is no guarantee that the GNP will grow.

Capital could be invested wisely, squandered on worthless consumption, or even destroyed by warfare; obviously the corresponding growth will be very different. In this paper we do not try to predict r_{gro} from first principles. Rather, we look for guidance in past history. Short term trends (decades or less) appear irrelevant, but averaged data since the industrial revolution may not be inappropriate for the next centuries, despite overpopulation, environmental degradation, and depletion of resources - Malthusian worries that were just as prevalent two hundred years ago. The last two centuries have seen appreciable growth, several horribly destructive wars notwithstanding. Is there reason to expect less growth now that we are on the threshold of the computer revolution, that family planning is ever more widely practised, and that the civilised world has learned the lessons of war and peace? Of course, there is no guarantee. Rome was overrun by barbarians - and yet a renaissance ensued.

In trying to predict future growth, one needs to distinguish population growth from other components of growth, because the population explosion since the industrial revolution is likely to come to an end. Even if technological progress allows the continued expansion of the supply of food and other resources, there is one limitation for which an expansion is difficult to imagine: the availability of land, in view of people's demand for "open space." For that reason we take the growth rate of GNP per capita as guide in choosing r_{gro}.

Some data for long term average GNP growth can be found in the classic textbook of Samuelson and Nordhaus (1985), as well as in Toutain (1987). We present their data in Table 1, together with data for population growth from McEvedy and Jones (1978) and USDOC (1992). The growth rate for GNP per capita is the difference between the first two columns.

The highest rates in this table are 4.2% for the Soviet Union and 3.5% for Japan, as well as 3.6% for France in the decades after World War II. These rates may reflect unique historical circumstances of catching up from a low position relative to richer countries. Furthermore, Japan may be unusual with an exceptionally high ratio of savings over consumption, and one may have doubts about the reliability of data from the Soviet Government. The long term average rates for France, Germany, the United Kingdom, and the USA are much lower, between 1.3 and 1.9%.

Now to the question whether GNP is an appropriate measure of wealth. It is obviously flawed, failing to account for such phenomena as the underground economy or the damage from pollution. Alternatives have been proposed, in particular the Net Economic Welfare index, but they are not yet universally accepted and data are much more limited. Here, too, an interesting data point is cited by Samuelson and Nordhaus (1985) which we show in Table 2: from 1929 to 1984 the Net Economic Welfare per capita of the US grew at an annual rate of 1.1% while the corresponding growth rate of the GNP per capita was 1.7%.

Table 1: Average growth rates in %/yr for GNP, population, and GNP per capita.

Country	Period	GNP	Population	GNP/cap
France	1789-1982			**1.5** [f]
	1949-1982			**3.6** [f]
Germany	1850-1984	2.8 [a]	0.9 [d]	**1.9**
Japan	1874-1929	4.5 [a]	1.0 [d]	**3.5**
Soviet Union	1885-1913	3.3 [a]	1.0 [d]	**2.3**
	1928-1983	4.9 [a]	0.7 [d]	**4.2**
United Kingdom	1855-1984	2.1 [a]	0.8 [d]	**1.3**
USA	1800-1984	3.5 [b]	2.1 [e]	**1.4**
	1890-1984			**1.8** [c]

Notes:

[a] Table 35-1, p.776, of Samuelson and Nordhaus, 1985.

[b] p.79, of Samuelson and Nordhaus, 1985.

[c] Fig.25-1, p.562, of Samuelson and Nordhaus, 1985.

[d] McEvedy and Jones, 1978.

[e] USDOC, 1992.

[f] Toutain, 1987.

If one disregards the high GNP growth rates for Japan and the Soviet Union as being less representative than the long term data for the other countries in Table 1, and if one takes the point of view that the relevant rate should perhaps be reduced towards the NEW/GNP ratio of Table 2, then one is lead to the conclusion that values of r_{gro} in the range of 1 to 2% may be a good guess. This range is clearly much lower than the values in the range of 3 to 8% usually recommended for the social discount rate (Lind, 1982; EPA, 1983; Howe, 1990; Quirk and Terasawa, 1991). To appreciate the significance of this difference in rates, one needs to consider the evolution of costs.

Table 2: Average growth rates in %/yr for GNP and NEW (Net Economic Welfare) per capita, for US from 1929 to 1984.[1]

GNP/cap	1.7
NEW/cap	1.1

Note:
[1] From Fig.6-4, p.119, of Samuelson and Nordhaus (1985)

5 The Evolution of Costs

Before discounting a cost one must estimate its value; this involves extrapolation from currently known costs. If one makes the customary assumption of constant growth rates, a cost C(t) that occurs at time t is calculated by applying a factor $\exp(r_{esc}\ t)$ with an escalation rate r_{esc} to the value C(0) that has been determined at time 0 (the present). Then one divides by a factor $\exp(r_{dis}\ t)$ with discount rate r_{dis} to obtain the present value:

$$\text{present value of } C(t) = C(0) \exp((r_{esc} - r_{dis})\ t). \tag{10}$$

Only the difference:

$$r_{eff} = r_{esc} - r_{dis} \tag{11}$$

matters. Even though this point is obvious and has been noted before, for instance by Fisher and Krutilla (1975), almost all of the environmental literature has been preoccupied with the discount rate while neglecting r_{esc}. Let us sketch some possibilities. For many intergenerational environmental costs, such as health impacts and global warming, a plausible value of the escalation rate is r_{esc} _ r_{gro}, at least in the intermediate term (the next few generations). The reason is that their valuation is based on willingness-to-pay (WTP), and as such it can also be expected to be approximately proportional to GNP/capita.

Quite generally WTP can be considered to evolve in two ways: firstly, it is proportional to per capita wealth, secondly, the constant of proportionality may change due to shifts in preferences or costs. The fraction of wealth spent on any one item can increase, but not sustainably beyond 100%. Since over the long term an exponentially growing term will exceed any constant, r_{esc} cannot be larger than r_{gro}. This implies that the net rate r_{eff} in Eq.11 should not be negative.

One also needs to consider the possibility of adaptive measures that allow us to reduce the impacts of environmental burdens. For example, if a cure for cancer is found, it would not be appropriate to apply today's valuation of cancers beyond that time. Thus in the very long run the costs may become small or negligible. A positive value of r_{eff} is compatible with this possibility, a value of zero is not.

These considerations lead us to recommend a simple working rule: forget about the escalation rate and calculate the intergenerational costs (beyond t_{short}) as if the discount rate were close to but slightly above zero, and in any case no larger than r_{gro}:

$$0 < r_{eff} < r_{gro} \approx 1 \text{ to } 2\%, \text{ probably } r_{eff} \text{ close to } 0 \text{ for next few generations.} \quad (12)$$

An attractive feature of this argument is that uncertainties due to the prediction of growth rates cancel approximately.

Now the break from conventional discounting becomes manifest: with conventional discounting the effective rate is several percent higher than zero while the discount procedure recommended here yields an effective rate close to zero. The consequence for the assessment of intergenerational costs can be dramatic: the present value of an infinite series discounted at rate r_{eff} is proportional to $1/r_{eff}$.

6 Conclusion

We have shown that the justification for the discounting of intergenerational costs of a project involves an inconsistency if the conventional social discount rate is applied beyond the generation that chooses the project. If one makes the public policy choice of accepting the cost-benefit perspective of future generations, one finds that costs (and benefits, by symmetry) beyond the first generation should be discounted at the growth rate of the economy. The transition time between short term and long term rate is determined by the duration of loans, and 30 years appears reasonable as first approximation. A more exact procedure would involve a gradual transition, corresponding to the spectrum of loans, but that is beyond the scope of this paper.

Contrary to the common belief of many people, discounting does not disadvantage future generations if this two-step discounting procedure is adopted. The cost-benefit criterion of Eqs. 8 and 9 assures intergenerational Pareto-optimality by giving each future generation a veto over our decisions. In fact, if we do not discount, future generations could chide us for decisions that make them worse off. But we have also pointed out that, when all relevant growth rates are taken into account, our procedure may often be quite close numerically to the position of people who are against discounting. The consequence for costs such as global warming or nuclear power can be very significant: the valuation can increase by an order of magnitude compared to values based on conventional discounting.

Acknowledgements

These ideas have evolved during discussions with many colleagues, both in the natural and in the social sciences. In particular I would like to thank Paul Bendt, Brigitte Desaigues, Jochen Diekmann, Mona Dreicer, Faye Duchin, Rainer Friedrich, Chuck Howe, Jan Kreider, Harald Ries, Tom Schneider, Bent Sørensen and Amos Zemel, as well as several referees. This work was made possible by a grant from the European Community, within the ExternE Program of the JOULE 2 Program of the DG XII.

References

Alcock, J. 1989. *Animal Behaviour: an Evolutionary Approach.* 4th edition. Sinauer Associates, Inc. Publishers. Sunderland, MA.

Burton, P. S. 1993. "Intertemporal preferences and intergenerational equity considerations in optimal resource harvesting". J. Environmental Economics and Management, vol. 24, 119-132.

Cline, W. R. 1992. *The Economics of Global Warming.* Institute for International Economics. 11 Dupont Circle, NW. Washington, DC 20036-1207.

Costanza, R. and L. Wainger (eds.) 1991. *Ecological Economics: the Science and Management of Sustainability.* Columbia University Press. New York.

Curtiss, P. S., B. Hernandez, A. Pons, A. Rabl, M. Dreicer, V. Tort, H. Margerie, G. Landrieu, B. Desaigues, and D. Proult 1995. "Environmental Impacts and Their Costs: the Nuclear and the Fossil Fuel Cycles." ARMINES (Ecole des Mines), 60 boul. St.-Michel, 75272 Paris CEDEX 06.

EC 1995. *Externalities of Fuel Cycles, "ExternE" Project: Summary Report.* Report N° 1. European Commission, Directorate-General XII, Science Research and Development. JOULE programme.

EdF 1994. "Les Chiffres Clés de 1993". Electricité de France.

EPA 1983. *Guidelines for Performing Regulatory Impact Analysis.* US Environmental Protection Agency, Office of Policy Analysis. Report EPA-230-01-84-003, reprinted March 1991. Washington, DC 20460.

Eurobarometer 1993. *"Les européens et l'environnement en 1992" (Europeans and the environment in 1992).* Opinion survey carried out in the framework Eurobaromètre 37.0. European Commission, DG XI/C/4. Brussels.

Fisher, A. C. and J. V. Krutilla 1975. "Resource conservation, environmental preservation and the rate of discount". *Quarterly Journal of Economics*, August 1975, p.364.

Howe, C. W. 1990. "The social discount rate". Introduction to special issue on social discount rate. *J. of Environmental Economics and Management*, vol. 18, N°2, p. S-1.

Krautkraemer, J., J. Pezzey, and M. A. Toman 1993. "Economic theory and 'sustainability.'" Discussion paper ENR93-14. Resources for the Future. 1616 P Street N.W., Washington, DC 20036.

Lind, R. C. (ed.) 1982. *Discounting for Time and Risk in Energy Policy.* Johns Hopkins Press for Resources for the Future. Baltimore, MD.

Markandya, A. and D. Pearce 1988. "Environmental considerations and the choice of the discount rate in developing countries." Environment Department Working Paper No.3. World Bank. Washington, DC.

Markandya, A. and B. Rhodes 1992. "External costs of fuel cycles; an impact pathway approach: economic valuation." EC/US fuel cycle study. Metroeconomica Ltd. London, UK.

McEvedy, C. and R. Jones 1978. *Atlas of World Population History.* Penguin Books. Harmondsworth, UK.

Ministère de l'Industrie 1993. "Les coûts de référence". DGEMP - DIGEC, Service de l'Electricité, Ministère de l'Industrie. 99 rue de Grenelle, F75353 Paris.

Nordhaus, W. D. 1994. *Managing the Global Commons: the Economics of Climate Change.* MIT Press. Cambridge, MA.

Norgaard, R. B. and R. B. Howarth 1992. "Economics, ethics, and the environment". Chapter 11, p.347-364, of *The Energy-Environment Connection*. J. M. Hollander (ed.) Island Press, Washington, DC.

Norgaard, R. B. and R. B. Howarth 1993. "Intergenerational transfers and the social discount rate". *J. of Environmental and Resource Economics*, vol. 3, N°2, p. 337-358.

OECD. 1992. *Proceedings of OECD Workshop on Life Cycle Analysis of Energy Systems*. Paris, 21-22 May 1992.

ORNL/RFF 1994a. Fuel Cycle Externalities: Analytical Methods and Issues. Report No.2 on the External Costs and Benefits of Fuel Cycles. July 1994. Prepared by Oak Ridge National Laboratory and Resources for the Future. Oak Ridge National Laboratory, Oak Ridge, TN 37831.

ORNL/RFF 1994b. *Estimating Externalities of Coal Fuel Cycles*. Report No.3 on the External Costs and Benefits of Fuel Cycles. September 1994. Prepared by Oak Ridge National Laboratory and Resources for the Future. Oak Ridge National Laboratory, Oak Ridge, TN 37831.

Portney, P. R. 1990. "Comments on Discounting Session." *J. of Environmental Economics and Management*, vol. 18, N°2, p. S-63.

Quirk, J. and K Terasawa 1991. "Choosing a government discount rate: an alternative approach". *J. of Environmental Economics and Management*, vol. 20, p. 16.

Ramsey, F. P. 1928. "A mathematical theory of saving". *Economics Journal*. Vol.38, p.543-559.

Rowe, R.D., C.M. Lang, L.G. Chestnut, D. Latimer, D. Rae, S.M. Bernow, and D.White 1995. *The New York Electricity Externality Study*. Oceana Publications, Dobbs Ferry, New York.

Samuelson, P. A. and W. D. Nordhaus 1985. *Economics*. 12th ed. McGraw-Hill. New York.

Solow, R. M. 1986. "On the intergenerational allocation of natural resources." Scand. J. of Economics, vol.88, p.141-149.

Toutain, J. C. 1987. *Le Produit Intérieur Brut de la France de 1989 à 1982*. Cahiers de l'I.S.M.E.A. Série Histoire Quantitative de l'Economie Française, No.15. Published by I.S.M.E.A., 11 rue Pierre-et-Marie-Curie (Institut Henri-Poincaré), 75005 Paris.

USDOC 1992. *Statistical Abstract of the United States*. 1992 edition. Published annually by the US Department of Commerce, Washington, D.C.

7 Is Adaptation Cheaper than Prevention? The Applicability of Cost-Benefit Analysis to Global Warming[1]

Peter Hennicke
Director of the Energy Department of
Wuppertal Institute for Climate, Environment, Energy GmbH
North-Rhine-Westphalian Science Centre
Wuppertal

Ralf Becker
Otto-von-Guericke-Universität
Magdeburg

1 Introduction

Is adaptation to climate change cheaper than prevention? To answer **this** question is a rather sophisticated job: More ecomomists, more **global** cost-benefit-analysis and heroic assumptions and calculations are needed. But is it the **right** question?

As a first answer to this question, we present some exercises with the modified DICE-Model which has been used by William Nordhaus. By changing some crucial assumptions the DICE-Model gives us the following results: **Reducing $C0_2$ by about 50% until 2050 is an optimal climate policy**. Maybe, William Nordhaus will not be happy with this conflicting view to his results. Perhaps he will argue that our assumptions are false, or he will change his model and will "prove" the opposite: Adaptation is cheaper than prevention.[2] Who is right at the end of the day? Decision-makers in industry and politics will be confused. Whose

---

[1] This paper is based on Hennicke, P./ Fischedick, M./ Thomas S.: Energy Efficiency in Europe: The "Economy of Prevention," Paper presented to the International Energy Efficiency Business Week in Prague, 5-6 November 1996.

[2] Perhaps William Nordhaus changed his mind, because he signed a statement together with more than 2000 US-economists saying that the US would be able to reduce its industrial emissions to slow global change without damaging its economy. Financial Times 14. 2. 1997.

results should they trust? We recommend: Don't believe us, as far as we argue only as pure cost-benefit economists, but don't believe William Nordhaus either, because he is an economist, too! Let us instead try to ask some other questions with more common sense:

First: Is climate change a problem? The IPCC[3] and most people of the scientific community will answer: Yes, it is! Second: Will there be damages of climate change? The IPCC-reports tell us: Yes, there will be many serious damages, but we do not know when, where and how much will happen, and it will never be possible to monetarise them all. Third: Are there costs of climate protection policy? The answer once more is: Yes, there are; but - in principle - we are able to make much better calculations **on costs of climate protection policy than on future damages.**

What can politicians learn from this? Our opinion is: Don't believe too much in global cost-benefit-analysis of climate change! It does not make sense to base fundamental decisions for mankind ("to adapt or not adapt"?) by using extremely speculative cost figures on **future damages** which will remain unknown or uncertain forever.

There is no doubt, that "no regret options" exist, though, up to now, we don't have enough empirical evidence how much climate protection policy really costs. Therefore, wouldn't it be much more intelligent to concentrate on **costs and benefits** of climate protecting policy? Wouldn't it be much more worthwhile to understand **why and how much "free lunches" or "no-regret options" do exist in real markets and how to overcome the barriers?**

The aim of this paper is, to answer **these** more common sense questions of cost-benefit-analysis of climate protection. This has to do with our understanding of the energy system. If the commodities on the markets were only kilowatt-hours and everything was in equilibrium (as many top-down models assume) there would only be **costs** of climate protection. But if we look at market failures and at existing barriers and if we ask what people really want from energy, namely **energy services,** kilowatt-hours are only a means to an end and the optimal allocation criterion is quite different. **Not cheap energy prices but cheap energy services** for the customers is then the goal of energy policy. The overall equilibrium of the energy system is only reached if the long run marginal costs of supply equal the long run marginal costs of energy efficiency. We will show in this paper: There is a broad range of options where the marginal costs of efficiency are less than the marginal costs of supply, that is the economic rational behind the "no-regret" options. Therefore, the crucial question remains **how many "free lunches"** could be derived from the technically feasible "efficiency revolution."

---

[3] Intergovernmental Panel on Climate Change (IPCC): Summary for Policymakers, Working Group I, II and III, Geneva 1995.

This has some fundamental implications to global cost-benefit-analysis: **If** there exists a large amount of "no regret-options" there is no need to compare **future** costs of damages with current costs of climate protection. On the contrary, we should systematically find out, how much "win-win"-options and what barriers do exist and how to overcome the barriers as quick as possible.

The underlying thesis of this paper is, that giving priority to energy efficiency and the concept of energy services is the key to minimise environmental, human, and geostrategic risks connected with energy consumption. On the basis of reported scenarios, this thesis is shortly summarised at the geographical levels of the world, the European Union and the Federal Republic of Germany. The analysis reveals that efficiency is not only the core of a risk-minimisation[4] strategy, but also of **a least-cost climate protection policy.**

Scenario analysis and case studies based on fostering efficiency show further more that a risk minimising strategy following the precautionary principle can be sound industrial politics and beneficial for the economy.[5] Therefore, in the last part of the paper we show: The prime strategy must be to apply an appropriate mix of instruments to reverse the perverse incentive structure that presently prevails in the energy system. It is not the additional consumption of energy and rising pollutant emissions which should, as is today the case, be worthwhile for suppliers and consumers, but rather the provision of additional energy services with less consumption of (non-renewable) energy. We term such a system "the Economy of Prevention."[6]

---

4 A risk-minimising energy strategy avoids both the risks inherent in harnessing nuclear power and those posed by climatic changes. In our understanding climate protection is a necessary, but not sufficient condition for sustainable development, because the reduction goals for reaching sustainability in a strict sense are far more comprehensive (concerning energy and material flows as well as area). See BUND / Misereor (Eds.): Zukunftsfähiges Deutschland, Berlin, Basel, Boston 1996 and Hennicke, P./ Seifried, D. 1996: Das Einsparkraftwerk. Eingesparte Energie neu nutzen, Berlin, Basel, Boston.

5 Some reasons why economists and politicians do not believe in these scenario results are shown below. As far as we know, up to now Denmark is the only country worldwide where climate protection is accepted by official policy as sound industrial policy.

6 "Prevention" does not exactly have the same positive connotation as the German term "Vermeidung". With "economy of prevention" ("Ökonomie des Vermeidens") we focus on market-oriented instruments and economic incentives to reach the physically necessary reduction of lead indicators (like CO_2) on the way to "sustainability." This way is connected with "new models of wealth" (E.U.v. Weizsäcker) and the questions of "sufficiency."

2 Global Scenarios: Wide Range of Conceivable 'Energy Futures'

A look at representative global energy scenarios published since the 1980s (cf. Figure 1)[7] shows a range of technically accessible energy futures with energy consumption levels for the year 2030 varying by a factor of 7 - despite comparable assumptions of economic and population growth! Energy consumption is not a matter of fate, but political resolution. In other words, energy policy suffers not so much from a technology deficit, but rather and above all from a policy and implementation deficit. Energy policy is failing to exploit the existing technically realisable and macroeconomically cost-effective options and leeway for decision-making. Grave risks could be avoided, but too little is being done. A first and urgent step for decision-makers and consultants in energy policy should therefore be, to systematically compare the underlying assumptions, the structure of the models and the results of "bottom-up" vs. "top-down"-analysis as well as of "use-oriented" vs. "supply-oriented" energy scenarios. This comparison will make it possible to answer the question whether the contradicting views on future energy pathways are caused by insufficient theory or insufficient policy.

Today we can definitively state: Without a substantial improvement in the efficiency with which every 'consumed' kilowatt-hour of energy is used (Ernst Ulrich von Weizsäcker speaks of a feasible enhancement of energy productivity by a "factor 4") it will neither be possible to contain the global risks of climate change and of nuclear energy, nor to peacefully resolve the distributional problems of scarce oil and gas resources in the future.

This pivotal realisation has now also been brought home to the world's major conference of energy suppliers, the World Energy Council (WEC). At the last WEC conference in Tokyo (October 1995), a long-term scenario (to 2050 and 2100) was presented[8] that intensively pursues - for the first time in the WEC context - the question of whether a risk-minimising 'sustainable' energy strategy is globally feasible (WEC/IIASA 1995). By WEC standards, the findings are sensational: In the 21st century, the challenging targets of a global climate protection policy can largely be achieved (CO_2 concentration rise less than 450 ppm, global temperature rise less than 2° C above preindustrial levels) while simultaneously phasing out nuclear energy worldwide. An adequate CO_2 reduction is, however, only achieved well after the year 2050: the German Climate Study Commission (Klima-Enquete) and the IPCC, however, demand a global reduction

---

[7] For references see: Schüssler, M. / Hennicke, P.: Potentiale und Kosten für eine risikoarme Energieversorgung, Übersicht über die Ergebnisse internationaler Studien, Wuppertal 1994 and WEC/IIASA: Global Energy Perspectives to 2050 and Beyond, WEC Report 1995, London 1995.

[8] Cf. WEC/IIASA: Global Energy Perspectives to 2050 and Beyond, op.cit.

of CO_2 emission levels by 50% by the year 2050 (Enquete 1995; IPCC 1994). The slower pace of the change of course in WEC Scenario C1 is primarily due to the assumed moderate average annual rise in energy productivity of 1.4%, although in 1992 the WEC still considered up to 2.4% p.a. to be feasible.

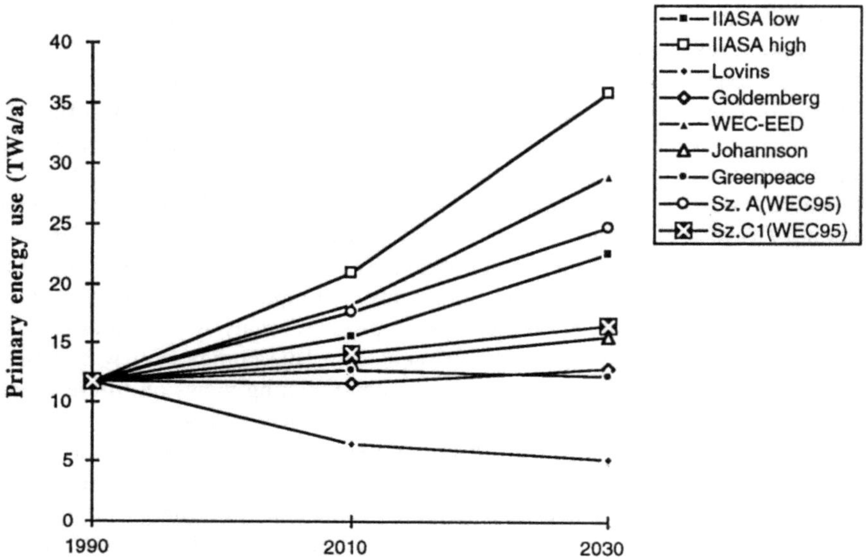

Figure 1: World Energy Scenarios - Overview

The risk-minimising WEC Scenario C1 'coexists' in the new WEC/IIASA analysis with five further variants (A1/2, B1/2, C2) that display more or less risk-cumulating effects. Thus, e.g., in Scenario A2 (A1), CO_2 emissions rise by the factor 2.5 (2.0) and nuclear capacity is more than doubled (quadrupled) by the year 2050. The most important new findings of WEC are, however, to be seen in the appraisal of the investment costs and realisation probabilities of the scenarios:

- For the period from 1990 to 2050, **the cumulative investment costs of Scenario C1 are 33% and 43% lower than the corresponding investment costs of the B and A1 Scenarios!**[9] Customer-side investments have albeit not been considered explicitly in any of the scenarios, and will be higher in C1 than in the others due to the higher efficiency growth. On the other hand, these efficiency improvements also avoid more running energy costs at the customer. Following the arguments of the WEC-authors it thus

9 See Grübler, A./ Jefferson, M./ Nakicenocic, N.: A Summary of the Joint IIASA and WEC Study on Long-Term Energy Perspectives, Laxenburg September 1995

appears justified to expect that the risk-minimising Scenario C1 will also be the more beneficial from an economic perspective.

- The WEC-authors stress that despite the long time horizon the basic decisions as to which path is to be taken must be made today, as the strategies modelled in the scenarios will become mutually exclusive within a few decades, and the capital tie-up periods in the energy system generally span several decades. Concerning the closeness to reality of the 6 scenarios, the authors expressly stress that: "All are held to be realisable. But none assumes that the developments will occur of their own accord." (WEC/IIASA 1995, p. 2). There is therefore at all events a need for energy policy action. Risk minimisation is possible and financeable, but: Energy managers, politicians, and the public at large - all of us must make up our minds soon.

If the WEC-C1-scenario is right there will not only be no trade-off between higher costs now and avoiding damages tomorrow. The contrary will be true: **Climate protection policy creates a double net benefit for societies:** First, for the living generation because it is cheaper than business-as-usual; and second for all next generations, because the risks and damages of future global change will be minimised. These results are in striking contrast to mainstream thinking in economics and to global cost-benefit-analysis of climate change. Therefore, let us make a short excursion. We will show with one typical example, the DICE-model, how the contradicting results of this kind of modelling could be reversed by only changing two or three strategic parameters.

3 The DICE-Model Revisited: An Excursion

William Nordhaus (1992a, 1994) combined a simple global economic growth model of Ramsey-type with a rudimentary model of climate costs and damages caused by economic output. He called it Dynamic Integrated Climate-Economy model or shortly DICE. The model has attracted considerably large attention. One reason for its popularity may be, that the about ten model equations – five for the climate impact chain and five for the economic growth model – are empirically interpreted despite the extreme simplification.

Figure 2 gives an optical overview of the model parts and the interdependencies of the endogenous and exogenous DICE-variables. The strategic variables are investment I and the control rate of GHG emissions. The latter effects the output and the emission rate of output in an opposite relation of magnitude. The world economy comprises the capital stock K, the world product Q and the world consumption C. The utility over generations has to be maximised. The effect chain of climate change starts with emissions E, which cause a higher GHG concentration in the atmosphere M. The induced radiative forcing F increases the temperature of the atmosphere T which interacts with the temperature increase of the ocean T*. In this chain we have two periods of time delay (decades). The transverse variables contain the final stock evaluation of M, K, and T using the corresponding shadow prices at the end of the model horizon.

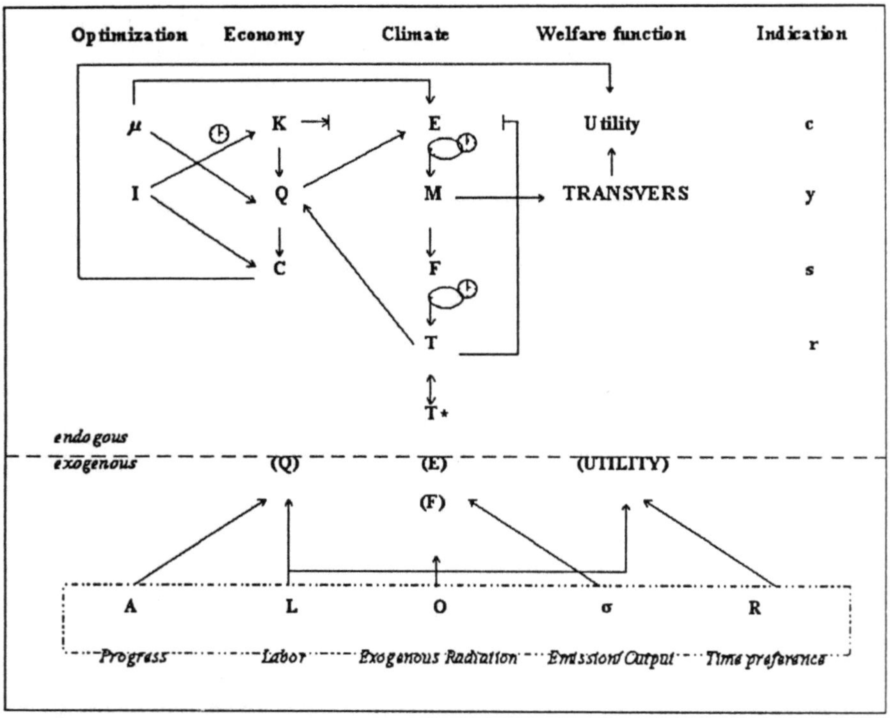

Figure 2: Interdependencies between the DICE variables

The seven approaches to climate-change policy in the following rank order of attractiveness from a pure economic point of view in DICE by Nordhaus (1994, p. 96) illustrate the tendency to favour adaptation instead of mitigation as the strategy for climate protection (see also Nordhaus 1994, Figure 5.6: GHG control rate, p. 94): the (DICE-) Optimal Policy strategy no. 2 recommends only very modest emission reductions of GHG starting at 8.8 percent reduction from baseline in 1995, a smooth increase leading to 13.4 percent in 2075 and an overall maximum rate lower than 17 percent:

1) **Geoengineering.**
   (Large-scale engineering to offset the warming effect; injecting particles, re-foresting, stimulating absorption of carbon in the oceans; shooting smart mirrors into space, seeding the oceans with iron to accelerate carbon sequestration)
2) **Optimal Policy.**
   (Optimal control rate and private investment rate to maximise discounted utility of generations; geoengineering can only be more attractive if this option is not completely included in the reduction cost function of optimal policy)
3) **Ten-Years Delay of Optimal Policy**.
   (Beginning in 2000)
4) **No Controls.**
   ("Baseline")
5) **Stabilising Emissions at 1990 Rates**.
   (Growing percentage reduction of GHGs in the future)
6) **Twenty Percent Emissions Reduction from 1990 Levels.**
   (As in the fourth position following Nordhaus there is no particular merit of this)
7) **Climate Stabilisation.**
   (Only feasible: to slow the GHG-induced global temperature increase to 0.2°C per decade after 1985 with a upper limit of total increase of 1.5°C from 1900)

These results have been criticised by economists in different regards. Barker (1996) outlines a general critique on the simplification in the DICE approach and illustrates shortly the bulk of reasons why the world economy cannot be pressed into a few equations, especially if the model shall be used empirically. In Hennicke and Becker (1996) it has been demonstrated, that DICE results are very sensible to variations of the effective discount rate (or more exactly the pure rate of social time preference), the utility function, the rate of energy efficiency, the CO_2 intensity, and economic growth assumptions. Becker (1996; 1997) shows, that the vote in favour of a relative high discount rate by Nordhaus (1994, p. 122-135) and others on the basis of the argument that the reality has to be described, does not hold; instead Becker explains the insisting on a high social rate of time preference through an insufficient interpretation in Lind (1982) of the criterion for public investment decisions by David Bradford (1975). The attempt to justify the differentiation of a relative high short-term and a low long-term discount rate conceals the conflict between sufficient opportunity cost implementation and "discounting away" long-term climate damages (see Becker, 1997). Richard Tol (1994) suggested to consider damages of climate change without a market price directly in the utility function in DICE. This already leads to an increase of the level of the (DICE-) optimal control rate of GHG between 5 percent and 15 percent.

In addition, Nordhaus himself concedes some methodical difficulties with the DICE calibration (1994, p. 78: "It should be emphasised, however, that there are many combinations of parameter values that could explain the historical data, so conformity to historical observation is but a partial validation of the model.") and a number of important qualifications (1994, p. 97: "The most important shortcoming is that the damage function, particularly the response of developing countries and natural ecosystems to climate change, is poorly understood at present; moreover, the potential for rapid or catastrophic climatic change, for which precise mechanisms and probabilities have not been determined, cannot currently be ruled out. Furthermore, the calculations omit other potential market failures, such as ozone depletion, air pollution, and research and development (R&D), which might reinforce the logic behind GHG reduction or carbon taxes. And finally, this study abstracts from issues of uncertainty, in which risk aversion and the possibility of learning may modify the stringency and timing of control strategies.") So there are innumerable directions of possible DICE extensions. Not only model extensions but a simple alteration of parameter assumptions can shift the DICE results decisively. Figure 3 demonstrates the high sensibility of the DICE results connected with the assumed discount rate.[10]

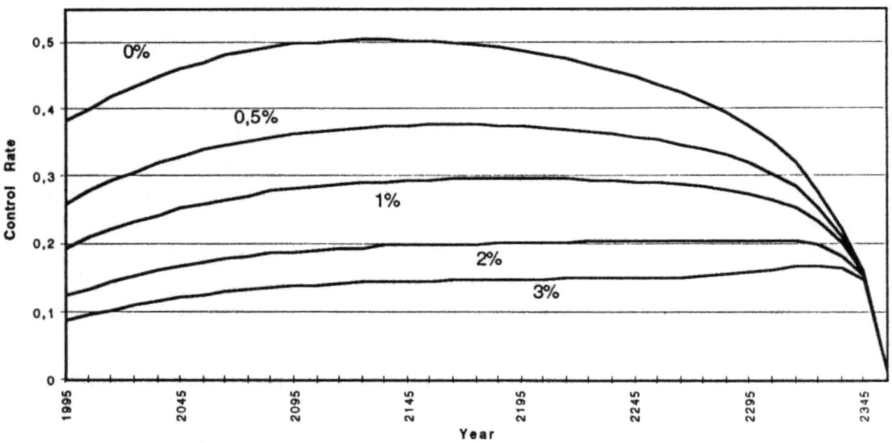

Figure 3: The dependence of the DICE-optimal control path from the pure rate of time preference

[10] In the view of the IPCC Working Group III (1995, p. 10-12) there is still no clear answer to the question of discounting: "The choice of the social discount rate for public investment projects is a matter of policy preference." This judgement should be unsatisfying for an economist regardless of his attitude towards discounting, and this especially, because it is behind the state-of-the-art of mainstream economic modelling (Becker, 1997).

We have computed three additional scenarios with the DICE-model in order to demonstrate the importance of the assumptions concerning the rate of efficiency increase and the market introduction of $C0_2$-free (renewable) energies. Figure 4 shows the significant effect of an increase of energy efficiency and the reduction of CO_2-intensity for the development of global atmospheric temperature growth.[11] If we start with an initial 3 percent growth of energy efficiency corresponding to scenario A2 the greenhouse gas problem would be "solved" because in DICE this leads to a stabilisation of T. In addition or to some extent alternatively, a larger reduction of CO_2 intensity could induce the same result (scenario B).

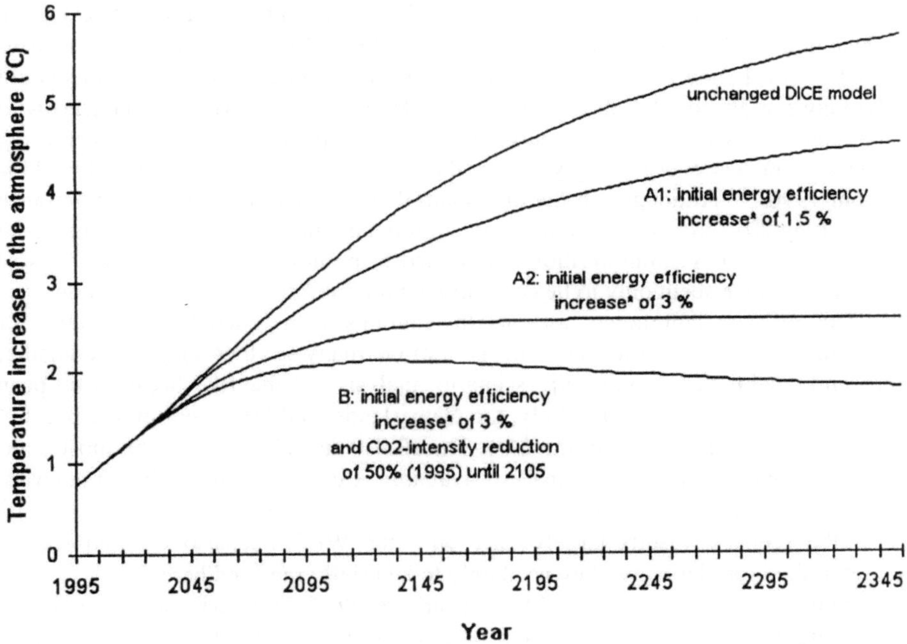

* Growth rate of energy efficiency halves every six decades (as in original DICE)

Figure 4: Sensitivity of results in DICE in relation to energy efficiency increase and reduction of CO_2-intensity

To summarise, first the DICE model recommends high control rates as well as low ones, if we change some strategic model parameters in a way that corresponds

[11] The initial rate of energy efficiency is reduced by 11% every decade, therefore in 2100 it would be about 1 %; this seems to be technically feasible.

to the scope of uncertainties of their values. Secondly, the model can be used to demonstrate the overwhelming importance of the market introduction of renewable energies and of the negawatt concept ("the efficiency revolution").

4 European Union: Risk Minimisation is Financeable

Under present conditions it is not probable that the EU will reach the target of stabilising $C0_2$-emissions by the year of 2000 "... and certainly not to ensure reductions in $C0_2$-emissions after the Year 2000."[12] Recently published scenarios by DG XVII show[13] that "...by 2020, European Community $C0_2$ emissions could be between 3% to 17% above 1990 levels. However, in "Forum", in concert with demand-side actions to keep energy consumption in check, there is an increasing share of nuclear energy supply. Under these circumstances, $C0_2$ could fall by 11% from 1990."[14] Although, on the one hand, these results underline the importance of energy efficiency for a climate protection policy, the message on the other hand is very clear: Risk minimisation is assumed to be not realisable - lower risks of climate change seem only to be compatible with more risks from nuclear power.

This message contradicts not only the findings of the cited WEC Scenario C1 and of some country studies (for the case of Germany see below), but also another recently published "bottom-up"-scenario analysis for the five largest European countries (France, Germany, Italy, the Netherlands, and UK, representing roughly 3/4 of the electricity use in the EU). The IPSEP study[15] presents the most comprehensive end-use-scenario analysis available at this moment for these countries, and an appraisal of the relative investment costs of a climate protection strategy. As the scenario results largely tally with the WEC Scenario C1 and studies undertaken for Germany, they need only to be briefly reported here.

Figure 5 gives an overview of the scenario results for primary energy use. In the two target scenarios, CO_2 reductions of almost 40% ("minimum costs") and almost 60% ("minimum risk"; including phase-out of nuclear power) are achieved by the year 2020. The "minimum risk"-scenario assumes that the available CO_2 reduction

[12] Commission of the European Communities: Report from the Commission. Second evaluation of national programmes under the monitoring mechanism of community $C0_2$ and other greenhouse gas emissions, Brussels 14.3.1996. Reducing $C0_2$ by 15% until 2010 (the common negotiation position of the EU for Kyoto!) therefore will never be reached under present conditions.

[13] Directorate Generale for Energy (DG XVII): Energy in Europe. European Energy to 2020, Special Issue, Brussels Spring 1996.

[14] op cit, p.20. "Forum" is one of four scenarios, the others are "Conventional Wisdom","Battlefield" and "Hypermarket."

[15] Krause, F. et al: Energy Policy in the Greenhouse. Volume II, Cutting Carbon Emissions: Burden or Benefit? The Economics of Energy Taxes and Regulatory Reforms on Climate, Growth and Jobs, Executive Summary, IPSEP 1995.

potentials provided by efficient energy use and generation are largely exploited. Both climate protection strategies generally give rise - this having been tested by extensive sensitivity analysis - to lower investment costs than the reference scenario, the "Conventional Wisdom Scenario" of the EU Commission (KEG 1990).

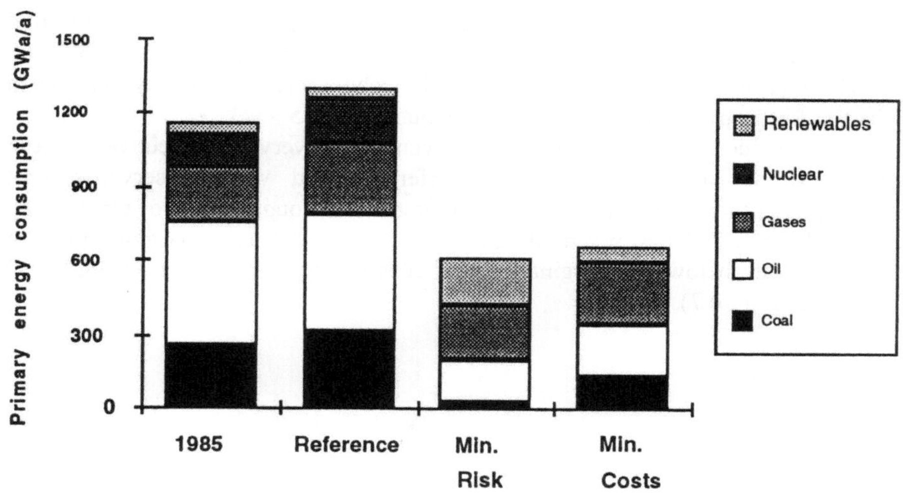

Figure 5: Primary energy use in EU-5 in the year 2020

Looking into the electricity sector in more detail the IPSEP-study arrives at the following results: the absolute savings in the Min-Costs-Scenario for the EU-5 region would be 20-50 billion ECU per year by 2020 (net of all investment costs). The net present value benefit of fully implementing the electrical efficiency resources in the 1995-2020 period is about 140-260 billion ECU: "These demand-side benefits are two to four times larger than the anticipated benefits from liberalising European electricity supply markets. If both reform opportunities are combined, economic benefits would be three to five times larger than those from the supply-side restructuring alone."[16] The main reasons for these striking differences compared to DG XVII-scenarios are due to different estimates about the potential and the cost-effectiveness of efficiency options. Following a "bottom-up"-approach the IPSEP-study looks deeply and in a disaggregated way into more than 90 end-uses and 240 efficiency techniques.

---

16 Krause, F. et al.: Energy Policy in the Greenhouse.Volume Two, Part 3B, NEGAWatt Power. The Cost and Potential of Electrical Efficiency Resources in Western Europe. Executive Summary, El Cerrito, p.iii.

Combining extensive "bottom-up"-analysis with in-depth socio-economic research on the implementation process, powerful arguments for intensified demand side policy engagement can be derived.

Integrated policy mixes based on target group-specific analysis of potentials and implementation conditions will contribute to a significantly higher diffusion and adoption of advanced energy technologies. Sensitivity analysis from the IPSEP project indicates the possible long-run impact of high implementation rates, which can be expected by socio-economic based policy strategies contributing to lasting social learning processes (see Fig. 6). Assuming implementation rates of 50% up to 100% (which can be expected for single technologies), electricity demand in the EU-5 region will equal or even be less than the 1985 level.

Demand-side efficiency policies are expected to be very cost-effective. Even in cases without any energy policy interference and with conservative cost assumptions (high DSE-costs, low generating costs), roughly 60% of the existing demand-side efficiency potentials are cost-effective, because marginal costs of efficiency are below the marginal cost of generating power (about 3.6 ECU/100 kWh)(see Figure 7).

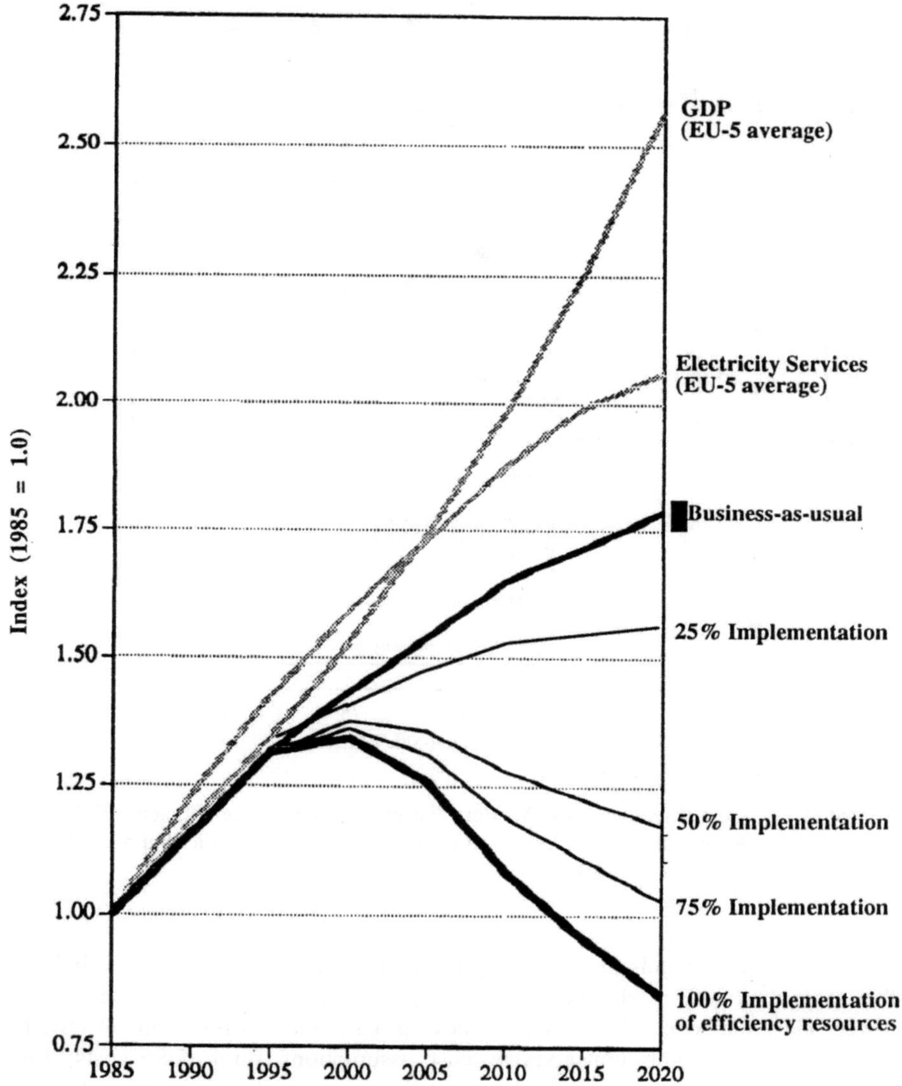

Figure 6: Electricity demand in the EU-5 region for the 1985-2020 period - business-as-usual and efficiency cases

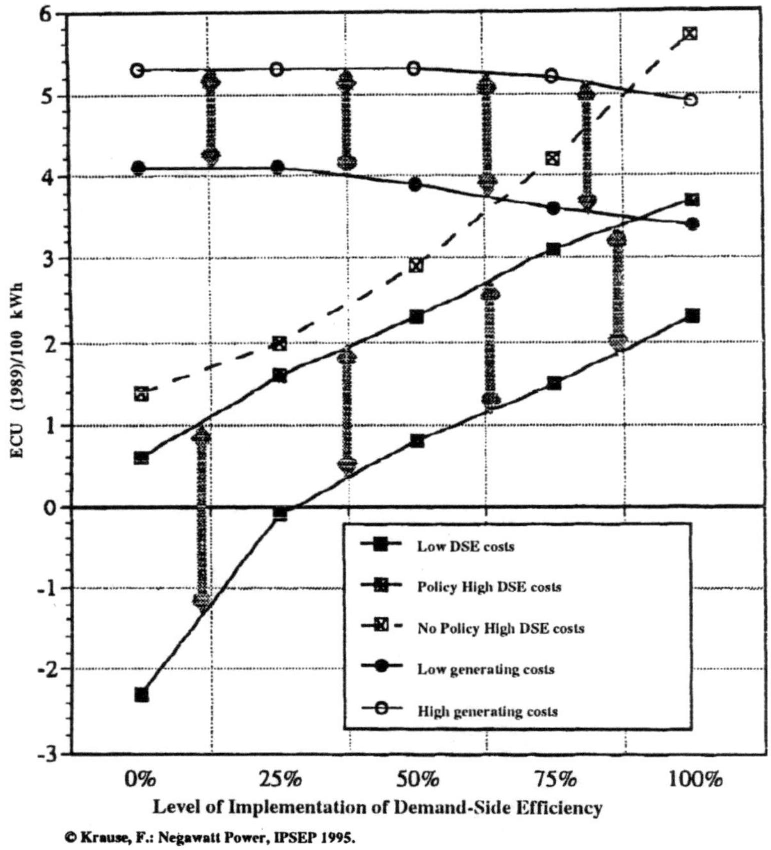

Figure 7: Cost-Effectiveness of Western Europe's Demand-Side Efficiency (DSE) Resources Against Levelised Costs of Marginal Electricity Supplies (EU-5 region, 1995 - 2020)

When taking demand-side efficiency policy into account, Krause et al. believe that more than 80% of all electric efficiency potentials can be implemented with lower marginal costs (including policy programme costs) than electricity generation, even with conservative cost assumptions (high DSE-costs, low generating costs). Assuming high generation costs, all efficiency potentials (100%) have lower marginal costs.

5 Germany: Room for Manoeuvre and Deficits in Implementation

5.1 The Scenarios of the German Bundestag's Climate Study Commission

The German Bundestag's Climate Study Commission has used the technique of scenario analysis to examine various future pathways of climate protection policy.[17] Despite a multitude of unresolved questions and methodological problems, the scenarios nonetheless offer a more comprehensive basis for energy economy studies than has previously been available in Germany, and thus improved orientation for medium-term climate protection policies.[18]

The Study Commission scenarios first illustrate the development that is to be expected if current trends continue ("Reference Scenario"). The Reference Scenario illustrates that business-as-usual policies will fall far short of realising environment and climate policy targets. Accordingly, two further scenarios were developed that lead to a CO_2 emission reduction of 45% by the year 2020 (with a 1987 baseline). The controversies surrounding the further use of nuclear energy in Germany were reflected by modelling two pathways: The first assumes nuclear capacity to remain constant over the medium term (i.e., only replacement investments). The second assumes a phase-out of nuclear energy by the year 2005. In the following, these two scenario variants are referred to as "Climate Protection" and "Climate Protection with Risk Minimisation".

In both climate protection scenario variants, CO_2 emissions are to be reduced by 27% by 2005 and 45% by 2020, taking 1987 as the baseline year. Because of time constraints in scenario preparation, the transport sector was not examined in detail, which is one of the greatest weaknesses of the Study Commission scenarios. Transport policy measures diverging only slightly from business-as-usual (changes in the modal split, reduction of the average consumption of new cars to 5 l/100 km by the year 2020) were assumed in the scenarios presented here

---

[17] Enquête-Kommission "Schutz der Erdatmosphäre" des 12. Deutschen Bundestages: Mehr Zukunft für die Erde - Nachhaltige Energiepolitik für dauerhaften Klimaschutz - Bundestagsdrucksache 12/8600, Bonn, 1994. The following section is based upon the article of Fischedick, M. / Hennicke, P., in: Greenpeace (Eds.): Der Preis der Energie, München 1995.

[18] Cf. also Nitsch, J. et al.: Bedingungen und Folgen von Aufbaustrategien für eine solare Wasserstoffwirtschaft, Bericht der Enquete-Kommission Technikfolgenabschätzung, Bonn, 1990; Masuhr, K. P. et al.: Konsistenzprüfung einer denkbaren zukünftigen Wasserstoffwirtschaft, Prognos AG, Basel 1991; Traube, K.: Perspektiven der Umstrukturierung des Westdeutschen Energiesystems angesichts des CO_2-Problems, Bremer Energie-Institut, Bremen 1992; Hennicke, P. (Ed.): Solarwasserstoff - Energieträger der Zukunft, Berlin, Basel, Boston 1995.

(R1V/"Climate Protection" and R2V/"Climate Protection with Risk Minimisation"). However, these measures fall far short of the necessary contribution of the transport sector to formulated national CO_2 reduction targets, so that all Study Commission scenarios assign an overproportional and thus also significantly more costly CO_2 reduction obligation to the energy sector. Moreover, the scenarios stipulate a minimum usage of German hard coal and eastern and western German brown coal.

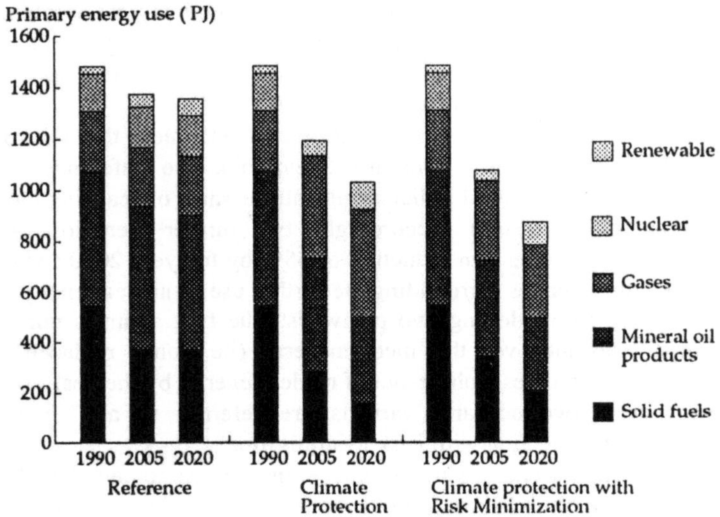

Figure 8: Primary energy consumption in the three Study Commission scenarios

Primary energy consumption drops in all three scenarios (see Figure 8), but to substantially varying extents. With the reference development, consumption drops from the 1990 baseline by 8.7% by the year 2020, leading to a reduction in CO_2 emissions of 14% by 2005, and 16% by 2020. This means that today's politically stipulated reduction targets would be far from being fulfilled with a business-as-usual development - the consumption reduction is mainly due to the restructuring of the (brown coal) energy system and general industrial restructuring in former East Germany, and can only marginally be viewed as a success of climate protection policy.

By contrast, the climate protection scenarios show a significantly larger reduction of primary energy consumption, namely by 24.3% and, respectively, 30.6%. This is based upon reduced energy consumption on the demand side, and improved efficiency on the supply side (e.g., through improved average conversion efficiency ratios, and an increased use of cogeneration). In all three scenarios, the consumption of solid fuels drops. Mineral oil consumption remains largely constant in the Reference Scenario but drops substantially in the climate protection scenarios, while gas consumption rises in the latter considerably.

The analysis shows that any climate protection strategy must be largely based upon the exploitation of the existing conservation and substitution potentials. It lies in the nature of the assumptions and the model used that this applies largely regardless of whether nuclear energy continues to be used or not. The two climate protection scenarios are **characterised by an annual average efficiency enhancement of 3.2%** ("Climate Protection") **and 3.6%** ("Climate Protection with Risk Minimisation") throughout the period from 1990 to 2020. By contrast, the **reference** development only entails an annual efficiency enhancement of some **2.2% p.a.**. When compared to the **actual figures** of the last 20 years, in which period the average annual efficiency improvement figured **1.75%** in former West Germany, these scenario figures imply substantially intensified conservation efforts. In the third, risk-minimised scenario, the compensation of the nuclear share of power supply is provided mainly by an increased use of gas for an interim period in the conversion sector, an exploitation of primary savings potentials through the increased use of cogeneration, and an increased use of renewables. The latter, however, is relatively moderate. Renewable sources of energy do albeit account in the third scenario for a substantially larger proportion of energy supply in the year 2020 (with some 10.5%) than they do today (about 2% in 1990), but compared to the reference development (about 5%) or the second scenario (about 6.3%), the contribution of renewables is at most doubled. To do so, besides hydroelectric power mainly the available nearly cost-effective wind and biomass potentials are partly exploited.

Both climate protection scenarios satisfy the stipulated CO_2 emission targets: CO_2 emissions are reduced by 27% by the year 2005, and 45% by the year 2020. On the basis of the selected assumptions (low CO_2 reduction contribution of the transport sector; excessive costs of energy savings; relatively favourable cost estimates for nuclear energy), the second scenario with a maintained nuclear capacity indicates lower macroeconomic expenditures than the reference development. By contrast, the third scenario with a simultaneous phase-out of nuclear energy indicates - again on the basis of these assumptions - cumulated, discounted extra costs over the period from 1990 to 2020 of some 150,000 million DM (or 5,000 million DM/a, i.e. about 60 DM per capita and year) compared to a development path following current trends. Compared to the second scenario with further use of nuclear energy, the assumptions favouring nuclear energy lead to extra costs of some 180,000 million DM, or about 75 DM per capita and year. We may assume for one thing that such a per capita burden would appear acceptable to most of the public as an 'insurance premium' for the avoidance of the risks of nuclear energy. For another, a more detailed analysis of the findings, taking into account the inconsistencies in the methodology and assumptions, the insufficient modelling of CO_2 reduction options in the transport sector, and the uncertainties in the price development of energy carriers, would come to the conclusion that a phase-out of nuclear energy in conjunction with a fulfilment of CO_2 reduction targets can be achieved entirely without additional macroeconomic burdens. Indeed, similarly detailed studies conducted elsewhere, in which comprehensive comparisons of supply- and demand-side resources were calculated within an integrated examination of energy supply and demand, come to the conclusion for Germany that **a nuclear phase-out** undertaken within the framework of effective climate protection measures **need not lead to extra costs,** but can even partly lead to reduced costs and thus positive macroeconomic effects.[19]

The rationale behind this kind of thinking can be best shown by using the **concept of supply curves of conserved energy**. Therefore, in the following chapter we sum up the results of different bottom-up-analysis[20] for the electricity sector Germany.

5.2 Supply Curves of Conserved Energy as a Screening Tool for Cost-Benefit-Analysis

Considering the wide range of efficient energy technologies already available on the markets, the insufficient degree of rational energy use seems to result rather from implementation deficits than from technology deficits. Because the market for energy service (instead of kilowatthours) does not function up to now, we should concentrate current cost-benefit-analysis on the question how to investigate least-cost options for energy services and how to overcome existing barriers.

From the analytical point of view, the observable degree of industrial energy efficiency represents a mix of different influence factors, interactions, and dependencies, incorporating all kinds of motivation and information deficits, barriers, or market imperfections.[21] In order to avoid a misinterpretation of the trend scenario and of the possibilities to deviate from this path, the analysis of theoretical potentials has to be enlarged by the analysis of real adoption processes and resulting policy options. Otherwise, a systematical under- or overestimation of the contribution of efficiency options and their impact on CO_2 mitigation would take place, causing wrong conclusions for energy policy-making.

In order to optimise the energy system in terms of emissions and costs, all different options both on the supply side and on the demand side should be systematically compared on the base of long-run marginal costs, representing the long term transformation costs of the energy system. Energy supply and energy conservation options both can be presented and compared by supply curves contai-

---

[19] Cf. pp. 441 ff. in: Enquête-Kommission "Schutz der Erdatmosphäre" des 12. Deutschen Bundestages: Mehr Zukunft für die Erde - Nachhaltige Energiepolitik für dauerhaften Klimaschutz - Bundestagsdrucksache 12/8600, Bonn 1994.

[20] See for example Wuppertal-Institute et al.: Interdisciplinary Analysis of Successful Implementation of Energy Efficiency in the Industrial, Commercial and Service Sector (InterSEE), Mid-Term Report (not published until now), January 1997.

[21] Cf. DeCanio 1993, Energie&Klima 1990, Gruber et al. 1995, Gruber/Brand 1991, IEA 1995, Sanstadt/Howarth 1994, Krause et al. 1993.

ning the absolute potential (e.g., in GWh) and the specific long-run marginal costs (e.g., in Pfg/kWh) (see Fig. 9 - 11).[22]

The supply curves of conservation illustrate the maximum cost-effective potential of savings under idealised market conditions. The average technology currently on the market has to be taken as the baseline for potential of savings. The specific savings costs of efficiency technology represent the additional investment and operation costs of the best available equipment as compared to the average on the market (e.g., compact fluorescent lamps vs. incandescent light bulbs). Due to the fact that all efficiency technologies considered are already available on the market, they have passed a certain commercialisation process and have left the stage of pure theoretical technical potentials without economic relevance. In analogy to the treatment of supply-side options, the levelised costs of conservation potentials have to be spread over the energy saved during the lifetime of the efficiency technology. The analysis is an idealisation in the sense that **a macroeconomic discount rate** (e.g., 4 % in real terms) is used to levelise the additional costs of energy-efficient technologies compared to business-as-usual technologies.

[22] For underlying data see Wuppertal Institute/Öko Institute 1995, further, e.g., A&W/Haas Consult 1994, Berg 1992, Brunner et al. 1988, Brunotte 1993, E-Source 1992, ebök 1988, Fischli 1993, Ravel 1992, Horbaty/Renggli 1993, Humm/Gasser/Bush 1991, Huser/Eisenhut/Bush 1992, Kaufmann/Ackermann/Pauli 1992, Münst 1992, Öko-Institute 1992, 1994, RAVEL 1991, 1992, 1993, Schäfer 1994, Sigg 1994.

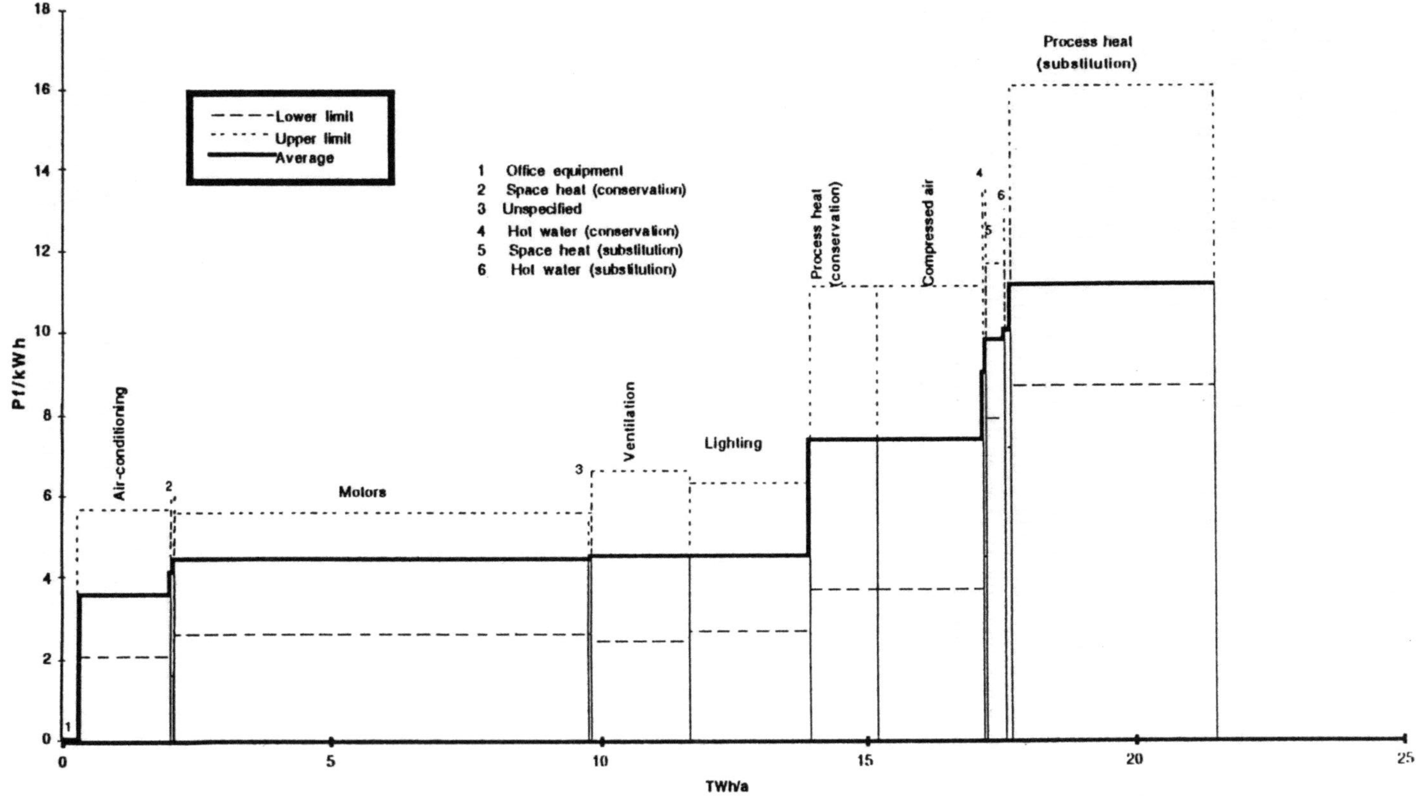

Figure 9: Supply curve for electricity conservation in the German industry (without basic industry; cumulated savings 22 TWh/a, total consumption 72 TWh in 1993)

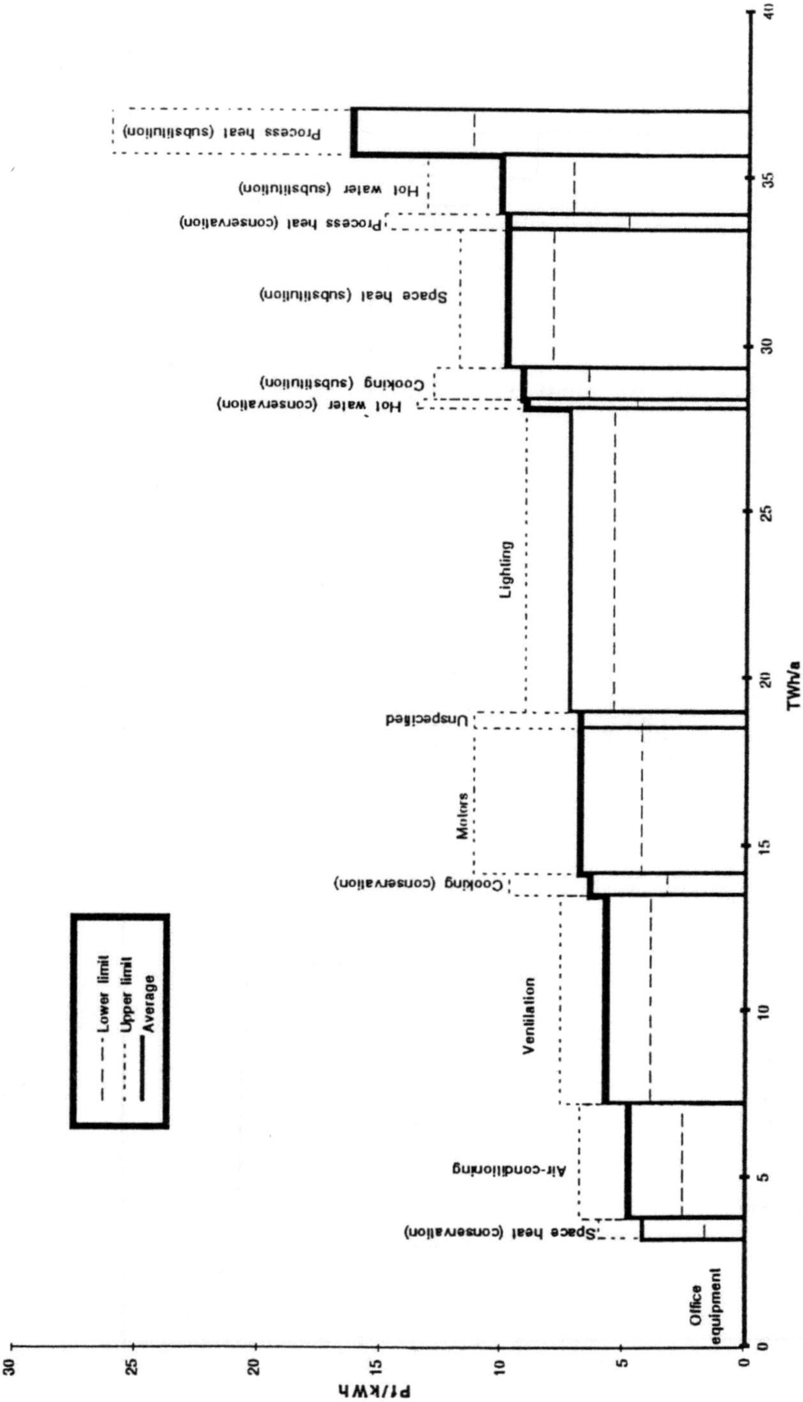

Figure 10: Supply curve for electricity conservation in the German commercial and service sector (total consumption 103.4 TWh in 1992)

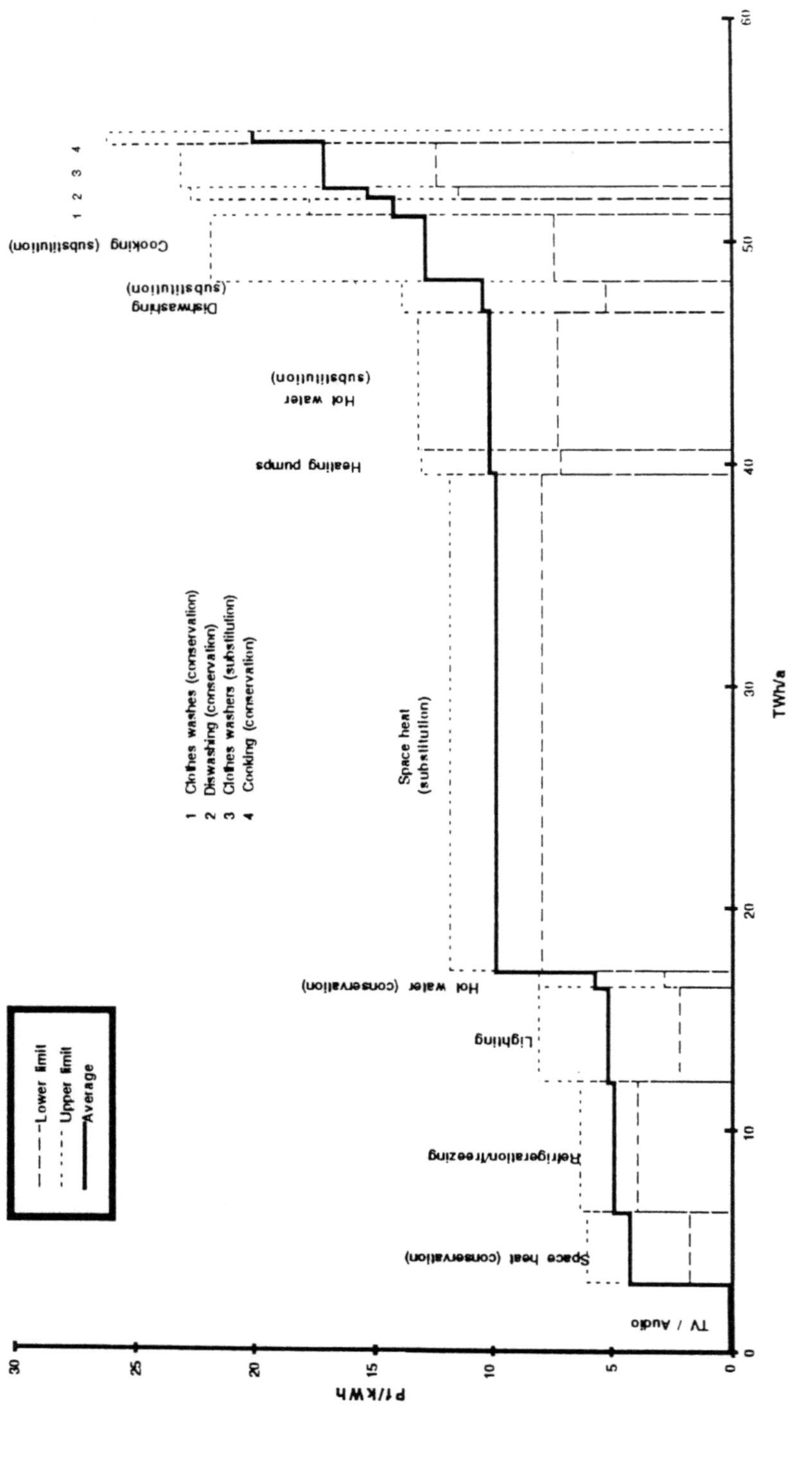

Figure 11: Supply curve for electricity conservation in the German residential sector (total consumption 122. 81 TWh in 1992)

From these supply curves, a (macroeconomically-defined) **least-cost mix of energy options satisfying a given demand of energy services can be derived.** For this purpose, transaction costs must be added, and the total costs of NEGAWatts must be compared with the total costs of MEGAWatts. Avoided long run cost of electricity supply in Germany has been calculated between 10 to 13 Pf/kWh (adding at least 2-3 Pf/kWh for avoided so called "external" costs).[23] Up to now, there are not too many hard figures for transactions costs in Germany. The main reason is that there are lot of pilot programs, but not so much evaluated full-scale programs in Germany. If you add on average 30% on techniques costs for transaction and marketing, there will be nevertheless a huge potential of NEGAWatts which should be cost-effective if the barriers for implementation could be overcome. Therefore, a very import conclusion can be derived: Within the boundaries of national and international CO_2 abatement targets, an unimpeded **competition between energy and capital** (efficiency technologies) is much more important than direct competition between energy suppliers. To reach a specified level of national energy services with least-costs (not only cost-effective energy supply) **should be the core of energy policy.** Within this concept, supply curves of conservation potentials represent a useful screening tool for defining least-cost options of energy policy and for analysing the size and the structure of a market for energy efficiency (NegaWatt market).

5.3 Closing the Implementation Gap: Instruments for Climate Protection

The scenarios presented above have shown - despite all their methodological deficiencies - that an adequate pool of **technological** options is at hand to provide a level of energy services that is climatically acceptable and entails a minimum of risks. The crucial question, however, concerns the issue which instruments can be applied to break down the barriers impeding market implementation of efficiency and of risk minimisation policies, and thus remove innovation and investment blockages currently standing in the way of more "soft energy paths."

Allocative processes in the market economy that are directed by competition and the private exploitation of capital have until now forced an astounding economic dynamism and - as long as economies and markets have **grown** - a certain degree of efficiency. But these 'acquisition techniques of the market' can neither set global quality and reduction targets, nor do they have self-regulating mechanisms that may solve distributional issues between rich and poor or between countries or generations. The German Climate Study Commission has shown

[23] See the agreed figures in the Round Table "Least-Cost Planning" of North-Rhine-Westphalia in: MWMTV, Abschlußbericht RT-Least-Cost Planning, Düsseldorf 1995; in this brochure together with RWE Energie AG, the biggest German utility, the external costs of electricity in Germany were pragmatically calculated with 2-3 Pf/kWh.

beyond doubt that the challenging CO_2 reduction targets set for Germany (30% by 2005, 80% by 2050) can only be achieved by a **comprehensive and differentiated policy mix**. For example policy mixes have been investigated for the important target group of the small and medium enterprises (SME) in more detail.[24] Promising policy mixes have to include measures concerning:

- Increase of general motivation (especially on the top management level) (e.g. Bush 1996, Institute for Energy Technology 1995).
- Supply of specific efficiency know-how and information including multiplicators, e.g., RAVEL Impuls programmes in Switzerland and NRW, Germany (Energieagentur NRW 1996, RAVEL 1996).
- Financial incentives, e.g., contributions to energy audits, utility rebate programmes (IRP/LCP), contracting (Wuppertal Institute/Öko Institute 1995, European Parliament 1996).
- Fostering of implementation know-how by technical assistance and dissemination of best practise cases, e.g., by utilities, energy agencies, and efficiency awards (VIK 1996, Lupi/Innocenti/Serani 1997).
- Increase of organisational know-how by energy and environmental management systems providing information on potentials and costs.
- Market transformation strategies increasing the (average) efficiency of equipment sold, e.g., by (combination of) standardisation, labelling, negotiated target values or technology procurement (Bundesamt für Energie 1996, Westling 1996).
- Establishment of supporting (local) social networks, e.g., energy management models (Bürki 1990).

In addition, a significant impact on energy efficiency can be expected from new multi-dimensional policy strategies combining stringent policy intervention with cooperative elements. Such a promising approach can be found in Denmark, where the recently introduced Danish CO_2 tax scheme opposes firms to a dynamic CO_2 tax, which can be lowered by a certain group of companies from energy intensive branches, if individual energy concepts are negotiated (MoF 1995).

In general, **strong political guidance** must be given and political decisions must be taken that put developments on the right track if climatically acceptable future markets are to be tapped in time and to a sufficient extent. The **concurrent shrinking of risk-burdened markets** (for fossil or nuclear energy) **and the politically induced emergence of 'soft' markets** (e.g., for energy-efficient universal technologies on the user side - in electric drives, ventilation, pressurised air, lighting; for renewable sources of energy; for low-energy houses; or for combined heat and power generation) demand long-term planning security for investors - something that can only be created through framework conditions set by the state. The "deregulation" and "liberalisation" of the energy markets need strong political

---

24 See Wuppertal Institut et al., Inter SEE, op.cit.

leadership and a framework, that markets, competition, and profits can lead in a sustainable direction. The guiding ideas of an economic policy for a new **combined regime of self-control and state regulation** in the energy system are: 'planned competition', 'economy of prevention', 'efficiency revolution', and 'new models of wealth'. The fundamental idea of a profit-driven energy-efficient economy is: The prevention of unnecessary consumption of energy and resources must not only provide benefit to the customers, but also to the suppliers - and to at least the same extent as an additional supply of energy and resources would. Failure to give the profit-orientation of individual economic actors a new, ecological market and development perspective must leave ecology and economy incompatible in a profit-driven market economy, and objectives such as climate protection and sustainability would remain unachievable in market economy systems.

Take the example of energy taxation (as a first module of ecological tax reform): If well designed, this market-oriented instrument acting via relative prices ('bads' becoming more expensive while 'goods' become cheaper) can accelerate the ecologically necessary structural change in a manner that is more conducive to a thriving economy and social equity than laissez-faire policies. **Macroeconomic** studies provide further evidence that a gradually rising and revenue-neutral energy tax will tend to trigger positive effects upon the **overall economy** (more employment, more qualitative growth). As long as no OECD-wide introduction of such a tax is on the horizon, it does remain purposeful, though, to use limited exemptions or transitional provisions to cushion the adjustment losses inevitably experienced by individual industries in any structural change.[25]

That these positive macroeconomic model results point in the right direction is verified by empirical evidence, and also by an overwhelming body of potential studies: The empirical evidence shows that the growth rates of demand-side and supply-side energy efficiency products have consistently been about twice as high as those of manufacturing over the past 10 years.[26] In the 1990s, this lead has increased.

Present environmental protection expenditure in Germany (44 billion DM in 1993) is, however, still largely made up of **end-of-pipe technologies**, which lead to new markets and returns for the manufacturers, but generally extra costs for the users (e.g., flue gas purification equipment for power plants). An 'Economy of Prevention', by contrast, relies primarily upon 'integrated pollution control', i.e., upon technical innovations that size and control processes and products in such a way that the user, too, receives material, energy and **costs** savings in general, and thus the basis for an improved competitive position.

---

[25] Cf. Görres, A. / Ehringhaus, H. / von Weizsäcker, E.U.: Der Weg zur ökologischen Steuerreform. Weniger Umweltbelastung und mehr Beschäftigung. Das Memorandum des Fördervereins ökologische Steuerreform, München 1994.

[26] Schmidt, A.: Ökonomische Auswirkungen rationeller Energieverwendung und erneuerbarer Energiequellen in den Bereichen Produktion und Außenhandel in Deutschland, 1976-1993, Darmstadt 1995.

This explains the great economic policy relevance of improved energy efficiency in conversion and use. It is certain that up to 45% of Germany's energy consumption could be saved with the technologies that are already known today.[27] With today's energy price level, this could cut the macroeconomic energy bill by some 100 billion DM per year. The tapping of this savings potential could create half a million secure jobs (net after deduction of losses suffered in the energy supply sector). There are still some doubts among energy experts as to the **economic** feasibility of this veritable 'efficiency revolution', but the **technical feasibility is no more debated**. One important point is that the "efficiency revolution" is a contribution to **the de-globalisation of energy service markets**, because CHP-systems for municipalities and industry, renewable energy, and efficiency equipment (mainly in buildings) create local and regional markets. It's an import substitution strategy: Imported oil, gas, or coal and the flow of money outside the economy are substituted by endogenous resources and local value added.

But only **strong political leadership** - the rediscovery and active exercise of the primacy of energy policy - can bridge the widening gap between knowledge and action. Unlike on the energy supply side, frameworks must be judiciously set out and vigorously supported by all agencies of the state, if the energy consumers are not to dictate a much too high future energy consumption through their daily undertaking of thousands of energy-inefficient investments.[28]

In contrast to the highly powered supply-side, energy conservation has no lobby. Up to now, energy users, efficiency manufacturers and NGOs have only a marginal influence on developing markets for energy services. Investments in more energy supply are calculated with payback periods of 15 and more years, while energy-saving investments must pay back from the perspective of the users within 3-5 years at most. As a consequence, without state intervention there is a steady flow of too much capital into the expansion of supply. Therefore, without appropriate flanking regulatory policies the implementation of NEGAWatts instead of MEGAWatts has little chance of succeeding in the struggle against the prevailing perverse incentive structure: At present, the more energy a utility sells and thus also damages the environment, the higher its profit is. The opposite should and could be the case: Earning more with less energy. This is the applied "Economy of Prevention."

The most promising instrument for implementing an "Economy of Prevention" and for changing the perverse incentive structure is Least-Cost Planning/Integrated Resource Planning. The 'building of conservation power plants' through LCP-energy efficiency programmes of utilities has high economic and

---

[27] Enquête-Kommission "Schutz der Erdatmosphäre" des 12. Deutschen Bundestages: Mehr Zukunft für die Erde - Nachhaltige Energiepolitik für dauerhaften Klimaschutz - Bundestagsdrucksache 12/8600, Bonn, 1994.

[28] Hennicke, P. / Richter, K. / Schlegelmilch, K.: Nutzen und Kosten von Energiesparmaßnahmen. Vorschläge für neue Förderinstrumente. Studie im Auftrag der deutschen Ausgleichsbank, Wuppertal 1994.

ecological benefits. A win-win strategy seems possible, as the following case-study for the Stadtwerke Hannover shows.

6 Integrated Resource Planning (IRP)[29]

During the last years, an increasing number of utilities in several European countries have discovered that their business can go beyond selling kilowatt-hours of electricity or gas. In a recent survey by VDEW, the association of German electric utilities, around 200 utilities in Germany from all levels[30] answered that they offer over 500 demand-side management (DSM) programmes to their customers (VDEW 1996).

In the following, we will shortly discuss the principle of Integrated Resource Planning. This will be clarified with the results of one of the most comprehensive IRP-projects completed so far in Europe, the LCP-case study which was prepared with co-funding from the EU's SAVE programme for Stadtwerke Hannover, a German municipal utility serving about 300 000 customers (Öko-Institute / Wuppertal Institute 1995).

6.1 The IRP Principle

The main differences between IRP and traditional utility planning are:

1) **Energy services** (e.g., properly lit and heated rooms, "hot pizza and cold beer"), not energy (i.e., kWh) have to be provided with as little damage to health and environment and as cheap as possible. Energy is no longer the end product, but only an intermediate product. This may include new, dispersed generation from small-scale cogeneration and renewables, but the most important new idea within IRP is that energy services, not energy should be delivered to the customer at the least-cost.
2) Utilities engage to **reduce energy consumption** whenever this is cheaper than providing additional energy ("demand-side management," "the business behind the meter"). Ideally, the utility integrates suppliers of energy-efficient

---

[29] The concept of IRP is a good example for applied cost-benefit-analysis for utilities. In this theoretical context we make no difference between the concepts "Demand-side management" (DSM), "Least-Cost Planning" (LCP), and "Integrated Resource Planning" (IRP).

[30] There are nine connected grid companies generating more than 80 % of all power, around 50 regional and about 800 municipal electricity distribution utilities in Germany; most of the municipal utilities offer also natural gas and sometimes district heat from cogeneration plants.

end-use technology and an appropriate form of energy supply into an efficient energy service package offered to the customer.

3) The **level playing field** between demand-side and supply-side resources means that both options should be analysed using the same economic methodology and parameters, especially regarding pay-back times and interest rates.

As explainded above Fig. 12 shows an aggregated supply curve of conservable or substitutable electricity in Hanover.

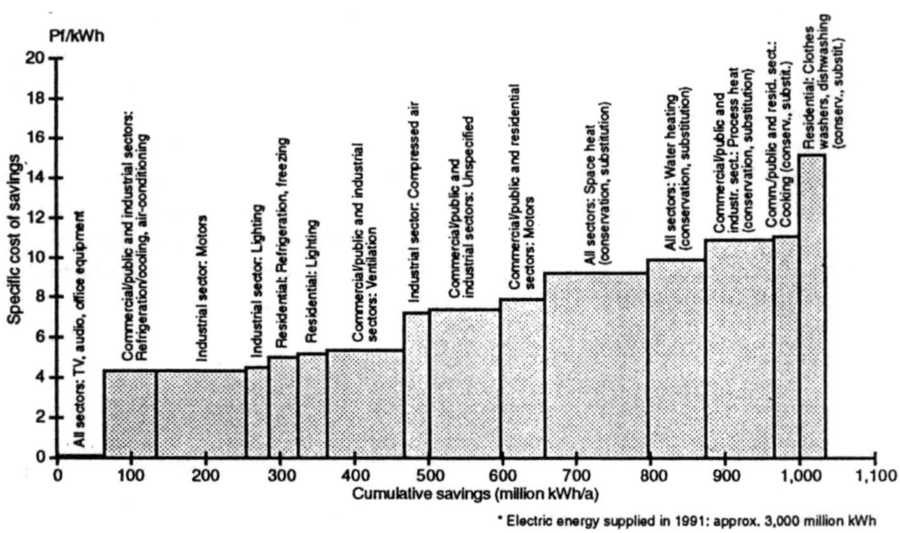

Figure 12: Supply curve of conservable or substitutable electricity for Stadtwerke Hannover (consumption in 1991: 3.027 GWh)

About one third of all end-use energy consumed in Hannover could be conserved (cf. Fig. 12) at lower costs than the long-run marginal costs of electricity supply (including avoided fixed and variable costs of generation, avoided losses, and avoidable costs of electricity transport, distribution, and sale), which under the specific conditions in Hanover are between 0,07 and 0,086 ECU/kWh (13 and 16 Pf/kWh without externalities).

6.2 Development of Nine Prototype DSM Programmes

For Stadtwerke Hannover, nine DSM programmes targeting all customer groups and those end-uses with the biggest savings potential where designed. Almost 90

% of the savings from these nine DSM programmes come from the commercial and industrial sector, and their benefit-to-cost ratio is predicted to be rather good (Fig. 13): The benefits, i.e., the long-term avoided marginal costs of electricity supply, are forecast to be higher than the costs, i.e., the sum of technology costs and programme delivery costs, for all nine programmes. Of course, only testing of these programmes under local conditions can provide the data to reconfirm this forecast.

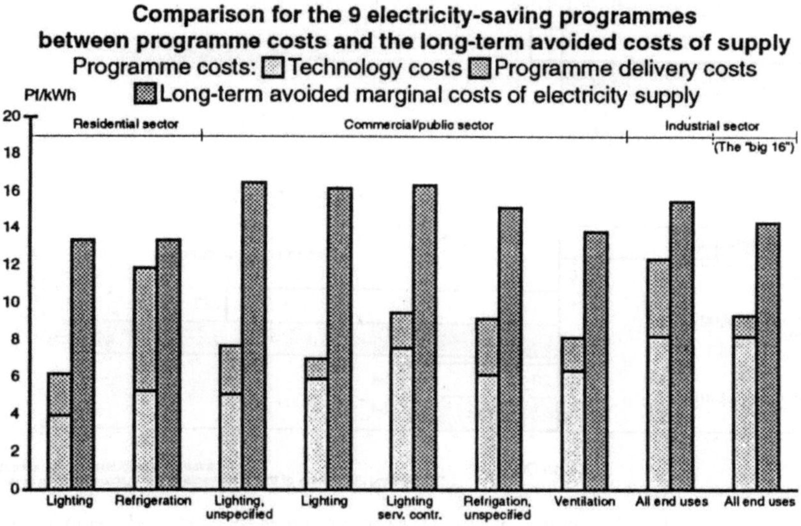

Figure 13: Costs and benefits from nine DSM programmes for Stadtwerke Hannover

Fig. 13 also displays the estimated costs for the utility to deliver the programmes, including programme design, marketing, administration, and evaluation costs. Incentive payments to customers are not included here since they are not real costs to society. They are only transfer payments from one part of society (the utility) to another (the participants of the programmes).

6.3 IRP and Economic Incentives

By implementing these nine programmes, Stadtwerke Hannover can reduce consumption by 200 million kWh per year and achieve a net benefit for the customers of about 5,5 million ECU (about 10 million DM) per year. These higher incomes, in turn, may generate higher consumption of all kinds of goods. Higher consumption will lead to better employment, which is also spurred directly by higher investment in energy-efficient equipment due to the programmes.

Simultaneously, emissions will be reduced, and an additional benefit from avoided externality costs will be created. This externality benefit is calculated at about 3.5 million ECU (about 6 million DM) per year using the data from the GEMIS model (Öko-Institute / GhK 1995).

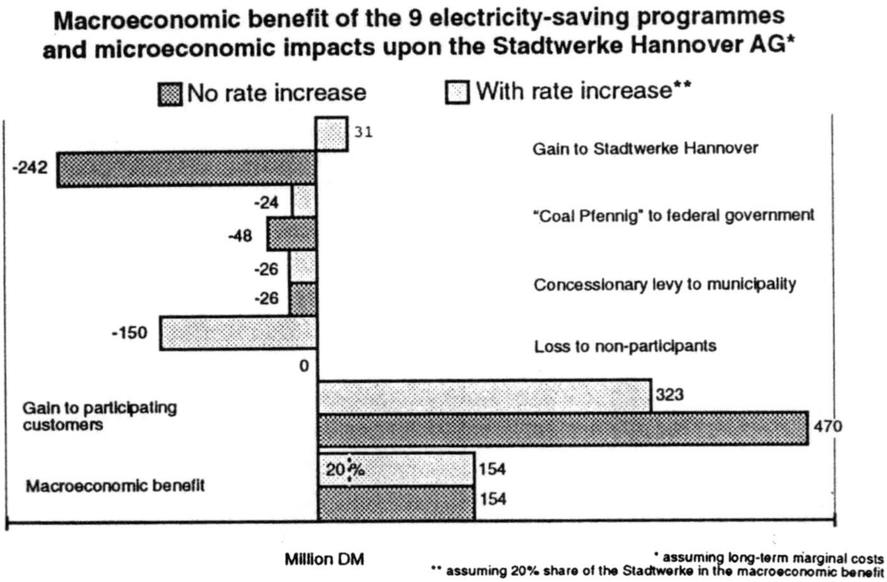

Figure 14: Cost-benefit analysis of nine DSM programmes for Stadtwerke Hannover without and with rate increase

However, if electricity rates and prices remain constant relative to the situation without the nine DSM programmes, these net benefits will be distributed rather unevenly between the utility and its customers (Fig. 14). The dark columns in Fig. 14 show that there is a cumulated net macroeconomic benefit over the useful life of the technologies of 154 million DM, but Stadtwerke Hannover will loose 242 million DM over this 15 year-period. The participating customers are the big winners, gaining 470 million DM from rebates plus bill reductions due to reduced electricity consumption. Another loser is the municipality of Hannover, since the concessionary levy which has to be paid by utilities in Germany is tied to kWh sales. This presents a disincentive to DSM programmes which should be removed by the Federal Government. The "Coal Pfennig", the national coal subsidy scheme, has been abolished with the beginning of 1996 (coal subsidies are now paid for by tax-payers, no longer by electric rate-payers).

But at constant prices, the disincentive to the utility due to the loss of 242 million DM is very high. Therefore, cost-benefit-analysis for utilities shows that, by implementing NEGAWatts, there is a striking contradiction between positive net benefits of the society and decreasing profits for utilities. The theoretical background for this real contradiction is that the market for energy services is not in an overall equilibrium. If cost-effective NEGAWatts are implemented there will be "stranded investments" at the supply-side and the costs could not be avoided at once. To overcome this extremely counterproductive negative incentive structure the utility should be allowed to raise its rates in order to recover DSM programmes costs just like costs for new power plants or transmission lines. Furthermore, the utility should be compensated for the difference between lost revenues and avoided marginal supply costs. And third, a moderate incentive for the utility to prefer DSM over expansion of supply is justified by the macroeconomic benefit of DSM. The regulatory authorities in Lower Saxony, which are responsible for regulating Stadtwerke Hannover, have said they will act in favour of such price increases if they are justified by cost-effective programmes.

Assuming a rate increase that compensates the net loss of 242 million DM and allows the utility to earn an additional 31 million DM, i.e., 20% of the net macroeconomic benefit, completely changes the picture. The utility now has a gain of 31 million DM (white columns in Fig. 10). Customers participating in the DSM programmes still have a net benefit of 323 million DM. Customers which do not participate in the DSM programmes loose 150 million DM.

Therefore, a specific feature of most cost-effective DSM programmes aiming at strategic reduction of electricity demand (kWh), not only capacity demand (kW), is that they lower total resource costs and bills but raise rates. This is because costs decrease but revenues decrease more, too. Therefore, lower costs have to be spread over fewer kWh, which normally leads to an increase in the price per kWh.

For the Hannover example, the examined price increases vary between 0.003 and 0.011 ECU/kWh (0.5 to 2.0 Pf/kWh) after ten years, compared to a situation without DSM. These increases are in most cases below the price decrease which occurred in 1996 in Germany due to the end of the national coal subsidy scheme.

Still, a rate increase due to DSM programmes causes a distributional effect between programme participants and non-participants. This should not be an argument against implementing DSM programmes. Rather, they should be designed in a way that allows all customers to participate in one part of a programme or another, e.g., through addressing a variety of technologies and end-uses, carrying on programmes long enough for every customer to have the opportunity to participate when a new appliance has to be bought anyway, and offering programmes for all customer groups.

It should be noted, however, that any costs for expanding electricity supply will also generate distributional effects between those customer groups which caused the expansion and those which did not or less than others.

However, it should be kept in mind that even after a possible rate increase, the bill of all customers for the energy services they require will be reduced by more than 80 million ECU (150 million DM) spread over the next 15 years in the Hannover case.

Extrapolated for the whole of Germany, over a period of 10 years, LCP electricity conservation programmes could save a good 20% (some 18,000 megawatts) of generating capacity. This could be financed with a moderate rate increase of 1-2 Pf/kWh (including an additional profit as incentive for the utilities) - the annual macroeconomic electricity bill would still drop by 10 billion DM p.a..[31] Despite a slight rise in rates in order to finance the conservation programmes, the aggregate customer bill for electricity and efficiency technologies would drop. **And it is after all energy bills that ultimately count for the competitiveness of industry and the pocket of the individual - not the energy rates!**

A bright future for IRP and DSM in Europe is justified by the emission reductions and the macroeconomic benefits that can be achieved. The possibility of rate-financed energy-efficiency programmes is open especially to grid-connected energies like electricity and gas, where there exist natural monopolies for the transport and distribution sectors. This is the main reason why the proposed EU-IRP directive on "Rationale Planning Techniques" (IRP-directive)[32] is correctly concentrating on the electricity and gas sectors.

It is necessary to implement an EU-**directive** for fostering the general market introduction of IRP. It is not sufficient to engage only in **voluntary** IRP activities: the IRP-directive is the important counterpart at the demand-side to the direct competition on the supply-side fostered by the EU-directive for the internal market for electricity.[33] An IRP-directive is necessary to harmonise the framework conditions for distribution/supply utilities in all EU countries to generate competition for quality of energy services instead of only a price competition for kWh. Transferring the costs of cost-effective IRP-programmes to the customers means a direct form of internalisation of the avoidance costs. Without harmonisation of the framework conditions of demand by the IRP-directive, utilities offering rate-financed energy-efficiency programmes would have to fear price competition from "cheap and dirty" utilities not offering such programmes. On the other hand, the direct internalisation of avoidance costs via IRP-programmes is complementary to a more moderate increasing energy tax. By directly removing market barriers through IRP-activities the tax rate could be lower though the same efficiency potential can be grasped by the combined effect of IRP and the energy tax.

---

[31] Hennicke, P. / Seifried, D.: Endbericht "Least-Cost Planning" im Auftrag der "Gruppe Energie 2010", Wuppertal, Freiburg 1994; Hennicke, P./ Seifried, D.: Das Einsparkraftwerk, Eingesparte Energie neu nutzen, Berlin, Basel, Boston 1996.

[32] Commission of the European Communities: Proposal for a Council Directive to introduce rational planning techniques in the electricity and gas distribution sectors. COM (95) 369/4, Brussels 1995.

[33] Wuppertal Institute: Rational Planning Techniques (RPT): A Tool to Enhance Energy Efficiency? Report to the European Parliament, Wuppertal 1996.

7 Outlook

The above scenario analysis for Germany and the EU refers to the period until the year 2020. The selection of instruments targets an even shorter, medium-term time horizon. Accordingly, the measures for further reduction of energy consumption and associated environmental impacts assumed in the development pathways that appear necessary by the middle of the next century can only be viewed as first steps. If we are to take seriously all the risks perceptible today in the global energy system, then we arrive for the future not only at the necessity of more far-reaching reductions of CO_2 emissions, but generally of a transition to a low-risk and **sustainable energy system**. This means above all an intensified use of renewables to satisfy energy demand, in addition to the complete exploitation of available energy savings potentials and the quest for new, environmentally sounder forms of production and life styles.

The design of a sustainable economic system must proceed with the formulation of a policy mix that aims, in addition to the reduction demands upon the energy system that are required from the present-day perspective, at a reduction of the consumption of 'environment' in general (energy, land use, materials consumption, and the derived environmental impacts), and of all risks associated with energy consumption.[34] The interactions arising between the individual measures instituted to achieve the various reduction targets or requirements are multitudinous. Thus, e.g., a reduction of heat consumption in buildings leads on the other side to an increase of material consumption. Similarly, the increased utilisation of renewable sources of energy (e.g., photovoltaics or biomass for power production) leads to increased consumption of materials and land. This is further compounded by a multitude of competing utilisation options (e.g., crops for energetic utilisation or for raw materials production).

This background in particular makes it clear that a development pathway that is to lead to far-reaching reductions of the environmental impacts associated with energy supply cannot, over the long term and with growing wealth expectations of the world's population, be based exclusively upon purely technical measures, e.g., the exploitation of savings potentials and the intensified use of renewables. Such a pathway must rather involve social innovations and societal structural changes leading to reduced demand for energy services.[35]

Particularly in view of the - also in industrialised countries - sharpening distributional problems, it cannot be our concern to normatively define certain 'sustainable' per-capita consumption levels. Instead, a process of societal research and quest must be organised, seeking for ways in which dematerialised and de-energised services and new models of wealth can generally contribute to an at-

---

[34] BUND / Misereor (Eds.): Zukunftsfähiges Deutschland, Berlin, Basel, Boston 1996.

[35] Cf. Norgard, J. / Viegand, J.: Low Electricity Europe - Sustainable Options, Lyngby 1992.

tenuation of unnecessary and environmentally damaging throwaway consumption. Catchphrases for such approaches are, e.g.: closed-loop economy and recycling, leasing ("use rather than own"), durability, multiple use, and reparability of products, exchange communities, flat sharing, car sharing, neighbourhood support structures, new forms of living together, communal day-nurseries, and launderettes. Furthermore, a greater degree of municipalisation and decentralisation of economic structures and infrastructure and social life in general can provide substantial reductions in consumption (e.g., through avoided traffic).

The drafting of such total-system scenarios that include aspects of social change and take competing or interacting reduction requirements into consideration requires more detailed research, and should come more to the forefront of energy policy discussion processes.

References

Barker, T. 1996: Essay Review (Buchbesprechung zu): NORDHAUS, 1994: Managing the Global Commons, The Economics of Climate Change, in: Energy & Environment, Vol. 7, Essay Review, 1996, p. 85 - 88.

Becker, R. 1997: *Economic-theoretical discussion of discounting and the implications for climate protection,* draft of an article, unpublished, Otto-von-Guericke-University Magdeburg, Germany (please contact: Fax 0049-391-11142).

Becker, R. 1996: Wirtschaftstheoretische Analyse der Annahmen und die Sensivität der Ergebnisse des DICE-Modells von William Nordhaus, diploma thesis, Bergische Universität Gesamthochschule Wuppertal, Wuppertal, Germany.

Bradford, D.F., 1975: *Constraints on Government Investment Opportunities and the Choice of Discount Rate,* The American Economic Review 65, Decembre 1975.

Bradke, H. et al.: Hauptbericht Industrie- Energieeffizienzpotentiale, Produktstrukturwandel und Produktionsentwicklungen bis 2020. IKARUS Projekt TP 6 Industrie, Karlsruhe/Jülich, 1996.

Brunner, C.U. et al. 1988: Elektrosparstudien - 16 Energieberater analysieren 22 öffentliche Gebäude, Studie im Auftrag des Kantons Basel-Landschaft und der Stadt Zürich, Zürich.

Brunotte, M. 1993: Energiekennzahlen für den Kleinverbrauchssektor, Freiburg.

BUND / Misereor (Eds.) 1996: Zukunftsfähiges Deutschland, Berlin, Basel, Boston.

Bundesamt (Hrsg.). 1996: Evaluation der Verbrauchs-Zielwerte für Elektrogeräte, Ausgearbeitet durch das Fraunhofer-Institut für Systemtechnik und Innovationsforschung (ISI), Karlsruhe.

Bürki, T.: Energie-Modell Zürich. Konzept einer Energieversorgung (insbesondere Elektrizität) für die Wirtschaft in der Stadt Zürich. Sozio-ökonomische Energieforschung in der Industrie, 1990

Bush, E.: Energie-Aperos - Energiesparmöglichkeiten lustvoll vermitteln, in: Impuls Nr. 19, Zeitschrift für IP BAU, RAVEL and PACER, Bundesamt für Konjunkturfragen, Bern, 1996.

Cline, W.R. 1992: Optimal Carbon Emissions over Time: Experiments with the Nordhaus DICE Model, Institute for International Economics, Washington, DC, USA.

Commission of the European Communities 1996: Report from the Commission. Second evaluation of national programmes under the monitoring mechanism of community $C0_2$ and other greenhouse gas emissions, Brussels, 14.3.1996.

De Beer et. al. 1994: ICARUS. Universiteit Utrecht, Department of Science, Technology and Society.

de Almeida, A.T. 1994: New Developments in Energy-Efficient Electric Motor Drive Systems, in: A. T. de Almeida et al. (eds.): Integrated Electricity Resource Planning, Dordrecht.

DeCanio St. 1993: Barriers within Firms to Energy-Efficient Investments. Energy Policy, Vol. 21, Nr. 9 (September 1993), S. 906-914.

Directorate Generale for Energy (DG XVII) 1996: Energy in Europe. European Energy to 2020, Special Issue, Brussels.

ELSAM (eds.) 1994: Integrated Resource Planning: From Concept to Practice, Integrated Resource Planning in the Danish Electric Utilities, Main Report, Fredericia.

Elser, M. 1994: Innovative Energiedienstleistungskonzepte für Industrie und Gewerbe in Dänemark, Norwegen und Schweden, Freiburg.

Energie und Klima, Band 2. 1990. "Energieeinsparung sowie rationelle Energienutzung und -umwandlungen" und Band 10 "Energiepolitische Handlungsmöglichkeiten und Forschungsbedarf". Bonn/Karlsruhe.

Energieagentur Nordrhein-Westfalen (NRW): REN Impuls Programm RAVEL NRW, Mehr Impulse für Umwelt, Wirtschaft, Beschäftigung durch rationelle Stromverwertung, Dokumentation Fachkongreß, 4.6.1996, Wuppertal.

Enquête-Kommission "Schutz der Erdatmosphäre" des 12. Deutschen Bundestages: Mehr Zukunft für die Erde - Nachhaltige Energiepolitik für dauerhaften Klimaschutz - Bundestagsdrucksache 12/8600, Bonn, 1994.

E-Source 1992: Switched Reluctance Motors, Snowmass.

European Parliament. 1996. Rational Planning Techniques (RPT): A tool to enhance energy efficiency ?, Working papers by the Wuppertal Institute, Wuppertal/Germany, Directorate-General for Research, Energy and Research Series W-21.

Fischedick, M. / Hennicke, P. in: Greenpeace (Eds.) 1995: Der Preis der Energie, München.

Fischli, U. u.a. 1993: Elektrizitätsbedarf von Industrielüftungen, Bern.

Goldman, Charles A./Kito, M. F. 1994: Review of Demand-Side Bidding Programs: Impacts, Costs, and Cost-Effectiveness, Berkeley.

Görres, A. / Ehringhaus, H. / von Weizsäcker, E.U. 1994: Der Weg zur ökologischen Steuerreform. Weniger Umweltbelastung und mehr Beschäftigung. Das Memorandum des Fördervereins ökologische Steuerreform, München.

Gruber E. et. al. 1995. Analyse von Hemmnissen und Maßnahmen für die Verwirklichung von CO_2-Minderungszielen, Fraunhofer Institut für Systemtechnik und Innovationsforschung, ifo-Institut für Wirtschaftsforschung, Gesellschaft für Energieanwendung und Umwelttechnik, Studienprogramm der Enquetekommision "Schutz der Erdatmosphäre" des Deutschen Bundestages, Band 3 Energie, Teilband II, Bonn, 1995.

Gruber E., Brand M. 1990. Rationelle Energienutzung in der mittelständischen Wirtschaft. Köln: TÜV Rheinland.

Gruber E., Brand M. 1991. Promoting energy conservation in small and medium-sized companies. Energy Policy 19(1991), Nr.3 S.279-287.

Grübler, A./ Jefferson, M./ Nakicenocic, N. 1995 : A Summary of the Joint IIASA and WEC Study on Long-Term Energy Perspectives, Laxenburg.

Hennicke, P./ Seifried, D. 1996: Das Einsparkraftwerk. Eingesparte Energie neu nutzen, Berlin, Basel, Boston.

Hennicke, P. / Becker, R. 1995: Ist Anpassen billiger als Vermeiden? Anmerkungen zur Ausagefähigkeit globaler Kosten/Nutzen-Analysen von Klimaänderungen, Zeitschrift für Energiewirtschaft 2/95, June 1995, p. 143-155; also in: HENNICKE, PETER (Hrsg.), 1995, Klimaschutz: Globale Kosten/Nutzen-Analysen, series "Wuppertal Texte", Berlin, Basel, Boston: Birkhäuser.

Hennicke, P. (Ed.) 1995: Solarwasserstoff - Energieträger der Zukunft, Berlin, Basel, Boston.

Hennicke, P. / Seifried, D. 1994: Endbericht "Least-Cost Planning" im Auftrag der "Gruppe Energie 2010", Wuppertal, Freiburg.

Hennicke, P. / Richter, K. / Schlegelmilch, K. 1994: Nutzen und Kosten von Energiesparmaßnahmen. Vorschläge für neue Förderinstrumente. Studie im Auftrag der deutschen Ausgleichsbank, Wuppertal.

Hofer, M.; H. Schnitzer: Energiesparen in Österreichs Industrie und Gewerbe – eine empirische Untersuchung. Graz: Technische Universität 1992.

Horbaty, R., Renggli U. 1992: Energieverbrauch in gewerblichen Küchen, Bern.

Humm, O., Gasser, S. Bush, E. 1991: Stromsparen im Dienstleistungsbereich, in: Technische Rundschau Heft 48 (1991), S. 50-54.

Huser, A., Eisenhut, H., Bush, E. 1992: Energieverbrauch von elektronischen Bürogeräten, Bern.

IEA (International Energy Agency), 1995: Market Barriers and the Deployment of New Energy Technologies. Working Party on Energy End-Use Technology, OECD, Paris March 1995.

Intergovernmental Panel on Climate Change (IPCC), 1995: Summary for Policymakers, Working Group I, II and III, Geneva.

Institute for Energy Technologies: Norwegian Industrial Energy Efficiency Network, Kjeller, 1995.

Intergovernmental Panel on Climate Change (IPCC), IPCC WG III, 1995: *Summary for Policymakers: Economic and Social Dimensions of Climate Change*, (IPCC WG III: Intergovernmental Panel on Climate Change, Working Group III), approved by the Working Group in October 1995 in Montreal, Canada.

Jaffe A.B., Stavins R.N. 1994. The energy efficiency gap. What does it mean? Energy Policy 22(1994)10 S. 804-810.

Johannson, T.B./Bodlund, B./Williams, R.H. (Hrsg.) 1989: Electricity - Efficient End-Use and New Generation Technologies, and Their Planning Implications, Lund University Press.

Kaufmann, U., Ackermann, R., Pauli, H. 1992: Kühlmöbel im Lebensmittelhandel, Bern

Krause, F. et al. 1995: Energy Policy in the Greenhouse. Volume II, Cutting Carbon Emissions,: Burden or Benefit? The Economics of Energy Taxes and regulatory Reforms on Climate, Growth and Jobs, Executive Summary, IPSEP, El Cerrito.

Krause, F. et al. 1995: Energy Policy in the Greenhouse.Volume II, Part 3B, NEGAWatt Power. The Cost and Potential of Electrical Efficiency Resources in Western Europe. Executive Summary, IPSEP, El Cerrito.

Lind, R.C., 1982: *A Primer on the Major Issues Relating to the Discount Rate for Evaluating National Energy Options*, in: LIND, ROBERT C. (Hrsg.), 1982, *Discounting for Time and Risk in Energy Policy*, Washington, Resources for the Future, p. 21 - 94.

Lupi, A., Innocenti, L., Serani, M.: Energy effectiveness and innovative electrotechnologies in industry - Messages from UNIPEDE'S ETA award, Joint Workshop Series "Electricity- the Bridge between energy and sustainable development? Eurelectric, Commission of the European Communities, General Workshop 22/23.1.1997.

Masuhr, K. P. et al. 1991: Konsistenzprüfung einer denkbaren zukünftigen Wasserstoffwirtschaft, Prognos AG, Basel.

Mills, Evan 1991: Evaluation of European lighting programmes, Utilities finance energy efficiency, Energy Policy, April 1991.

MoF [Ministery of Finance]. 1995: Energy Tax on Industry, Copenhagen, December 1995.

Müller, M. / Hennicke, P. 1995: Mehr Wohlstand mit weniger Energie, Darmstadt.

Münst, F. 1992: Wirkungsgradoptimierung der Drucklufterzeugung und -verteilung, Bern.

Nitsch, J. et al. 1990: Bedingungen und Folgen von Aufbaustrategien für eine solare Wasserstoffwirtschaft, Bericht der Enquete-Kommission Technikfolgenabschätzung, Bonn.

Nordhaus, W.D., 1994: Managing the Global Commons, The Economics of Climate Change, MIT Press, Cambridge, Massachusetts; London, England.

Nordhaus, W.D. 1992: Explaining the "DICE": Background Paper on a Dynamic Integrated Model of Climate Change and the Economy, Yale University, New Haven, CT, USA.

Norgard, J. / Viegand, J. 1992: Low Electricity Europe - Sustainable Options, Lyngby.

Öko-Institute/GhK (Gesamthochschule Kassel) 1995: Umweltanalyse integrierter Energie-, Transport- und Stoffsysteme: Gesamt-Emissions-Modell Integrierter Systeme (GEMIS) Version 2.1, Aktualisierter und erweiterter Endbericht im Auftrag des Hessischen Ministeriums für Umwelt, Energie und Bundesangelegenheiten, U. R. Fritsche u.a., Darmstadt/ Freiburg/ Berlin/ Kassel (veröffentlicht durch HMUEB, Wiesbaden).

Öko-Institute/Wuppertal Institute 1995: Integrated Resource Planning, The Hannover LCP Case Study, Final Report, Hannover/Freiburg/Wuppertal.

Öko-Institut 1992a: Datenanalyse zur westdeutschen Industrie: Potentiale zur rationellen Energienutzung, überarbeiteter Endbericht, Gotelind Alber/Uwe Fritsche/Stefan Thomas, Darmstadt/Freiburg.

Öko-Institut 1994a: Stromeinsparung bei Beleuchtungsanlagen in Industrie, Gewerbe und Dienstleistungsunternehmen, Walter Orlik, Freiburg.

Öko-Institut 1994b: Stromeinsparung bei Lüftungsanlagen (Entwurf), Freiburg.

RAVEL (Bundesamt für Konjunkturfragen, Bern, Impulsprogramm RAVEL) 1991: Umwälzpumpen, Auslegung und Betriebsoptimierung, Bern.

RAVEL (Bundesamt für Konjunkturfragen, Bern, Impulsprogramm RAVEL) 1992: Strom rationell nutzen, RAVEL-Handbuch, Zürich.

RAVEL (Bundesamt für Konjunkturfragen, Bern, Impulsprogramm RAVEL) 1993a: Elektrische Antriebe energie-optimal auslegen und betreiben, Bern.

RAVEL (Bundesamt für Konjunkturfragen, Bern, Impulsprogramm RAVEL) 1993b: Elektroantriebe.

RAVEL (Bundesamt für Konjunkturfragen, Bern, Impulsprogramm RAVEL) 1996: RAVEL Kompetenz: Schrittmacher für nachhaltigen Erfolg. Schlußbericht.

Sanstadt A.H., Howarth R.B. 1994: Normal markets – market imperfections and energy efficiency. Energy Policy 22(1994)10 S. 811-818.

Schäfer, H.: Stromsparen und Energiesparen mit Strom in der Industrie, in: UTECH '94 - Seminar "Stromsparen" am 24./25.2.1994, S. 141 - 159.

Schmidt, A. 1995: Ökonomische Auswirkungen rationeller Energieverwendung und erneuerbarer Energiequellen in den Bereichen Produktion und Außenhandel in Deutschland, 1976-1993, Darmstadt.

Schüssler, M. / Hennicke, P. 1994: Potentiale und Kosten für eine risikoarme Energieversorgung, Übersicht über die Ergebnisse internationaler Studien, Wuppertal.

Sigg, R. 1994: Stromsparen bei Heizungs- und Klimaanlagen, in: UTECH '94 - Seminar "Stromsparen" am 24./25.2.1994, S. 123 - 133.

Thomas, Stefan 1995: Evaluation of RWE's KesS Rebate Programme for Efficient Residential Appliances, in: ECEEE, Proceedings of the 1995 ECEEE Conference, Stockholm.

Togeby, M., Larsen, A.: The Potential for Electricity Conservation in Industry - From Theory to Practice, AKF-Institute for Local Government Studies, Copenhagen, Paper for 18th IAEE International Conference, Washington D.C., July 1997.

Togeby, M., Pedersen, T., Lorenzen, J. 1990: Electricity Saving in Danish Industry, Kopenhagen (AKF).

Tol, R. 1994: The damage costs of climate change: a note on tangibles and intangibles, applied to DICE, Energy Policy, 1994, 22 (5), p. 436-348.

Traube, K. 1992: Perspektiven der Umstrukturierung des Westdeutschen Energiesystems angesichts des CO_2-Problems, Bremer Energie-Institut, Bremen.

VDEW (Vereinigung Deutscher Elektrizitätswerke) 1994: DSM-Projekte Deutscher EVU, Stand Mai 1994, Frankfurt/M.

Verband der Industriellen Kraftwirtschaft (VIK): VIK Beratungsinitiative zur Förderung der rationellen Energieverwendung, Essen 1996.

WEC/IIASA 1995: Global Energy Perspectives to 2050 and Beyond, WEC Report 1995, London.

Westling, H. 1996: Cooperative Procurement - Market acceptance for innovative energy efficient technologies, NUTEK, Stockholm.

World Energy Council. 1995. Global Energy Perspectives to 2050 and beyond, Working Group, Session 7, No. 7.1, 16th WEC Congress, Tokyo, 1995.

Wuppertal Institute / BEM (Beratungsgruppe Energie+Marketing) 1995: Evaluierung des KesS-Programms der RWE Energie AG, Peter Hennicke u. a., Wuppertal/Icking.

Wuppertal Institute 1996: Rational Planning Techniques (RPT): A Tool to Enhance Energy Efficiency? Report to the European Parliament, Wuppertal.

8 Some Critical Remarks on Today's Bottom-up Energy Models

Eberhard Jochem, Fraunhofer Institute for Systems and Innovation Research (ISI)
Breslauerstr. 48, D-76139 Karlsruhe

0 Introduction

Working Group III of the Intergovernmental Panel on Climate Change has discussed the significant differences in the costs of mitigating greenhouse gases in OECD countries in great detail (IPCC, 1996). Most of the analyses and cost projections have been made on limiting or reducing fossil fuel-based CO_2 emissions. The report of the Working Group III concludes that "the results of top-down and bottom-up analyses differ because of such factors as higher estimates of no-regrets potential and technological progress or earlier saturation in energy services per unit GDP." Unfortunately, the report does not attempt to clarify the discrepancies due to differences in the methodological approach in its summary for policy-makers, but simply states that "in the most favourable assessments, savings of 10-20 % in the total cost of energy services can be achieved" (IPCC, 1995, p.22).

This oracle-like statement implicitly suggests that interdisciplinary research seems to be unable to achieve more precise results and allows politicians all kinds of arguments within the broad range of substantial additional costs of mitigating greenhouse gases by any policy intervention on the one side, and high net benefit ("no-regrets", "free lunches") to the economy by climate change policy on the other (see Figure 1). Given this range of subjective interpretation of the differences in the calculated results, the non-expert will be very likely to conclude that "the truth may be somewhere between the engineers' optimism and the economists' pessimism" (Grubb et al., 1993).

A closer look at the differences in the results of top-down and bottom-up models, however, reveals that "differences in results are driven less by differences in model structures than by differences in input assumptions about the way the economy functions, the relative efficiency of energy markets, and the cost of improving the efficiency of these markets" (Hourcade/Robinson, 1996). This conclusion will be elaborated in more detail in this paper, taking the model structure and used cost and energy data of the LP model of the IKARUS project as an example of today's bottom-up energy models (STE, 1997).

Before going into the deficiencies of bottom-up energy and cost data, a short overview of the deficiencies of top-down models seems to be appropriate.

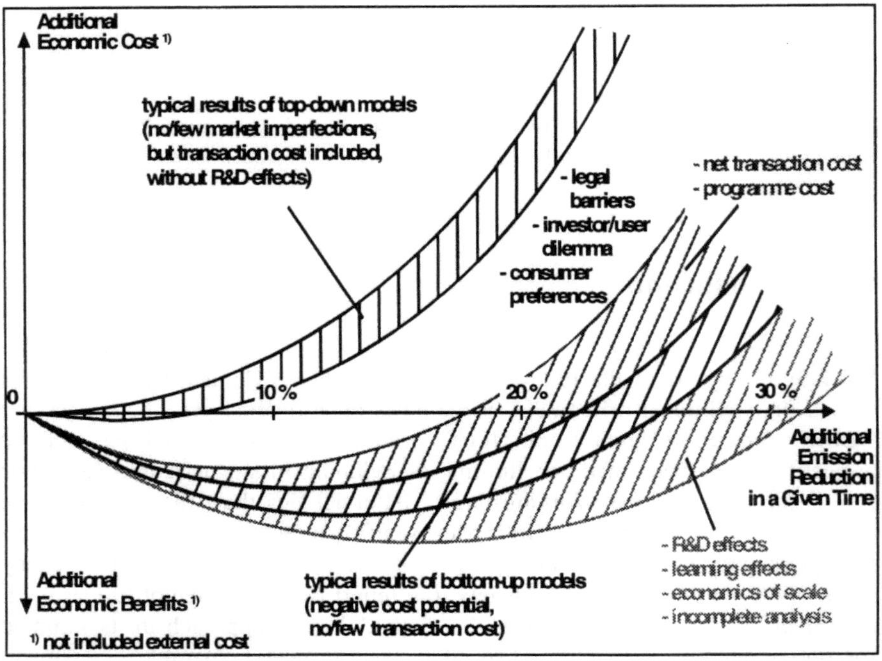

Fig. 1: Differences in greenhouse gas mitigation costs and the perception of top-down and bottom-up models

1 Understanding the Gap Between the Engineers' Optimism and the Economists' Pessimism

Many top-down models are based on the existence of a competitive stationary equilibrium assuming a few restrictive simplifications (Ayres, 1996):

- The supply and demand side of a competitive equilibrium consist of a large number of independent producers and independent consumers who are unable to influence prices or aggregate production levels (no monopolies or oligopolies).

- Every commodity or service in the economy is produced from labour or from other commodities and services produced by the economy and sold in the market (no exhaustible resources are taken from the environment, no externalities).
- Each actor in the market is a perfectly rational utility maximiser and continuously aware of his own preferences (no investor/user dilemma, no legal obstacles to rational energy use, no group-specific traditional decision patterns, no distrust of external consultants).
- Each actor in the market is perfectly informed about the prices and characteristics of all products and services offered for sale at all times (no lack of knowledge, no need for professional training).

Besides these essential simplifications, the most important theorem of competitive equilibrium models is: once a state of equilibrium has been reached, no transaction can improve the position of one actor without hurting that of another or others; at best, all losses and gains add up to zero.

Looking at these assumptions and theorems of many top-down models, the discrepancy with reality observed by consulting engineers, salesmen of energy efficiency equipment and by energy services (Gillissen et al, 1995; Gruber/Jochem, 1990; IEA, 1987) becomes quite understandable. The model's causal relationships lack any explicit consideration of legal barriers to energy efficiency, the investor/user dilemma or consumer preferences, all of which could be changed by specific policy measures (see Figure 1), which would open up new potentials of cost-effective efficiency investments.

2 The Optimistic Side of Bottom-up Analyses, Particularly in the Short Term

Having worked with **bottom-up** models in the energy field for more than 25 years, the author is well aware of the deficiencies of this second approach:

- As most model applications do not consider full net **transaction costs** of the investor, e.g., information gathering and evaluation, cost for planning, ordering, managing, and start-up of the investment, it seems quite plausible that less energy-efficient equipment is realised in real life than sophisticated LP models suggest (e.g., STE, 1997). But, of course, this discrepancy is mainly due to the fact that bottom-up models generally do not consider obstacles and market imperfections (see below).
- Secondly, the **programme costs of mitigation policies** to overcome existing market imperfections and some obstacles have not been taken into account or

tend to be underestimated in target groups of small energy consumers. Generally, they are only explicitly considered in policy evaluation of utility demand-side management activities or national programmes, but not in national bottom-up energy strategy models.

Although these two cost elements tend to contribute to an underestimation of short-term mitigation cost, it is important to say that these two categories of cost are not necessarily additive, but they can influence each other (see Figure 1). A good example of the interaction between programme cost and transaction cost is publicly funded information, professional training and first financial incentives for contracting activities: the new energy service companies are given some necessary assistance to specialise in the particular contracting market and to build up their specific know-how and routine in order to reduce the transaction costs of their customers by taking over all the cost of planning up to financing and maintenance and doing it more professionally and cost-effectively than the customer could do. In any case, however, these aspects give rise to the conclusion that the 'free lunches' may be somewhat smaller than bottom-up models suggest, at least for short-term considerations (see below).

3 Are the Results of Bottom-up Models too Pessimistic?

Krause (1996), however, states that even bottom-up studies overestimate the costs of saved energy, even if the neglected transaction and programme costs are taken into account. Some observations of the author from his own experience can confirm this view, particularly for the industrial and tertiary sectors and long-term perspectives:

- In very heterogeneous subsectors such as capital goods or consumer goods industries, the **analysis** of end uses is **incomplete,** because it is too cumbersome due to lack of data on the individual technologies which are too numerous to be analysed in a thoroughly technical and empirical manner; instead, the specific energy consumption of top-down analyses is taken as a substitute and cost estimates are based on top-down approaches, but not on micro-economic analysis (Jochem/Bradke, 1996, p. 29-36). This means, however, that the energy efficiency potential is likely to be too small as no information about the market imperfections and existing obstacles is available. The full economic efficiency potential is, therefore, unknown, which may be substantial in many sectors (see Figure 2, point (1)), as several empirical findings suggest (e.g., Gruber/Brand, 1991; Jaffe/Stavins, 1994; Sanstadt/Howarth, 1994).

In addition to this uncertainty of the full economic efficiency potential, the corresponding cost data are likely to be too high, as they cannot be derived from individual technical investments, but have to be calculated on the basis of the expected energy savings, branch-specific energy prices and profitability expectations. For reference scenarios, the cost for achieving "reference efficiency improvements" is assumed to be zero in the calculations of the bottom-up models. In many cases, however, they may be negative as many empirical cases and efficiency price competitions suggest (see Table 1 and the Eta competition of the Association of the German Utilities in 1996 and 1993). For policy scenarios, the additional costs for achieving "extra efficiency improvements" calculated on the basis of the rule mentioned above seem to be often too high when compared with empirical findings, as they do not consider "free lunches" at all nor do they include efficiency-relevant side effects of new production, building or transportation technologies which are introduced anyway for other reasons (e.g., higher productivity, quality, comfort, or improved logistics).

Table 1: Empirical cases of energy efficiency improvements and their profitability, identified by ISI in 1996

Type of off-site		Efficiency potential	Internal rate of investment[1]
11 MW	Air compressor station	30 %	20 %
25 MW	Heat and vapour generation	35 %	25 %
4 MW	Heat recovery for preheating glass raw material and cullets	40 %	18 %
90 kW	Water circulation pumps	25-35 %	20 %
75 kW	Illumination of a storage and production hall	20-30 %	15 %
100 kW	Air compressor station (valves and control by staff)	25 %	20 %
1 MW	Waste heat use from plastic manufacturing machines	40 %	18 %

[1] including planning costs and today's energy prices

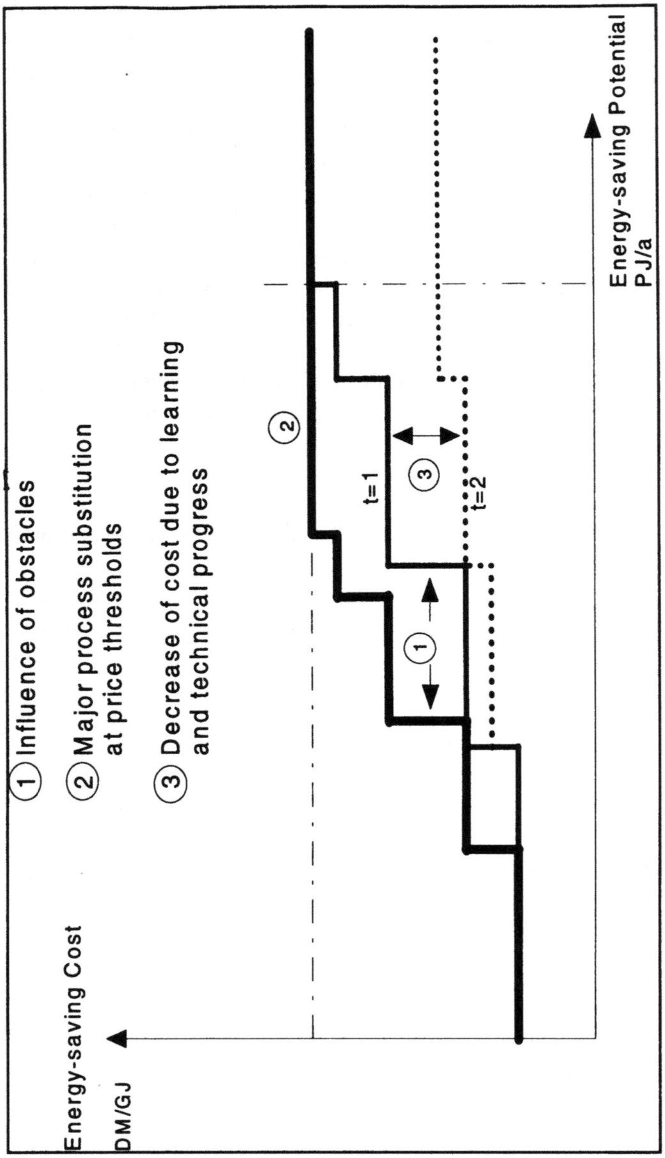

Source: Jochem and Bradke, 1996

Fig. 2: Limitations of energy-saving / cost curves, particularly in the industrial and tertiary sectors

- In many cases, analysts are **not aware of demonstrated near commercial and advanced technologies**. A detailed study on future energy efficiency in three industrial branches made in 1978 demonstrates that experts interviewed in 1977 did not know about substantial innovations their branch would undergo within the next 10 to 15 years (Jochem et al., 1978). This observation draws the attention to a major deficiency in energy cost modelling: very often, cost-efficiency potential curves have a progressive shape according to traditional economic theory (e.g., Schipper et al., 1992). In practice, however, major substitutions of technologies occur at a given cost level which implies degressive cost-efficiency cost curves or constant curve segments over some 20 or 30 percentage points (see Figure 2, point (2)). This degressive part of cost curves is often not considered because the particular level of cost the model chooses in a reduction run had not been foreseen during the data collection phase, or because the cost data of the substitution technology was unknown or difficult to estimate. There is some evidence that the energy-saving cost of new production technologies substituting traditional technologies is marginal or even negative because substituting production processes are based on new concepts and unit operations, e.g., membranes instead of thermal separation, thin slab casting of steel instead of rolling (Smit et al., 1994), biochemical bleaching of pulp instead of chemical bleaching at high pressures and temperatures, or the use of oxygen in glass melting instead of air with high heat losses by the exhaust gases.
- But a major source of substantial "pessimism" seems to be the **costs** of new energy-efficient technologies **based on prototypes**. In many cases the effects of future learning in production and mass markets are not sufficiently taken into account if considered at all (see figure 2, point (3)). Here, again, the cost reduction potential depends on the speed of innovation and market diffusion and indirectly on the intensity and efficiency of policy packages. Higher programme costs, for instance of simultaneous procurement, labelling, minimum standards, and financial incentives for the application of electric motors (Almeida, 1997) may reduce total investments and operating costs to a net gain due to intensive mass production of highly efficient motors, variable speed drives and more efficient pumps, ventilator and compressor systems.

  But even in cases of very traditional technologies being used since decades further cost reduction, even unexpected cost reduction may be possible. The cost reduction of big steam and gas turbines by 40 % within three to four years realised by Western European manufacturers is a typical example in case.
- In quite a few bottom-up models (e. g., simulation or I/0 models), **secondary benefits** due to increased productivity and consumer amenity or **credits** for avoided transmission and distribution investments in energy networks are not considered. However, these omissions may not be particularly important in any case. But given the trend to higher uses of grid-based energies with high marginal and peak load costs, credits for avoided energy demand may be substantial.

The above described deficiencies in bottom-up models are of varying significance: they have small relevance in monofunctional applications and traditional energy converting technologies like boilers or power plants; but the - often forgotten - negative cost potential increases with new technologies at the beginning of the learning curve (e.g., condensing boilers, fuel cells) and with increasing heterogeneity of subbranches (see Figure 1).

The deficiencies listed above are one major explanation why industry and the tertiary sector seem to be rather inflexible with increasing CO_2 emission targets as compared to the residential or conversion sector where monofunctional and expensive technical solutions are described with their technical and economic parameters - and often without mentioning their obstacles and market imperfections (see Table 2). The deficiencies, which are more pronounced for long-term perspectives (more than 10 to 15 years), tend to overestimate the cost of energy efficiency improvements and to underestimate their potentials. The deficiencies essentially depend on the bottom-up approach which tries to take individual technologies as its basis.

But due to the lack of statistically relevant empirical data on obstacles and individual technologies, the **bottom-up approach generates its own dilemma:** The analyst could try to overcome this data dilemma by speculative assumptions and technological visions, but scientific criteria and lack of empirical evidence do not sufficiently support these assumptions or visions.

Table 2: CO_2-emission reduction of the reference and reduction cases, Western Germany 2005

Sector	1995 share of CO_2 emissions	CO_2 emission reduction in 2005 in Mill t/a						Type of analysis technology mainly
		compared to 1989 "Reference"		by additional -15 % in total as compared to "Reference"		by additional -25 % in total as compared to -15%-case		
Electricity generation	30	12	(46%)	12,6	(20%)	29,3	(35%)	individual
Other energy conversion	5	5	(19%)	16,7	(27%)	6,7	(8%)	individual
Industry	21	29[1)]	(111%)	1,3	(2%)	3,2	(4%)	mixed
Transport	23	-25[2)]	(-96%)	8,8	(14%)	10,6	(13%)	individual
Commercial/public	7	8	(30%)	6,8	(11%)	4,1	(5%)	mixed
Residential	14	-4[3)]	(-15%)	15,1	(25%)	29,5	(35%)	individual
Sum of final energy sectors	65	8	(30%)	32,0	(52%)	47,4	(57%)	+
All sectors	100	26	(100%)	61,3	(100%)	83,4	(100%)	+

[1)] including important shares of structural change [2)] due to further motorisation [3)] due to no major obstacles of energy-efficient building stock

Source: STE, Dec. 17th, 1996

4 Policy-relevant Consequences of Sector-specific Pessimism

If these deficiencies and dilemmata are not considered while looking at results derived from bottom-up models, it is unavoidable that wrong conclusions will be drawn. Typical cases of misinterpretation and misuse of results are:

- Industrial and tertiary sectors are considered to have very little flexibility in contributing in addition to the reference case to targets of climate change policy (Gerster, 1995). Consulting engineers and manufacturers of capital goods, however, strongly disagree with this view given the fact of substantial potential of 'no-regrets' and new energy-efficient technologies at some point of early market diffusion, or given the fact that the economic efficiency potential at some 20 to 25 % within the next 15 years has been constant over several decades in almost all industrialised countries (Schipper et al., 1992) like the oil reserve to consumption ratio at some 30 to 40 years during decades.
- Representatives of energy-intensive industries of the Federal Association of German Industry take the results as supporting arguments that additional energy or CO_2-taxes will not change energy intensity but only increase the tax burden on industry and reduce the competitiveness of German industry (BDI, 1996). This position is quite controversial, at least for long-term considerations, because investments and innovations in industry and the commercial sectors are considered to be energy price sensitive.
- Macroeconomists take the cost figures of the bottom-up models as valid information for determining specific CO_2 mitigation costs; in some cases, they may even use them for calculating additional CO_2 mitigation costs (e.g. for the voluntary agreements of the German industry) although the bottom-up analysts point to the fact that these reduction potentials cover mostly 'no-regrets', i.e., that these potentials can be realised with net negative cost, even taking into account programme and transaction costs of the voluntary agreements. Instead of large economic benefits by realising 'no-regrets' (including the transaction and programme costs), study results present economic cost of climate change policies (e.g., RWI/Ifo, 1996).

In order to avoid such misinterpretations, specific attention should be given to

- research on obstacles and market imperfections which would clarify the potentials of "no-regrets" or "free lunches" and their relevant negative costs, very important in the statistical distribution of the no-regrets in branches and individual technological fields such as electric motors, boilers, air compression systems or cooling systems;

- cost estimates of new commercial and advanced technologies with their cost structure and learning curve potentials; this field is of major importance, but hardly known as a field of determining political processes;
- cost developments of prototypes where mass production and learning curves may drastically reduce today's cost in the long term.

5 A Long-term Example

One conclusion of this situation could be to distrust results of these models and rely on conventional desk analyses and intelligent thinking for the decision making process. This position, however, does not take into account the possibility of systematic sensitivity analyses in order to identify sensitive input data or causal relationships of the models used. The reader may also be aware of the fact that manufacturers and operators of long-lasting capital goods or transportation infrastructure or governments have to make their decisions despite the lack of valid information.

A recent bottom-up analysis aimed to identify the least-cost energy supply and demand system with CO_2 reduction targets varying between 60% and 80% by 2050 for West Germany (Tönsing et al., 1997). The technical and cost data were estimated by a team of researchers at ISI and experts from various business areas of Siemens. The influence of data uncertainties was considered by varying technical and cost data by a systematic lower and upper end of two scenarios. The higher the CO_2 reduction target, the less important were the uncertainties of cost estimates for the technical structure of the energy system and the more important the specific CO_2 emission per energy unit. One of the interesting results was that, in the reference scenario, the total energy system costs in Germany increased by almost 50 %, but their share of GDP decreased from 8.2 % to 4.5 % with some deviation of the technical structure due to the uncertainties of the cost estimates (see Figure 3). But even in the -75% CO_2- reduction scenario, total annual energy system costs only increased by 90% with minor deviations of the technical structure influenced by uncertainties of data.

Of course, the authors distrusted the results of such a long-term analysis which, however, was felt to be necessary to discover the latest point in time to start R&D on long-lasting investments which have to fit the reinvestment cycle of buildings, gas pipelines or electricity supply systems, railways, and other transportation systems. After all the results had been reviewed, a simple test of calculating the long-term average change in energy intensity was made; the figures were -1.5 % per year for the reference case and -1.9 % per year for the -75 % CO_2 reduction case. This range corresponds to the long-term trend of the past (see Figure 4) and theoretical considerations of efficiency potentials in the next century (Jochem, 1991).

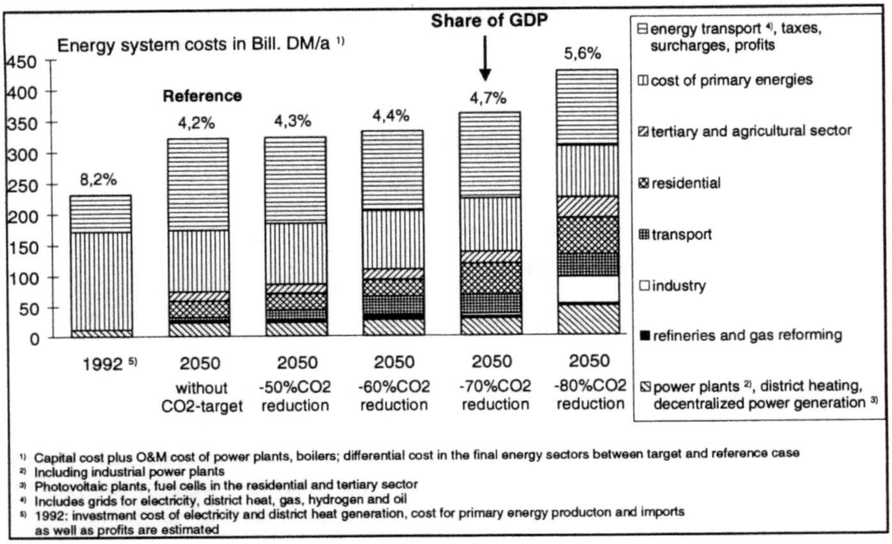

Fig. 3: Energy system cost in Germany in 1992 and 2050 with different reduction targets of CO_2 emissions

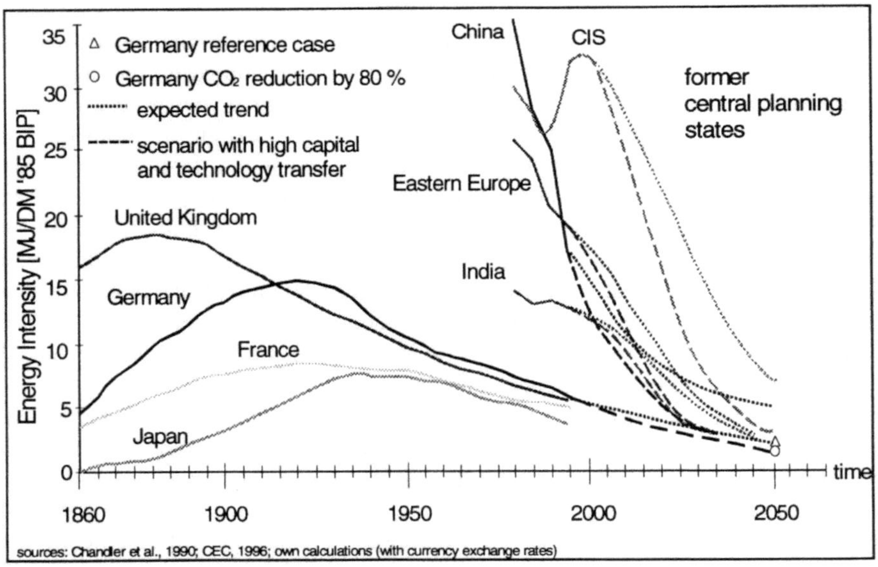

Fig. 4: Energy intensities of some OECD countries, Eastern Europe, CIS, China and India, past trends and future prospects until 2050

The results of this long-term analysis (until 2050) were also quite useful to discuss results of bottom-up analyses with a time horizon of 2010 or 2020. The deficiencies discussed in Chapter 3 and Figure 2 became evident from the point of view of new technologies which are likely to penetrate their markets after 2020 or of potential cost reductions expected for new technologies in energy and end use sectors.

6 Conclusions and Recommendations

Despite the fact that most analysts expect the "real truth of costs" to be somewhere between the cost figures of top-down and those of bottom-up analysis, there seems to be some valid evidence that the net transaction costs and additional programme costs to mobilise efficiency potentials in the order of 15 to 30% will be outweighed by additional microeconomic benefits which have been neglected so far by bottom-up analyses at least for the industrial and tertiary sector; major arguments are:

- existing top-down and bottom-up analyses do not adequately consider technological innovations, because
  - they are not explicitly included in the models, or
  - it would be too cumbersome to analyse so many technologies;
- existing bottom-up models are still partially top-down models because of a lack of data, particularly in industry and the commercial and services sectors;
- cost effects of learning, of professional specialisation (e.g., contracting) and economies of scale are not sufficiently taken up by bottom-up models.

On looking more closely at existing deficiencies of bottom-up analyses, there is no doubt that these models are not yet ready for immediate use in energy or climate change policy. Today's results need careful analysis and qualitative conclusions may be drawn, still with some uncertainty, and a sceptical attitude seems a good basis for periodic data reviews and sensitive decisions.

The **next steps of research and development** should cover aspects of branch-specific obstacles, the size and distribution of no-regrets and their related transaction cost, particularly of off-sites and postponed reinvestments of energy-intensive processes. It should also cover major technical innovations and potential cost-reductions of new energy-efficient technologies by increasing mass production and technological learning. Finally, empirical data on costs of programmes to overcome existing obstacles should be collected in order to calculate full CO_2 mitigation costs and to build an information basis reliable for energy and climate change policies.

The paper did not touch upon the cost to society, the ecological rebound effect of additional economic growth due to saved energy costs and the energy intensity of marginal consumption. The paper also omitted aspects of the social cost of energy use, which, again, is likely to compensate the ecological rebound effects by almost one order of magnitude.

Literature

de Almeida, A. (ed.): Energy Efficiency Improvements in Electric Motors and Drives, Proceedings of the International Conference, Lisbon, 29-31 October 1996, Springer 1997

Ayres, R. U.: Disequilibrium, Double Dividends and No-Regrets Policies. INSEAD, Fontainebleau, France November 1996

BDI (Bundesverband der Deutschen Industrie): Aktualisierte Erklärung der deutschen Wirtschaft zur Klimavorsorge. Dokumentation. Köln, März 1996

Gerster, H.J.: IKARUS: Erste Ergebnisse einer CO_2-Reduktionsstrategie für das Jahr 2005. Potentiale und gesamtwirtschaftliche Mehrkosten für die alten Bundesländer. In: Energiewirtschaftliche Tagesfragen 46. Jg. (1995) Heft 4

Gillissen, M.; Opschoor, H.; Farla, J.; Blok, K.: Energy Conservation and Investment Behaviour of Firms. Dep. Science, Technology and Society, Univ. Utrecht, NL, Oct. 1995

Grubb, M. et al. (1993): The costs of limiting fossil fuel CO_2 emissions: a survey and analysis. Annual Review Energy Environment (18) p. 397-478

Gruber, E., Brand, M.: Promoting energy conservation in small and medium-sized companies. In: Energy Policy, April 1991, p. 279-287

Gruber, E., Jochem, E.: Obstacles to rational electricity use and measures to alleviate them. Energy Policy 18(1990)5, p. 340-350

Hourcade, J.-Ch.; Robinson, J.: Mitigating factors. Assessing the costs of reducing GHG emissions. Energy Policy 24(1996)10/11, p. 863/873

IEA (Internat. Energie-Agentur). Energy Conservation in IEA-Countries. OECD. Paris 1987

IPCC (Intergov. Panel on Climate Change): Summary for Policymakers: Economic and Social Dimensions of Climate Change. Montreal, Canada October 1995

IPCC (Intergov. Panel on Climate Change): Economic and Social Dimensions of Climate Change. Cambridge University Press 1996

Jochem, E.; Gruber, E.: Obstacles to rational electricity use and measures to alleviate them. Energy Policy 18(1990)5 p. 340-350

Jochem, E.: Long term potentials of rational energy use – the unknown possibilities of reducing greenhouse gas emissions. Energy & Environment 2 (1991) 1

Jochem, E.; Bradke, H.: Energieeffizienz, Strukturwandel und Produktionsentwicklung der deutschen Industrie. IKARUS-Monographien, Forschungszentrum Jülich, Band 19/1996

Jochem, E. et al.: Rationelle Energienutzung. Energieverbrauch und Möglichkeiten rationeller Energienutzung in der Verarbeitenden Energie: Steine und Erden, Zellstoff und Papier, Textilindustrie. In: Informationen zur Energiepolitik, Heft 11, Karlsruhe, Mai 1978

Krause, F.: The costs of mitigating carbon emissions. A review of methods and findings from European studies. Energy Policy 24(1996)10/11, p. 899/915

Tönsing, E. et al.:Szenarien der Energienutzung und -versorgung mit reduzierten CO_2-Emissionen bis 2050. Energiewirtschaftliche Tagesfragen (to be published 1997)

Schipper, L.: et al.: Energy Efficiency and Human Activity: Past Trends, Future Prospects. Cambridge Univ. Press, 1992, p. 281

Smit, R. et al.: Long-term potentials of energy efficiency improvement: Technology Descriptions. Univ. Utrecht, NOVEM 1994

Jaffe, A.B.; Stavins, R.N.: The energy efficiency gap. What does it mean? Energy Policy 22(1994)10 p. 804-810

Sanstadt, A.H.; Howarth, R.B.: "Normal markets" – market imperfections and energy efficiency. Energy Policy 22 (1994)10 p. 811-818

STE (Programmgruppe Systemforschung und technologische Entwicklung): Bericht des Teilvorhabens 1 – Modelle. Jülich (to be published 1997)

RWI/Ifo (Rheinisch-Westfälisches Institut für Wirtschaftsforschung/ ifo Institut für Wirtschaftsforschung): Gesamtwirtschaftliche Beurteilung von CO_2-Minderungsstrategien. Untersuchungen Heft 19, Essen 1996

9 Double Dividend of Climate Protection and the Role of International Policy Coordination in the EU - An Applied General Equilibrium Analysis with the GEM-E3 Model

Klaus Conrad, Mannheim University
Tobias F.N. Schmidt, ZEW
Mannheim, Germany

Abstract. While there is some hope that the ongoing climate change negotiations will soon come up with concrete, time scheduled and binding emission reduction commitments, the question of how to achieve these targets is still unsolved.

The objective of this paper is to analyse alternative settings of an environmental tax reform and its economic and environmental impacts on the EU. The methodological framework used is based on a multi-country and multi-sectoral computable general equilibrium model for eleven EU-member states. The emphasis of the analysis lies on the institutional setting of a carbon dioxide reduction policy and on the specification of the labour market. The institutional settings analysed are related to the degree of environmental policy coordination. As standard neo-classics neglect the problem of involuntary unemployment, we relax this restriction in the second part of the analysis in order to test alternative (more rigid) labour market specifications. The major findings of the paper can be summarised as follows: 1) There is some potential for a double dividend in the EU. 2) Coordination beats not always unilateral actions. 3) Labour market rigidities play a crucial role to both, the double dividend and the coordination issue.

Acknowledgements
This research is based on two modelling projects financed by the JOULE-II programme of the European Commission (DGXII). We are indebted to Pantelis Capros, Takis Georgakopoulos (NTUA-Athens) Stef Proost and Denise Van Regemorter (CES at the Katholic University of Leuven), who are co-developers of the model GEM-E3. Nevertheless, we take responsibility for all errors and omissions, that might have remained.

1 Introduction

While there is some hope that the ongoing climate change negotiations come up soon with concrete, time scheduled and binding emission reduction commitments, the question of how to achieve these targets is still unsolved. This irresolution holds for both, the type and/or mix of instruments and the degree of international cooperation with respect to joint or unilateral implementation. In the pre-Kyoto period the European Commission claimed for all industrialised countries a 15% reduction of greenhouse gases by 2010 based on the emissions of 1990 not knowing exactly how such a promise could be achieved.

Even those emission projections that take into account some political action (conventional wisdom scenarios) estimate an EU-wide increase of greenhouse gas emissions up to 8% by 2010. Based on these growth estimates the actual reduction accounts for 23% - a rather ambitious goal that will be hardly reached if efficient instruments are not launched within the next few years. Unfortunately many EU member states are currently sharing the same economic problems: high unemployment, high taxes that are hard to reduce because of the public deficit criterions given by the Maastricht Treaty and the Monetary Union and increasing social security rates because of increasing unemplyoment and the aging of societies.

Obviously, the implementation of environmental policy instruments should not hinder other economic or political goals like GDP growth, international competitiveness or employment. The discussion whether there exists a 'no regrets' policy, that improves environmental quality without negative impacts on the economy, has been raised a long time ago but has not been solved yet. While the initial discussion has addressed mainly the longer term which ranks around the Porter-Hypothesis, the recent debate revolves more around the short term and is linked to optimal taxation. The hope that advocates have in mind is to obtain a so called double dividend (one for the environment and one for employment and/or economic welfare) by generating a revenue neutral tax reform that is more efficient than a given pre-existing one. Considering both, the climate change issue and the problem of unemployment, it seems straightforward to tax emissions or energy and reduce labour costs by recycling the tax receipts to employers' social security contribution in a revenue neutral way. If one could prove the existence of a double dividend that is linked to such an environmental tax reform, its acceptance could be improved considerably.

The objective of this paper is to analyse such a tax reform and its impact on the EU economy. For this purpose we first present an analytical framework that enables the analysis of alternative policy instruments on the national and the EU-wide level. This part of the paper deals with the description of a computable general equilibrium model GEM-E3 for eleven EU member states. A brief outline of the model will explain the general scope, the assumptions made, the agents considered and their underlying behaviour.

Subsequently we apply this framework to identify and assess the economic and environmental impacts of alternative tax reforms. The emphasis of the analysis lies on the institutional setting of a carbon dioxide reduction policy and on the behaviour of agents in the labour market. Both aspects seem to be rather crucial for the economic impact of environmental tax proposals. The institutional settings analysed are related to the degree of environmental policy coordination. In the first setting Germany decides to be a forerunner in combating climate change by reducing the emissions by 10% within ten years. In the second setting the EU member states decide to combat climate change in a coordinated way. The target (10% emission reduction) is set at the EU level and all countries try to reach this goal jointly. As the policies differ considerably with respect to the environmental improvement obtained, comparing the two implicitly assumes that the (German) policy maker is indifferent to the actual environmental benefit achieved by the policy. One could think about such a behaviour if one finds that the policy maker reacts mainly to international political pressure than to eventually occuring national environmental damages (that might be very low). A reduction of 10% could offset the international pressure on whatever level (national or EU-wide) this reduction will be. Hence, what the policy maker minds is the national economic impact of the policy implemented and not the overall environmental benefit achieved. In this respect, the comparison undertaken is more a concern of political opportunity than of economic efficiency.

A standard criticism to the neo-classical framework of CGE-modelling is that it neglects the problem of unemployment. To grasp this criticism the assumption of a market clearing wage rate is relaxed during the course of the analysis. Imposing the same policy (environmental tax reform) under alternative wage rules enables a comparison of impacts with respect to alternative labour market regimes.

A reconsideration of our findings with respect to the results of similar analyses that can be found in the literature closes the paper.

2 The Specification of the Model

2.1 The State of the Art

The global nature of economic activity and environmental consequences and the increasing understanding of the interdependence of these human-related systems have led to the development of a model generation that tries to address this linkage to improve the knowledge about the current and future potential effects of human behaviour. These integrated assessment models usually incorporated the know-how of several disciplines and vary in the presentation of complexity according to the question they intend to address.

Models with economic origin often use the computable general equilibrium (CGE) approach. These models have been used quite extensively to study energy-economy-environment interactions, but they cover usually only a part of the policy relevant problems. Most effort has been done with respect to global warming, some models being also used within the IPCC[1] activity. Examples of global or multi-regional models in this field are the Nordhaus-models DICE (Nordhaus (1994)) and RICE (Nordhaus/Yang (1996)), the MERGE model of Manne et al. (1995), the OECD model GREEN of Burniaux et al. (1991), the model G-CUBED of McKibbin/Wilcoxen (1992), the model of Perroni/Wigle (1994) and the EU-model WARM of Carraro/Galeotti (1996). Welsch (1996) uses an aggregated CGE model for the European Community to study the effect of joint versus unilateral carbon taxation in a two-region framework.

Furthermore there are a couple of one-country applications like the models of Boyd et al. (1995), Glomsrød et al. (1992) and Proost/Van Regemorter (1995) which try an integrated assessment of alternative policy options. The acidification problem was addressed mainly by one-country approaches like the model of Bergman et al. (1995), Glomsrød et al. (1992), Boyd et al. (1995) and Conrad/Schröder (1991a). There is a wide range of model developments that do not handle the environmental consequences explicitly but deal with the topic in purely economic terms. Examples of this group of applications are the models of Jorgenson/Wilcoxen (1993), Hazilla/Kopp (1990) or Stephan (1989).

2.2 Cost Function and Input Coefficients and the Price System

We characterise the technology of a cost minimising industry by nested CES cost functions. $C(X, PK, PLEM, g)$ is the cost function at the first stage for producing output X given the input prices for the capital, PK, and the labour/energy/material aggregate, $PLEM$ respectively. The price diminishing technical progress g is specified by exponential rates of diminuition. This is the dual approach of incorporating input using or input saving technical change. This type of technical change considers autonomous (costless) energy efficiency improvements.

Profit maximisation under constant returns to scale implies marginal revenues equal to marginal costs which explains the output price PX of domestic production in terms of a CES unit cost function:

(1) $$PX = PX(PK, PLEM, g)$$

From Shephard's Lemma we derive the factor demand functions as variable (price dependent) input coefficients:

(2) $$\frac{K}{X} = a_K, \qquad \text{with } a_K = a_K\left(\frac{PX}{PK}, g\right);$$

---

1 Intergovernmental Panel on Climate Change

(3) $$\frac{LEM}{X} = a_{LEM}, \quad \text{with} \quad a_{LEM} = a_{LEM}\left(\frac{PX}{PLEM}, g\right).$$

In principle, one could include all the input prices of the model in one CES unit cost function. This, however, would imply the assumption, that the elasticity of substitution between all inputs is the same. Nesting of inputs into several aggregates allows for different elasticities of substitution. Hence, the production structure presented in Figure 1 follows conceptually the assessment of substitutability of inputs.

Figure 1: The nested production and factor price scheme

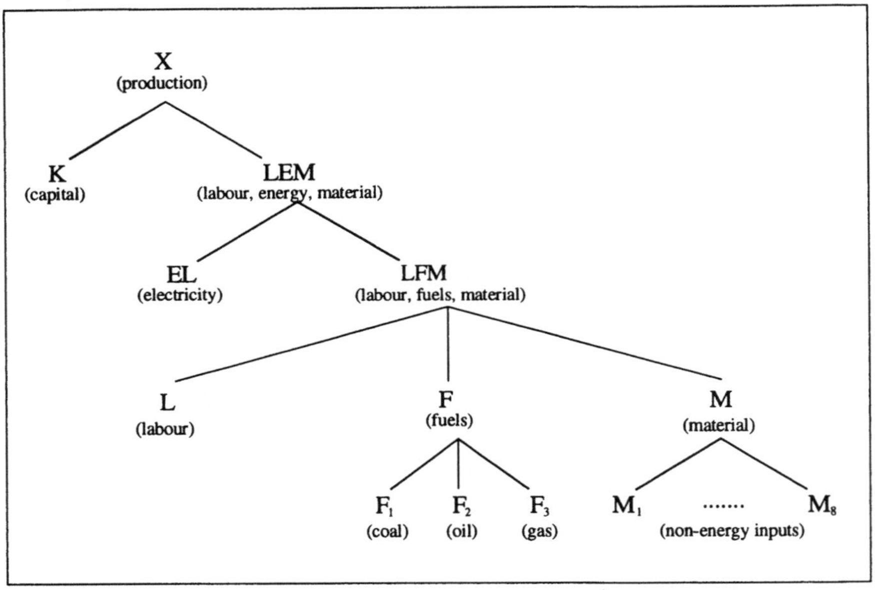

Capital input as derived from (2) is the desired capital stock, say K_{des}. In the GEM-E3 model we treat, however, capital as a quasi-fixed stock over the current year at a level from the end of the previous year, say K_{-1}. We therefore use (2) to determine an endogenous ex-post price of capital based on a rate of return which the industry has earned ex-post. For that purpose we solve (2) for PK_{post}:

(4) $$PK_{post} = PX \cdot f\left(\frac{X}{K_{-1}}, g\right)$$

PK_{post} is the endogenous shadow price of capital which clears the market for fixed K_{-1}. It will be used to calculate capital income $PK_{post} \cdot K_{-1}$ in period t.[2]

If we determine an exogenous ex ante price of capital PK_{ante}, then (2) can be employed to determine the desired stock of capital K_{des}. Let this ex ante price be the standard user cost of capital formula $PK_{ante} = PI \cdot (r + \delta)$ where PI is the price of investment goods, r is the rate of return on risk-free government bonds (exogenously given or determined by the closure rule) and δ is the rate of replacement. Then the desired capital stock is

$$\text{(5)} \qquad K_{des} = X \cdot a_K, \quad \text{with } a_K = a_K\left(\frac{PX}{PK_{ante}}, g\right).$$

Net investment I_{net} with 'adjustment' is given by

$$\text{(6)} \qquad I_{net} = m(K_{des} - K_{-1})$$

Finally, the capital stock for the next period is

$$\text{(7)} \qquad K = I_{br} + (1 - \delta) K_{-1},$$

where $I_{br} = I_{net} + \delta \cdot K_{-1}$ (gross investment).

We next specify a CES price function for the aggregate *LEM*

$$\text{(8)} \qquad PLEM = PLEM(PEL, PLFM, g),$$

where PEL is the price of electricity (EL) and $PLFM$ an aggregation of the prices of labour (L), other fuels (F) and material (M).

One level further down of the nesting the unit cost function for the *LFM* aggregate has to be specified:

$$\text{(9)} \qquad PLFM = PLFM(PL, PF, PM, g)$$

Again, we derive the *LFM* and *EL* aggregates as well as the price-dependent composition of the *LFM* aggregate from Shephard's Lemma.[3] This yields the input coefficients for labour, the material aggregate and the fuels aggregate.

The final level is given by a CES-composition of these aggregates. The fuels aggregate consists of three fuel inputs (1,2,3: coal, gas, oil) whereas the material aggregate considers seven non-energy inputs (5-11: agriculture/forestry/fishery products, energy intensive products, equipment goods, consumer goods, transport, services and non-market services). Sector 4 is the electricity sector specified already by equation (8).

$$\text{(10)} \qquad PF = PF(\tilde{PY}_1, \tilde{PY}_2, \tilde{PY}_3, t)$$

---

[2] It is easy to check that our calculation of PK_{post} is equivalent to calculating it from the zero profit condition.

[3] i.e., similar to equations (2) and (3).

(11) $PM = PM(\tilde{PY}_5, \ldots, \tilde{PY}_{11}, t)$,

where $\tilde{PY}_i$ denotes the price of domestic supply PY_i plus indirect taxes.[4]

Applying again Shephard's Lemma yields the derivation of the input coefficients. If we multiply the input coefficient of the aggregates by the coefficients of their sub-inputs we obtain the overall input coefficients a_i with respect to the domestically produced supply:
For fuel inputs

$$(12) \qquad a_{i,X} = \frac{F_i}{X} = \frac{F_i}{F}(\cdot) \cdot \frac{F}{LFM}(\cdot) \cdot \frac{LFM}{LEM}(\cdot) \cdot \frac{LEM}{X}(\cdot) \qquad i = 1,\ldots,3$$

for material inputs

$$(13) \qquad a_{i,X} = \frac{M_i}{X} = \frac{M_i}{M}(\cdot) \cdot \frac{M}{LFM}(\cdot) \cdot \frac{LFM}{LEM}(\cdot) \cdot \frac{LEM}{X}(\cdot) \qquad i = 5,\ldots,n$$

for electricity

$$(14) \qquad a_{EL,X} = \frac{EL}{LEM}(\cdot) \cdot \frac{LEM}{X}(\cdot)$$

and for labour

$$(15) \qquad a_{L,X} = \frac{L}{X} = \frac{L}{LFM}(\cdot) \cdot \frac{LFM}{LEM}(\cdot) \cdot \frac{LEM}{X}(\cdot)$$

The dot in the parentheses indicates that the coefficients depend on relative prices.

The parameterisation is based on calibration using country specific social accounting matrices which link input-output tables with common national accounting. This procedure permits to calculate most parameters from observed input coefficients, given a range of other parameters like the elasticities of substitution.

2.3 The Foreign Trade Specification

For modeling intra-industry foreign trade between the EU member countries, the Armington approach is widely accepted: domestically produced goods and imports from different countries are imperfect substitutes. Thus dual to a CES production function $Y_c=f(XD_c,IM_c)$, giving domestic supply Y_c in country c as an aggregate of domestic production (for domestically demanded goods) XD_c and imports IM_c, is a CES unit cost function.[5]

---

4 see also equation (16).

5 Note that XD_c is the production for the domestic market only.

$$(16) \qquad PY_c = \left(c_x \cdot PXD_c^{1-\sigma_x} + (1-c_x) \cdot PIM_c^{1-\sigma_x}\right)^{\frac{1}{1-\sigma_x}},$$

where $\sigma_x > 0$ is the elasticity of substitution, c_x a distribution parameter and PY_c, PXD_c, and PIM_c are the corresponding prices of Y_c, XD_c, and IM_c (price of aggregated imports is in national currency of country c). From this cost function we derive both the share of domestic production (for domestically demanded goods) and the share of imports in total supply:

$$(17) \qquad \frac{XD_c}{Y_c} = c_x \cdot \left(\frac{PY_c}{PXD_c}\right)^{\sigma_x}$$

$$(18) \qquad \frac{IM_c}{Y_c} = (1-c_x) \cdot \left(\frac{PY_c}{PIM_c}\right)^{\sigma_x}.$$

If the model determines total domestic supply, then we have to allocate aggregate import demand, derived from (18), to the 11 EU member state countries and to the rest of the world (*row*) who contribute to this aggregate import demand. Thus, considering a specific good, the imports of country c consist of the exports of the other 11 countries in that good. Therefore, the demands for the 11 goods are also distinguished by place of production. There will be French import demand of consumer goods produced in the United Kingdom and produced in Spain. To obtain such a trade matrix with 11 x 12 import demand functions by good and place of production we specify a CES import unit cost or price function (12 = *row*):

$$(19) \qquad PIM_c = \left(\sum_{k=1}^{12} c_{m,k} \cdot PIMP_{c,k}{}^{1-\sigma_m}\right)^{\frac{1}{1-\sigma_m}} \qquad c = 1, \dots, 11,$$

where $PIMP_{c,k}$ is the price of imports coming from country k in currency of country c. As there are import taxes and duties (t_{dut}), it is $PIMP_{c,k} = (1 + t_{c,k,dut}) \cdot PEX_{k,c} \cdot e_c / e_k$, where all exchanges rates are in national currency per ECU and $PEX_{k,c}$ is derived from equations (25) and (26).[6] Given the price index PIM_{row} and the exchange rates $e_{row,c}$, the eleven prices PIM_c can be calculated.

Again, a cost minimising composition of the import aggregate is the objective of the importing country. Shephard's Lemma, applied to the cost function PIM_c in (19) yields:

---

6 Since we distinguish 11 goods, we would have to write
$PIMP_{i,c,k} = (1 + t_{i,c,k,dut}) \cdot PEX_{k,c} \cdot e_c / e_k$

$$(20) \qquad \frac{IMP_{c,k}}{IM_c} = c_{m,k} \cdot \left(\frac{PIM_c}{PIMP_{c,k}} \right)^{\sigma_m}, \qquad k = 1,\ldots,11,\ row$$

where $IMP_{c,k}$ is the import by country c from country k in currency of country c.[7] Multiplying equation (20) by IM_c, derived from equation (18), yields a trade matrix for each of the 11 commodities.

The country specific imports $IMP_{c,k}$ have to match the export demand by definition, i.e.,

$$(21) \qquad EXP_{k,c} = \frac{IMP_{c,k} \cdot e_k}{e_c}.$$

In contrast to many other model specifications the version of the GEM-E3 model used here differentiates export supply by place of destination: Domestic production X_c is supplied on the domestic market XD_c, on the EU market $EX_{c,eu}$ and on the rest of the world market $EX_{c,row}$. Firms maximise revenue, given the production level X_c determined by the input-output solution:

$$\max_{XD,EX,EX_{row}} \quad R = PXD_c \cdot XD_c + PEX_{c,eu} \cdot EX_c + PEX_{c,row} \cdot EX_{c,row}$$

$$\text{s.t.} \qquad X_c = X_c(XD_c, EX_c, EX_{c,row}),$$

where $X_c(\cdot)$ is a transformation function with a constant elasticity of transformation (CET). We choose again the dual approach in terms of a revenue function $R(X, PXD, PEX_{eu}, PEX_{row})$, dual to the CET transformation function. Since the latter is assumed to be homogeneous of degree one, the revenue function can be written as:

$$(22) \qquad R(X,\ldots) = X \cdot \left(\gamma_c \cdot PXD^{1-\sigma_T} + \gamma_{eu} \cdot PEX_{c,eu}^{1-\sigma_T} + \gamma_{row}\ PEX_{c,row}^{1-\sigma_T} \right)^{\frac{1}{1-\sigma_T}}$$

where $\sigma_T < 0$ is the elasticity of transformation. The supply functions are derived by using Hotellings Lemma, i.e., by differentiating the revenue function with respect to the product prices.

Since the sales identity $PX \cdot X = R(X,\ldots)$ has to hold, the product prices have to satisfy the price equation:

$$(23) \qquad PX_c = \left(\gamma_c \cdot PXD_c^{1-\sigma_T} + \gamma_{eu} \cdot PEX_{c,eu}^{1-\sigma_T} + \gamma_{row} \cdot PEX_{c,row}^{1-\sigma_T} \right)^{\frac{1}{1-\sigma_T}}$$

Therefore, the export supply is given by the following equations:

---

[7] Because of $\sum_k c_{m,k} = 1$, the adding up condition $\sum_k PIMP_{c,k} \cdot IMP_{c,k} = PIM_c \cdot IM_c$ is automatically satisfied.

domestic market

$$(24) \qquad XD_c = X_c \cdot \gamma_c \cdot \left(\frac{PX_c}{PXD_c} \right)^{\sigma_T}$$

EU-market

$$(25) \qquad EX_{c,eu} = X_c \cdot \gamma_{eu} \cdot \left(\frac{PX_c}{PEX_{c,eu}} \right)^{\sigma_T}$$

world market

$$(26) \qquad EX_{c,row} = X_c \cdot \gamma_{row} \cdot \left(\frac{PX_c}{PEX_{c,row}} \right)^{\sigma_T}$$

As export demand is determined by (21), equation (25) can be used to find the equilibrium price $PEX_{c,eu}$ which matches demand and supply of exports. Since the price PX_c is the unit cost, determined in (1), (23) can be solved for PXD_c. Then XD_c in (24) can be calculated. Finally, given the import demand of the rest of the world ($IMP_{c,row}$, see (27) below), the price $PEX_{c,row}$ is derived from the export supply function (26).

To determine the import demand of the rest of the world (*row*) we proceed as in equations (16) to (19) by choosing the index *row* instead of *c*. Dividing (18) by (17) yields the import demand function:

$$(27) \qquad IM_{row} = X_{row} \cdot f\left(\frac{PX_{row}}{PIM_{row}} \right)$$

The world market prices PX_{row} and the output of the rest of the world X_{row} are exogenous for all eleven commodities. One of these prices serves as the numeraire. Finally, there is a price function for PIM_{row} similar to (19) and import demand functions $IMP_{row,k}$ analogous to (20).

This yields a trade surplus or deficit (TS):

$$(28) \qquad TS = PEX_{c,eu} \cdot EX_{c,eu} + PEX_{c,row} \cdot EX_{c,row} - PIM_c \cdot IM_c \,.$$

As the nominal exchange rates are assumed to be exogenous, the current account is not balanced. Depending on the scenario to be simulated, the current account might increase or decrease and even change the sign.

2.4 Consumer Demand and Labour Supply

The behaviour of consumers is assumed to perform a two-stage budgeting procedure: an intertemporal allocation of lifetime wealth endowment between present and future consumption of goods and leisure, and an intratemporal allocation of total consumption of goods into durable and non-durable goods. The durables are distinguished by three (cars, heating systems, electrical appliances), the non-durables by nine categories of consumption. Demand for non-durables like gasoline or electricity is linked to the use of durables. Hence, demand for linked non-durables and the demand for services from durables has to be reconciled with investment demand for modifying the stocks of durables towards their optimal levels. We therefore employ a restricted expenditure function with stocks of durables as quasi-fixed goods. The expenditure function is derived from the Stone-Geary utility function which underlies the linear expenditure system. Figure 2 shows the household's allocation problem.

Figure 2: The household allocation scheme

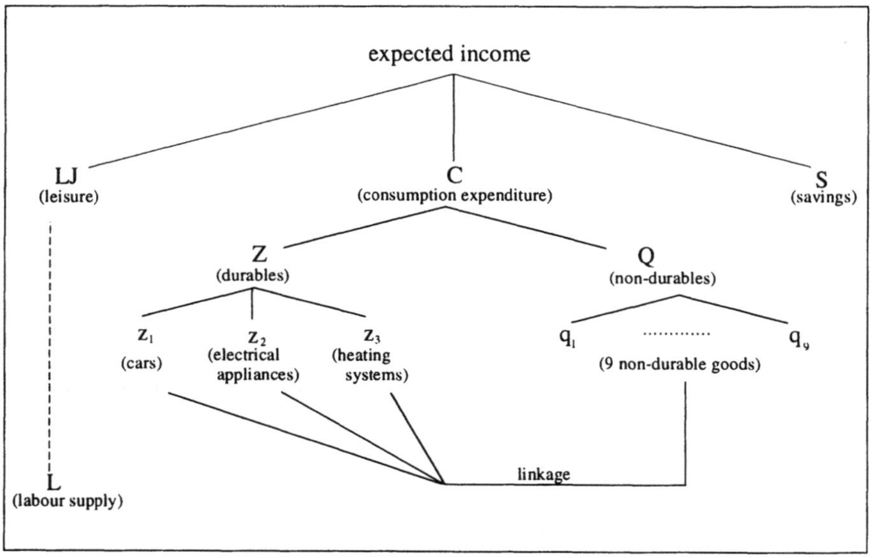

At the first stage a representative household determines an allocation of his resources between present and future consumption by maximising an intertemporal utility function subject to an intertemporal budget constraint:[8]

8 β_C and β_{LJ} are normalised so as to sum up to one.

$$\text{(29)} \qquad \max_{C_t, LJ_t} \Sigma_t (1 + s)^{-t} (\beta_c \ln(C_t - C_0) + \beta_{LJ} \ln(LJ_t - LJ_0))$$

$$\text{s.t. } WT = \Sigma_t (1+r)^{-t} (PC_t \cdot C_t + PLJ_t \cdot LJ_t)$$

where WT is present value of wealth. C_t is private consumption in volume, C_0 its subsistence level, LJ_t is leisure (in volume) and LJ_0 its subsistence level, s is the subjective discount rate and r is the nominal interest rate. An initial commitment for leisure could be $LJ_0 = 12$ hrs/day multiplied by the average working days per year. The price of leisure is $PLJ = (1 - t_{hss} - t_{fss}) \cdot (1 - t_{hdir}) \cdot w^{nom}$ where t_{hdir} is the marginal direct tax rate for labour income and t_{hss} and t_{fss} are the contributions of employers and employees to social security. Under myopic expectations and the assumption of constant and equal growth rates for inflation and the nominal wage rate, the Fisher relation can be used to reduce the demand functions for consumption and leisure to the following formulas:[9]

$$\text{(30)} \qquad C = C_0 + \frac{s}{r_r} \left(\beta_c \frac{1}{PC} \right) (Y_{disp} + PLJ \cdot LJ - PC \cdot C_0 - PLJ \cdot LJ_0)$$

$$\text{(31)} \qquad LJ = LJ_0 + \frac{s}{r_r} \left(\beta_{LJ} \frac{1}{PLJ} \right) (Y_{disp} + PLJ \cdot LJ - PC \cdot C_0 - PLJ \cdot LJ_0),$$

where r_r is the real long term interest rate which is assumed to be constant in the standard version of the model.[10]

The last equation is implicit in LJ and has to be solved for LJ. Labour supply is given by the remaining time resources, i.e., total time resources minus leisure demand. In the standard neo-classical version of the model, the wage rate serves to match labour demand of firms and leisure demand of households. The savings of households can then be determined by $S = Y_{disp} - PC \cdot C$.

Two types of consumption expenditure are distinguished: the expenditure for non-linked non-durable goods (e) which is allocated on the second stage of the consumer decision problem and the expenditure for the use of durables (this covers capital costs and demand for linked non-durables associated with the use of durables).

The expenditure function for the variable non-durables with three quasi-fixed durable goods (Z_1, Z_2, Z_3 for cars, electric appliances, and heating, respectively) is

$$\text{(32)} \qquad e(p_1, \ldots, p_m; u; Z_1, Z_2, Z_3) = \sum_{i=1}^{m} p_i \cdot Q_{o,i} + u \cdot \prod_{j=1}^{3} (Z_j - Z_{o,j})^{-\gamma_j} \cdot \prod_{i=1}^{n} \left(\frac{p_i}{\beta_i} \right)^{\beta_i}$$

---

9 see Schmidt (1997) for a complete representation of the derivation.

10 The long term interest rate is endogenised if the constraint of a balanced current account is imposed (other version of the model).

where we denote the prices of consumption categories p_i by small letters to indicate that they do not match directly to prices of products $P\tilde{Y}_l$. The variable u is the utility level, $Q_{0,i}$ is the minimum required quantity of good i, $Z_{0,j}$ is the minimum required quantity of a durable good j, and $\sum p_i \cdot Q_{0,i}$ is 'subsistence expenditure'. The expenditure minimising demand for non-durable goods, given utility u and the stocks of the three durables, can be derived by partial differentiation of the expenditure function with respect to the prices:

$$(33) \qquad Q_i = Q_{0,i} + \frac{\beta_i}{p_i}\left(e - \sum_{i=1}^{m} p_i \cdot Q_{0,i}\right), \qquad i = 1,\ldots,m$$

The desired stocks of durables and the ex-post service prices of durables can be derived in an analogous way as used for the restricted cost function approach. With an exogenous ex-ante user cost of durables p_{Z_j}, the desired stock follows from

$$\frac{\partial e(\cdot, \tilde{Z}_j)}{\partial Z_j} = -p_{Z_j}, \qquad i.e.$$

$$(34) \qquad \tilde{Z}_j = Z_{0,j} + \frac{\gamma_j}{p_{Z_j}}\left(e - \sum_{i=1}^{m} p_i \cdot Q_{0,i}\right)$$

Purchases of new durables under partial adjustment restrictions ($0 < m_j \leq 1$) are:

$$(35) \qquad I_{Z_j}^{net} = \tilde{m}_j \cdot (\tilde{Z}_j - Z_{-1,j}) \qquad j = 1, 2, 3.$$

We finally obtain the total consumer expenditures $PCE \cdot CE$ including both, non-durables and the services from durables.

$$(36) \qquad PCE \cdot CE = \sum_{i=1}^{m} p_i \cdot Q_i + \sum_{j=1}^{3} p_{Z_j} \cdot (Z_{-1,j} + I_{Z_j}^{net})$$

We should finally say some words about the specification of the user cost of a durable, p_{Z_j}. In principle, we could set p_Z equal to $p_Z = PI \cdot (r + \delta)$, where PI is the price of the durable, δ is the rate of replacement and r is the interest rate. However, as some non-durable goods as gasoline, electricity, and heating are linked to the stock of durables, we used a composition of these goods into a linked part and into a disposable part. The idea behind such a composition is that demand for gasoline (Q_G) is linked to the use of the stock of automobiles (Z). Or, in algebraic terms, $Q_G = \alpha_{G,Z} \cdot Z + \tilde{Q}_G$, where $\alpha_{G,Z}$ is yearly gasoline consumption per unit of purchase price of the car and $\tilde{Q}_G$ is gasoline consumption from fast driving or bad maintenance of the car.[11] The latter is considered to be a part of the non-linked non-durables and is therefore not further specified. The user concept for durables implies a cost price p_Z of the services of, e.g., an automobile which includes the user cost of capital plus the cost of gasoline, i.e., $\tilde{p}_Z = PI(r+\delta)+\alpha_{G,Z}\cdot p_G$. The introduction of a tax on CO_2 or NO_x will increase the price of gasoline. Hence, the user cost of an already purchased car will increase while the demand for new cars will decline. Under a carbon dioxide tax, for instance, the cost price of a car is $p_Z = PI\,(r+\delta)+\alpha_{G,Z}\cdot(p_G+t_{CO_2}\cdot e_{CO_2})$, where t_{CO_2} is the tax rate and e_{CO_2} is the emission coefficient for gasoline. If we incorporate furthermore a property tax or motor vehicle tax rate τ, then the user cost of a car is

$$(37) \qquad p_Z = PI\cdot(r\cdot(1+\tau)+\delta+\tau)+\alpha_{G,Z}\cdot p_G$$

For guess-estimation of the parameters $Q_{0,i}$, β_i and γ_j we make use of the properties of a linear expenditure system, i.e., from guess-estimates of n income elasticities one obtains the n parameters β_i and from guess-estimates of n direct price elasticities one obtains the n parameters $Q_{0,i}$, given the β_i' s (and similarly for the parameters of the durables).

2.5 Demand, Supply and Model Closure

The standard system of input-output accounting is given by

$$(38) \qquad Y_i = \sum_{j=1}^{n} a_{i,j}\cdot X_j + F_i$$

where F_i is final demand with $F_i = C_i + CG_i + I_i + IG_i + EX_i$. C_i is private consumption of good i, CG_i and IG_i are government consumption or investment, respectively (exogenous), I_i is gross investment by origin and EX_i are exports.

Since our demand system determines consumption goods by categories and our system of investment functions investment demand by destination, transition matrices are required transforming demand into deliveries from the industries. Therefore, the C_i's in final demand have to be seen as the result of the transition matrix of the type (branches x categories) multiplied by the consumption categories. Similar to the matching of consumption categories to products, an investment matrix with fixed technical coefficients serves to compute investment demand by origin (products) from investment demand by destination (branches) as evaluated from investment behaviour in (5) and (6), together with investment for replacement and decay, i.e., $\delta\cdot K_{-1}$. The system (38) can be written as a system in the unknown variables Y_i if we rewrite it as

---

[11] For more details see Conrad/Schröder (1991b).

(39) $$Y_i = \sum_{j=1}^{n} a_{ij} \cdot \left(\frac{X_j}{Y_j} \right) \cdot Y_j + F_i$$

with X_j / Y_j determined by (17).

In value terms, demand has to be equal to supply:

(40) $$PZ_i \cdot Y_i = \sum_{j=1}^{n} \tilde{PY_j} \cdot X_{ij} + (1+t_i) \cdot \tilde{PY_i} \cdot (C_i + CG_i) + \tilde{PY_i} \cdot (I_i + IG_i) + PEX_i \cdot EX_i$$

where PZ_i is the market price including indirect taxes, PY_i is the price of domestic supply including indirect taxes on production and t_i indicates the value added tax rate on consumption. $X_{i,j}$ represents the intermediate demand of good i . The accounting identity from the input side is:

(41) $$PY_j \cdot Y_j = \sum_{i=1}^{n} \tilde{PY_i} X_{ij} + PL_j \cdot L_j + PK_j \cdot K_{-1,j} + PIM_j \cdot IM_j$$

If we sum (40) over i and (41) over j and then subtract (41) from (40) we obtain the national accounting identity saying that the private gross domestic production from both the flow of cost approach and from the flow of product approach should be equal, i.e.,

(42) $$\sum_j (PL_j \cdot L_j + PK_j \cdot K_{-1,j}) = \sum PY_i \cdot F_i - \sum_j PIM_j \cdot IM_j$$

This identity is satisfied if and only if total saving, involving income distribution and fiscal policy relationships, equals total investment. Following Walras' Law, this market $(n+1)$ is in equilibrium if an equilibrium price vector has been found for the other n markets (supposed that the demand, supply and price functions are specified according to the needs of an Arrow-Debreu economy). Therefore, the saving-investment identity $(I=S)$ and the corresponding global shadow price of capital (mobility of (new) capital between sectors but not across countries is assumed) is automatically given.

2.6 The Environmental Module in GEM-E3

The scope of the environmental issue is limited to three pollutants: nitrooxides (NO_X), sulfur dioxide (SO_2) and carbon dioxide (CO_2). For SO_2 and NO_X we specify abatement costs which will increase the cost price of using pollution intensive inputs. To derive such a cost price, we start with the primal production function approach where material input M consists of material input M_1 for production and of material input, M_2 required for complying with environmental regulation, i.e., $M=M_1+M_2$. Let us assume that the environmentally related input M_2 is proportional to the flow of pollutants, which in turn depends on the input of fossil fuel F, i.e., $M_2 = \alpha \cdot e \cdot F$, where e is an emission or waste coefficient in terms of tons of an air pollutant per unit of the energy input. Combining the two equations yields: $M = M_1 + \alpha \cdot e \cdot F$. Using this composition we will rewrite the standard cost minimisation approach given by:

$$C(X, PK, PL, PEL, PF, PM) = \min_{K,L,EL,M,F} (PK \cdot K + PL \cdot L + PEL \cdot EL + PM \cdot M + PF \cdot F)$$

$$s.t.\ X = f(K, L, EL, F, M),$$

replacing M by $M_1 + \alpha \cdot e \cdot F$ one obtains

$$C(X, PK, PL, PEL, PF, PM) = \min_{K,L,F,M} \left(PK \cdot K + PL \cdot L + PEL \cdot EL + PM \cdot M_1 + \widetilde{PF} \cdot F\right)$$

$$s.t.\ X = f(K L, EL, F, M_1 + \alpha \cdot F),$$

where $\widetilde{PF} = PF + PM \cdot \alpha \cdot e$ is the cost price of energy. It consists of the energy price and the additional costs due to environmental regulation when using one unit of energy input. As the cost of regulation increases with the enforcement intensity, α should depend on the degree of abatement. We therefore specify α as a function in the degree of abatement a, which represents the enforcement in pollution control:

$$\alpha = c(a) \cdot a$$

The degree of abatement is defined as the ratio of abated emission over potential emissions ($0 \leq a \leq 1$) and $c(a)$ are the costs of abatement measures per unit of emission or waste, measured in base year prices. They depend on the degree of abatement with $c'(a) > 0$ and $c''(a) > 0$. Finally, e is an emission or waste coefficient in terms of tons of an air pollutant per unit of energy input. With this interpretation of α we obtain the following cost price for energy

(43) $$\widetilde{PF} = PF + PM \cdot \alpha \cdot e = PF + PM \cdot c(a) \cdot a \cdot e.$$

This cost price of energy increases over-proportional by an enforcement in environmental regulation. On the production side this implies an increasing share of complementary material inputs. The change of the cost price of energy will also cause the firm to alter its input choices. A stricter environmental policy will have a substitution effect which will result in a reduced demand for energy and its price complements and in an increased use of its substitutes. This integration of abatement costs in a cost-price concept will be used for modeling the impact of regulation on household and firm behaviour; for the latter each sector will be treated separately.

The cost price approach can be extended for the case of several pollutants. Then $\widetilde{PF}$ is

(44) $$\widetilde{PF} = PF + PM(c_{SO_2}(a_{SO2}) \cdot a_{SO_2} \cdot e_{SO_2} + c_{NO_x}(a_{NO_x}) \cdot a_{NO_x} \cdot e_{NO_x})$$

with abatement costs for SO_2 and for NO_X. If there is a tax on a pollutant, then there is also a cost price component for the actual emissions, i.e.,

(45) $$\widetilde{PF} = PF + PM \cdot c\,(a) \cdot a \cdot e + t \cdot (1 - a) \cdot e\,.$$

Finally, if there is an energy tax (t_F) and/or an emission tax on carbon dioxide, t_{CO_2}, where no convenient end-of-pipe measures exist, then a turns out to be zero in the equation above.

The cost price approach can be embedded in the CES price function by replacing in equation (10) the price of the energy components $P\widetilde{Y}_i$ by $\widetilde{PF}_i$. This also increases the overall price index of energy, $\widetilde{PF}$. Environmental regulation will then have an impact on the composition of the energy aggregate *F*. It will also increase the price of the product according to equation (1), and it will reduce the demand for energy derived from equation (10).

2.7 Welfare Measure

The welfare change used for the evaluation of policy scenarios is represented by Hick's measure of equivalent income variation (*EV*). The EV is based on the intertemporal utility maximisation problem and has to be derived from equations (29) - (31). In a single period t we have

(46) $$EV_t = FE(PC_t^0, PLJ_t^0, U_t^1) - FE\,(PC_t^0, PLJ_t^0, U_t^0)\,,$$

where *FE* is the expenditure function corresponding to (29) - (31). *EV* gives the change in expenditure at base case prices PC^0 and PLJ^0 that would be equivalent to the policy-implied change in utility. In order to derive the expenditure function from the utility function, we insert the demand functions (30) and (31) into the utility function (29), and solve for the level of utility, say U_t:

(47) $$U_t = \frac{s}{r_r} \cdot \left(\frac{\beta_c}{PC_t}\right)^{\beta_c} \cdot \left(\frac{\beta_{LJ}}{PLJ_t}\right)^{\beta_{LJ}} \cdot (FE_t - PC_t \cdot C_0 - PLJ_t \cdot LJ_0)$$

where FE_t is total expenditure:

$$FE_t = Y_{disp,t} + PLJ_t \cdot LJ_t$$

(47) solved for FE_t gives the expenditure function used in (46) to determine *EV*:

$$FE_t(PC_t, PLJ_t, U_t) = U_t \cdot \left(\frac{\beta_c}{PC_t}\right)^{-\beta_c} \cdot \left(\frac{\beta_{LJ}}{PLJ_t}\right)^{-\beta_{LJ}} \cdot \left(\frac{s}{r_r}\right)^{-1} + PC_t \cdot C_0 + PLJ_t \cdot LJ_0$$

The utility level U_t *is* calculated from the tth element of the sum of utilities in (29).

If $EV<0$, welfare after the policy measure is lower than in the base case. The consumer would be willing to pay the maximum amount EV at the fixed budget level $FE^0=FE(PC^0,PLJ^0,U^0)$ to avoid the decline of utility from U^0 to U^1. Similarly, if $EV>0$, the consumer would be willing to pay the maximum amount EV to see the change in environmental policy implemented.

The standard model version considers full competitive equilibrium in all markets, including the labour market. At this stage it covers eleven countries (Belgium, Germany, Denmark, France, Greece, Ireland, Italy, The Netherlands, Portugal, Spain and United Kingdom) and eleven products or sectors (agriculture, coal, crude oil and refined oil products, gas, electric power, energy intensive industries, equipment goods industries, consumer goods industries, transport, services, non-market services).

3 Empirical Analysis of the Double Dividend Issue

3.1 Policy Options and Scenario Definition

As an application of the GEM-E3 model we have estimated the welfare gain of alternative CO_2-reduction policies. Each policy is linked to a reduction target that is achieved by a CO_2 tax with a rate just high enough to achieve the given reduction goal. The revenue from this tax is used to reduce the employers' contribution to social security. This is the so-called "double dividend" analysis. The carbon tax should affect the substitution of energy for other inputs and reduce global warming (first dividend). This substitution effect could have already a positive impact on the demand for labour. However, the recycling of the tax money to social security as a partial compensation for employers' contribution should increase the demand for labour. The hope, the advocates of the double dividend have in mind is that the substitution effect towards more labour outweighs the output effect in terms of lower growth. Some authors describe this positive effect on the labour market as the second dividend, others define the second dividend in terms of economic welfare. Our analysis deals with both definitions as we are in general not convinced that the double dividend criterion is sufficient to be 'the one and only' measure that should be used to accept or neglect a policy. From a policy maker's view one might accept a loss in economic welfare if the policy measure supports more employment significantly.

In our empirical analysis we focus on the double dividend issue under two institutional settings. The settings are related to the degree of environmental policy coordination. In the first setting Germany decides to be a forerunner in combating climate change by reducing the emissions (of the base year) by 10% within ten years. The other EU countries do not apply particular measures for reduction. In the second setting the EU member states decide to combat climate change in a coordinated way. The target of 10% emission reduction is set at the EU level and all countries try to reach this goal together. Hence, in this setting, the marginal costs are equalised not only across sectors but across countries as well.

One could think about other schemes of coordination, e.g., every country reduces 10% of it's base year emissions or a 'full' coordination policy where the winners compensate the losers via side payments. These schemes are not discussed here. We rather view the climate topic from a German policy maker's perspective. Hence, the issue of concern is: 'What can Germany gain from an EU-wide coordinated environmental tax reform?.'

It is evident that the overall EU-wide emission reductions of the two scenarios specified above are different. Hence, we do not analyse solely an efficiency gain due to a better allocation from coordination.[12] The policies differ considerably with respect to the environmental improvement obtained and the economic burden on the economy. Comparing the two policies implicitly assumes that the German policy maker is indifferent to the actual environmental benefit achieved by the policy. One could think about such a behaviour if one finds that the policy maker reacts mainly to international requests given for example by the ongoing negotiations of the Climate Convention than to eventually occuring national environmental damages (that might be very low). Advertising a reduction of 10% could offset the international pressure on whatever level (national or EU-wide) this reduction will be. Hence, what the policy maker minds is then the national economic impact of the policy implemented and not the environmental benefit achieved. To this concern, the comparison undertaken is more a question of political opportunity than of economic efficiency.

For both policy options the emission reductions are obtained by an (endogenously computed) tax on CO_2 emissions. This tax is levied on producers and consumers. The tax reform is supposed to take place in a revenue neutral way, i.e., the receipts are kept in the country that imposes the tax and are used to reduce the employers' contribution to social security. The revenue neutrality is achieved by keeping the share of public deficit on GDP constant. As the tax and the recycling is endogenous, the problem of tax erosion induced by the decrease in emissions is implicitly solved. Hence, the reduction of wage costs through recycling of tax receipts is limited by the public deficit constraint.

The time horizon of target realisation is set to ten years. The reduction within these ten years is assumed to take place linearly, i.e., in each year the level of emissions (in terms of the base year) is reduced by 1%.

As the 10% target is expressed in terms of emissions of the base year, economic growth will increase the actual reduction effort. To take this aspect into account, we consider the expected emission growth in the eleven countries. The rates assumed for this study are based on an estimate of the European Commission. It is

[12] This was done, e.g., by Conrad/Schmidt (1996a or 1996b).

considered to be a 'conventional wisdom' scenario[13] and covers both economic growth and growth in energy efficiency. The country specific emission growth rates are depicted in the last column of table 1.

One way to consider the growth aspect in the simulations could be to calibrate energy efficiency, emission coefficients and economic growth in such a way that the model generates ('ex-post') the expected economic and environmental development. This procedure turned out to be very difficult in the complex world of our multi-country and multi-sector model.

Another way is to consider the growth aspect in the reduction plan of the policy. Then, the targets imposed include the 10% reduction of the base year emissions plus the expected emission growth.[14] Former sensitivity analyses have shown, that it is sufficient to limit the growth consideration to the policy targets if the structural changes due to the expected growth are reasonable and if one looks at the results in relative terms only. In the tables below, the policy impacts of the counterfactual equilibrium (policy simulation) are presented as percentage change of the reference equilibrium (reference simulation). What is shown then is the pure policy effect; assumptions that are kept unchanged in both scenarios are ruled out.

Furthermore, we limit the presentation of results to the impacts that are obtained at the end of the implementation horizon. In the tables below yearly values of the tenth year are depicted by a '10' in the period line while figures that concern the entire time horizon are marked by a '1-10' in the period line.[15]

3.2 Simulation Results

3.2.1 Unilateral Action Versus Coordination

We start with the unilateral case where Germany decides to reduce its emissions solely. The last column of table 1 shows the underlying CO_2 emission growth rates assumed. The column eleven depicts an actual (net) emission reduction of -7%.[16] Together with a negative expected growth rate for Germany of -3%, the 10% goal is reached. The increase of emissions in the other countries indicates the EU internal carbon leakage. The emission growth in the EU is lowered (almost 2%) but the absolute increase remains positive.

---

[13] i.e., it is not very optimistic but also not very pessimistic and includes some measures that will probably take place in the countries without imposing a 'major' policy instrument (like an emission tax).

[14] As we had no better data, we assumed, that all sectors grow at the same rate.

[15] Further details on the periodical values can be obtained from the authors on request.

[16] Throughout the discussion of results the sign of the numbers in the tables are kept whatever formulation is used. For example we will not change the sign in the following expressions: 'the emission growth rate is -3%' and 'the emissions fall by -3%'.

As table 1 shows, the unilateral 10% reduction of Germany requires an emission tax of 8.24 ECU per tonne of CO_2 at the end of the reduction plan. The welfare effect measured in equivalent variation and expressed in percent of GDP is positive (0.02%). As both environment and welfare are improved, the so called double dividend is obtained. The welfare gain is achieved due to the increase in consumption (0.13%) which outweighs the 'negative' effect linked to the increase in employment (0.26%) (more employment means less leisure as population and therefore total time endowment is fixed). As employment and the real wage rate (0.47%) go up, there is more (labour) income for consumption. Other incomes (i.e., non-labour income)[17] decrease which reinforces the effect on employment (increase in labour supply). Since gross domestic production falls by -0.20%, labour productivity declines. The investment of firms falls by -0.18%. Therefore, the positive effect in GDP is linked solely to the increase in consumption and the changes in the current account. Imports decrease (-0.24%) more than exports (-0.13%) which leads to an 'improvement' of the current account. The other countries gain or lose slightly according to the trade relations. The interested reader is referred to the figures in table 1.

The effect on employers' rate to social security is depicted in figure 3. Recycling the receipts of the emission tax in a revenue neutral way leads to a reduction of this rate by 1.67 percent points (see the bar for DE-nc). As Germany acts unilaterally there is no reduction in the other countries.

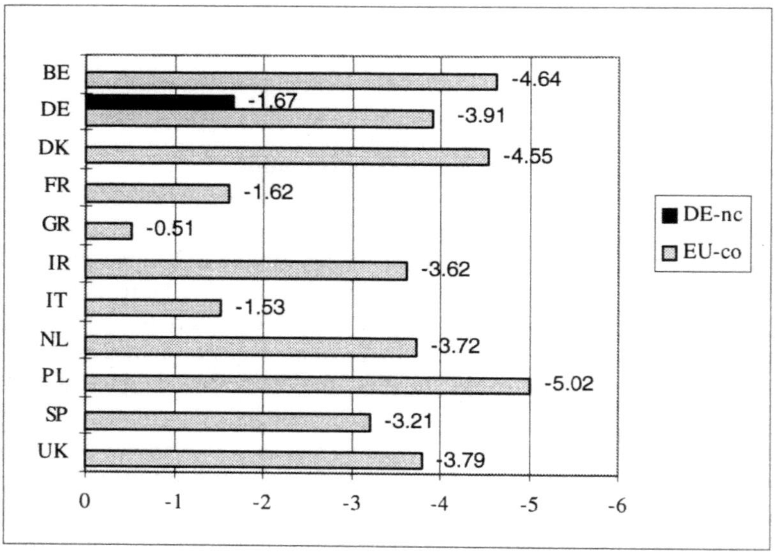

Figure 3: Employers' contribution to Social Security (change in %-points)

---

[17] The changes in other income are not depicted in the tables.

The simulation results of the coordinated policy are presented in table 2 and figure 3. The country specific reductions of employer's social security contribution due to the CO_2-tax receipts are depicted in figure 3 (see bars of EU-co). The cuts are between 0.51 (Greece) and 5.02 (Portugal) percent points. Most countries (Belgium, Germany, Denmark, Ireland, Netherlands, Spain, United Kingdom) find a reduction between 3.5 and 4.5 percent points.

Real wage rates and employment increase in all countries with the exception of Greece. Greece faces a very high actual CO_2 reduction of -24.06%.

The EU-wide actual emission reduction (including the growth effect) is -13.74% while the contribution of countries to the common target differs according to their ability to adjust towards lower emissions (see table 2a and b). The target requires a uniform tax rate of 22.01 ECU per tonne of CO_2.

The positive effect of wage cost reduction is outweighed by the decline in production (-0.78%). Hence, there is a decrease in employment (-0.06%) in Greece. Private consumption rises between 0.06% (Great Britain) and 1.47% (Belgium). Exceptions are Greece (-0.92%, due to the decrease in real income) and Italy (-0.04%, due to a drop in other incomes that compensates the positive effect in labour income). Gross domestic production falls in all countries as the relief of labour taxes (i.e., rate of social security) is not fully compensating the increase in domestic production costs due to emission taxation. Again with the exception of Greece, GDP rises everywhere in the EU. The welfare measure EV in percent of GDP is negative for Greece (-1.71%), Great Britain (-0.38%), Germany (-0.22%), and Italy (-0.21%). For the other countries the EV shows up as a a (positive) welfare gain.

What are the conclusions that can be drawn at this stage of the analysis:

1. Based on the parameter set used for the above two simulations, and the assumption of a perfect competitive world, we always (with the exception of Greece) find a dividend for the labour market, i.e., employment increases. This is the case irrespective of the policy imposed, i.e., coordinated/non coordinated or high/low emission reduction effort (for Germany: -7.25% in the unilateral case and -15.47% in the coordinated case).
2. Obtaining a second dividend in terms of economic welfare is crucially linked to the design of the policy. This is shown by the German example: while the unilateral 'low' emission tax gives a positive effect on EV, the coordinated but higher emission tax gives an overall negative effect, i.e., the EV of Germany is negative. Amongst others there are particularly two effects of importance: First, the increasing distortion of a higher emission tax outweighs the positive effect that is linked to the decrease in labour tax and its distortionary effect. Secondly, a coordinated policy leads to a drop in exports as the demand in other EU-countries is now affected by own policy measures.
3. Within our model framework and its parameterisation, the strong double dividend hypothesis with respect to economic welfare has to be rejected because

of its lack of stability. The second dividend is linked not only to the tax level but to country specific characteristics as well. In the coordinated case, Denmark and Italy face approximately the same actual reduction effort under the unique tax rate, while they differ in the sign of the welfare gain: The EV is positive for Denmark whereas it is negative for Italy.

3.2.2 Labour Market Regimes

It is intuitively clear, that the imposed flexibility on the labour market plays an important role for the dividend of employment as well as for a welfare dividend.

To get an idea how sensitive impacts are due to the assumption of a perfect competitive labour market, alternative labour market regimes are tested. Imposing labour market rigidities implicitly assumes another utility function of households. For the purpose of our sensitivity analysis we assume that unemployment is completely compulsory, i.e., leisure due to unemployment is not utility increasing. For reasons of simplicity leisure demand of households is kept fixed (this is equivalent to a fixed labour supply). As an explicitly specified wage rule describes the potential development of the real wage rate, unemployment becomes a residual, i.e., supply of labour minus demand of labour.

Two imperfect labour market regimes are examined: In one variation the real wage rate is kept constant. Unions are supposed to have some bargaining power as the decrease in labour productivity linked to an environmental tax reform has no impact on the real wage rate. The nominal wage rate covers inflationary effects, i.e., it changes according to the consumer price index.

In a second analysis we assume that unions are not able to keep the level of the real wage rate but have to accept an adjustment according to the (lagged) changes in labour productivity.

Table 3 shows the findings of this sensitivity analysis with respect to equivalent variation (EV) as percentage of GDP, consumption and employment. Note that the EV is an ordinal measure which can be used to evaluate policy simulations within one model variation only. As the rigid labour market regimes implicitly assume a different utility evaluation, the EV can show which policy is preferred under a particular labour market regime but it is not appropriate to assess policies under different labour market regimes:[18] While there is no unemployment in a classical labour market, leisure of the unemployed is supposed to be involuntary in the rigid labour market regimes. As in the latter more employment is matched by former unemployed, there is no loss in utility because of a decrease in leisure. Hence, to compare the results across model variations one has to take the two components of utility, i.e., consumption and employment (respectively leisure) into account.

[18] The main reason for this inconvenience is the fact that we have to compute a separate reference run for each labour market regime. As the computations of EVs refer to different bases they are not comparable.

In the classical labour market regime, Germany would prefer the national policy. There is a welfare gain for the households of 0.02% of GDP, whereas the coordinated policy yields a welfare loss of -0.22% of GDP. If real wages are assumed to be inflexible (fixed real wage rate) German households are indifferent to the institutional setting of the policy but they would like to see one of the two implemented: the EV of both is equal and positive. For a wage rule with labour productivity adjustment, Germany is better off to choose the national policy, but German consumers would like to see none of the two policy options to be implemented.

To enable a comparison across model variations changes in consumption and employment have to be taken into account. We will exemplary discuss the impact of different labour market specifications for the unilateral policy[19]: If real wage rates are fixed, demand for labour is higher than in the flexible labour market as there is no price mechanism that adjusts demand and supply (which would lead to higher wage costs). The additional labour demand is met by the formerly unemployed. Production costs and the fall in total output are lower. At the same time private consumption rises less than in the flexible labour market case as the increase in real income is lower. Hence, households have to work more for an additional unit of consumption than in the classical labour market case. However, the distortionary effect of the pre-existing labour tax (rate of social security contributions) is higher in the case of unemployment,[20] which is another source of the positive effect in employment. In the third labour market specification the decrease in labour productivity lowers real wage rate. The effect on employment is even a bit higher than in the case with fixed real wage rate. But private consumption is almost unchanged as the level of real income is not affected considerably; the employment effect is outweighed by the decrease in the real wage rate.

How do the results of the coordinated policy differ with respect to the different assumptions for the labour market? In principle the findings are similar to those for the unilateral policy: there is more employment under a rigid labour market but real income and therefore consumption increases less. The marginal rate of consumption to employment is highest for the flexible labour market, decreases in a market with fixed wage rate and is lowest for a labour market where the real wage is adjusted according to labour productivity. Hence, in terms of households' efficiency the rigid labour market regimes lowers the positive effect of the policy.

If the goal of the policy is to increase employment without lowering the level of consumption, the coordinated policy is appropriate for most countries. Under the labour market regimes 'perfect competition' and 'fixed real wage rate' only Greece and Italy are affected negatively, whereas in the third regime Germany, Greece, Italy and United Kingdom are worse off. Under all three market regimes, there is

---

[19] For reasons of clarity we suppress a more detailed presentation of the figures of these simulations.

[20] This is in line with the findings of Bovenberg/de Mooij (1994b) and Bovenberg/van der Ploeg (1994)

an EU-wide positive effect for labour without lowering the total level of consumption.

What are the conclusions of the labour market sensitivity analysis?

Labour market rigidities alter the results found in a perfect competitive world. But again everything depends: Labour market rigidities will worsen the impact of the policy on employment in those model applications, which find no dividend for employment in a competitive world, as less flexibility in the labour market will increase adjustment costs. Whereas a rigid labour market improves the (positive) effect on labour in applications that find a dividend for employment in the perfect world already. This is obvious as a positive effect on employment in a competitive world is linked to the increase in the after tax real wage rate. A wage rule that keeps the real wage rate fixed enables firms to demand the formerly unemployed labour force at unchanged unit costs. Hence, production costs are lower than in the competitive case.

It is beyond the scope of this paper to discuss the appropriateness of the labour market regimes imposed and to judge about the most realistic one. Nevertheless in undertaking a medium term analysis one might expect some inflexibility in real wages. A decrease in real wages or the assumption of a clearing labour market seems at least in the short term less plausible. If one believes in the inflexibility of wages in the short and medium term, a coordinated policy that taxes emissions and reduces wage costs would be favourable for all countries with the exception of Greece. The other labour market regimes traces more losers, i.e., there are more countries who refuse the mitigation policy because an economic dividend is not obtained. Another aspect is the question of compensating the losers. As the figures in the tables show, there is an EU-wide benefit of coordination under the flexible labour market and under the regime with fixed real wage rate. If one would find a compensation mechanism that makes the losers better off (at least indifferent to the policy), one might find a strategy where all countries are willing to cooperate. This aspect is a topic of future research.

3.3 Exploring the Deviancy

Why are our empirical results not fully compatible with some other theoretical and empirical findings of, e.g., Bovenberg/de Mooij (1994a, 1994b), Pethig (1996) or Goulder (1995b), who reject the double dividend hypothesis more or less systematically? In summary the differences are linked to the following issues:

- full general equilibrium effects versus comparative static analysis,
- agents considered,
- endogeneity of factors and number of inputs considered,
- optimality of the pre-existing tax system,
- assumed factor mobility,

- foreign trade specification and its parameterisation.

Analytical studies done, e.g., by Bovenberg/de Mooij (1994a, 1994b), Bovenberg/van der Ploeg (1994) or Pethig (1996) are based on more simple static general equilibrium models. Their work focus on an analytical comparative static evaluation of these models. As Pethig stresses, *comparative static analysis* becomes "... very messy even in simple models..." if one releases the very restrictive assumptions on preferences and technologies that enable a full characterisation of conditions under which a double dividend might be obtained. Applied general equilibrium analysis allows for more complex specifications which cannot be solved analytically any more. Hence, it is a priori not clear, why the results obtained by comparative static analysis in simplified models should hold in a much more complex world that uses real world data and traces *full general equilibrium effects.*

Another explanation is the *consideration of agents.* In contrast to the theoretical analysis of Bovenberg and others, where only three agents are distinguished (households and firms are not separated), the model used here takes four agents into account (household, firms, government and foreign). Therefore firms keep a part of capital income for investment. Hence, a policy-induced decrease in capital income affects both, firms and households, which in turn lowers the (negative) impact on private consumption.

Furthermore, the theoretical result in Bovenberg/de Mooij (1994a), that pollution taxes reduce the incentive to supply labour, is not in contradiction to our model results because their proof is based on the *assumption of a single input* (labour) and on a constant labour productivity. In Bovenberg/Van der Ploeg (1994) three inputs are used (labour, energy, and capital), however the prices of capital and energy are determined on global competitive markets, i.e., they are exogenous. In their factor price frontier $\frac{(w+t_L)}{p} = \phi\left(\frac{\overline{PK}}{P}, \frac{\overline{PF}+t_F}{P}\right)$,

a given tax on fossil fuel (t_F) uniquely determines the producer wage $\frac{(w+t_L)}{P}$.

Hence, the energy tax is fully born by the immobile factor labour and thus amounts to an implicit labour tax. The factor price frontier in our model, derived from the unit cost function, is (in a shorter version):

$$1 = \phi\left(\frac{PL}{P}, \frac{PK}{P}, \frac{(PF+t_F)}{P}, \frac{PM}{P}\right),$$

where prices of capital and of energy are endogenous. The carbon tax is therefore not an implicit labour tax, i.e., the effect of a lower tax on wages is not fully offset by the carbon tax. This explains why in Bovenberg/Van der Ploeg an increase in the energy tax harms employment as the higher energy tax is shifted onto the immobile factor (i.e., labour) only. Hence, the higher cost of energy depresses the market wage $\frac{w}{P}$. In the GEM-E3 model it depresses both factor prices, i.e., labour and capital, as the user cost of capital is endogenous. In an open economy framework the endogenous PF absorbs also some of the tax impact (depending on the 'openness' assumed).

The GEM-E3 model includes a range of *pre-existing taxes* and transfers, e.g., direct taxation of households and firms, employers and employees contribution to social security, import duties, subsidies, value added tax and social benefits. In contrast to the assumptions of, e.g., Bovenberg/de Mooij, the pre-existing tax systems in the countries considered are not optimal. Hence, one could argue that a tax reform could enable efficiency gains which are not linked solely to a new environmental tax but to a general change in (effective) factor tax rates.[21] But this is still not in line with the findings of Bovenberg/Goulder (1996), who show that both, the analytical and the empirical analysis coincidence even if one considers pre-existing taxes: While they find analytically that the prospects of a double dividend are enhanced if "... a revenue neutral tax reform shifts the burden of taxation to the less efficient (undertaxed) factor...", there is no empirical evidence obtaining such a situation: no double dividend is obtained in their numerical analysis for a wide range of parameters.

Again these findings are crucially linked to the *factor mobility assumptions* they make. In both, the analytical and the numerical analysis capital is fully mobile. While in a comparative static analysis capital mobility is usually assumed to cover the long term effects of a policy, the meaning is different if it is imposed for an intertemporal (periodically computed) numerical model. There the assumption of full capital mobility holds for both, the short (which seems to be rather unrealistic) and the long run. It is worthwhile to stress this aspect as the mobility issue has important impacts on the excess burden of taxes. Böhringer et al. (1996) find for their CGE-model for Germany a marginal excess burden (*meb*) of 36.5 for labour taxes and a *meb* of 82.8 for capital taxes under the assumption of full capital mobility. This situation changes significantly if existing capital is assumed to be sector specific, i.e., sectorally and internationally immobile. The *meb* for labour taxes is then 14.2 while the *meb* for capital taxes drops to -12.5. Hence, the marginal excess burden of the overall tax system decreases sharply if the assumption of full capital mobility is removed. As Bovenberg/Goulder (1996) concede, a highly suboptimal pre-existing tax system (i.e., a tax system where the *meb* of different taxes vary quite a lot) enhances the chance of obtaining a double dividend, if the burden of the new environmental tax falls mainly on the undertaxed factor (which is capital in our case) and relieves the burden of the overtaxed factor (labour).

The putty-clay approach used in our recursively dynamic model assumes sectoral fixed capital stocks within a single period. In such a situation the burden of the environmental tax falls partly on capital as stocks can adjust only gradually

[21] Goulder (1995b)

over time by depreciation and gross investment. This specification is in line with the empirical evidence that can be found at least in the short and medium term and which denies that firms can easily grab their facilities and go somewhere else if production conditions become a bit worse than they were before.[22]

Another explanation for the mismatch of our results to those of others can be drawn from the *foreign trade specification and its parameterisation*. There are three interrelated subjects of importance:

First, in countries with little or without own primary energy resources, an energy or emission tax affects mainly imported goods. Hence, the (direct) burden due to substitution falls to those sectors that produce the taxed good, i.e., it is shifted abroad. Second, as exports are produced with taxed energy, the demanding country (the rest of the world in particular) contributes to the reduction of domestic wage costs. And third, the impact of an environmental tax reform is also crucially linked to the trade specification and trade closure that is imposed.

While most single country model applications often use closed economy and/or small country assumptions, our model specifies open economies and relieves the small country assumption partly. In particular, there is no constraint on the trade balance in the standard version of the model, i.e., the current account is not balanced through an endogenous exchange rate. Furthermore, adopting the Armington assumption in the foreign trade specification is - at least in its operation - equivalent to a specification that gives countries some monopsony power to influence the prices on the world market.[23] In a multi-country trade framework, the imports of country *a* delivered by country *b* have to match the exports of country *b* to country *a*. If imports and domestically produced goods are imperfect substitutes an increase in export prices of country *a* will not cut imports in country *b* coming from country *a* to zero. Hence, a part of the burden of the environmental tax reform undertaken in the exporting country *a* falls on the importing country *b*. Country *b* pays the tax through the export price of *a* without gaining anything from revenue recycling.

This aspect seems to be crucial for the results: A higher elasticity of substitution in the Armington function will decrease the chance of obtaining a second dividend. Many studies imply elasticities which do not correspond to the empirical evidence. The GREEN-model of Burniaux et al. or Böhringer et al. use elasticities of substitution around four for both levels, the domestic supply (nest of domestically produced goods and the import aggregate) and the import aggregate (nest of imports by origin, i.e., exporting country). Even if the literature on the empirical estimation of foreign trade elasticities is very limited, assuming two times a substitution elasticity around four seems rather unrealistic. Reconsidering the findings of these studies most estimations of own price elasticities are between 0.5 and 2.

---

22 see, e.g., Dean (1992), Jaffe et al. (1995)

23 This is independent of the use of an (CET) export supply function.

Jorgenson/Wilcoxen, e.g., have estimated import demand price elasticities for the U.S. economy within a range of 1.5 to 2.0.[24]

Deriving the relation between the price elasticity of demand ($el_{i,i}$) and the elasticity of substitution (σ) for our model specification under ceteris paribus assumptions gives the following expression:

$$el_{i,i} = \sigma \cdot \left(1 - \frac{PIM_i \cdot IMP_i}{PY \cdot Y}\right).$$

It indicates that the elasticity of substitution σ should be close to the own price elasticity, as long as the import shares of domestic demand are small. A similar relationship can be found for the cross price elasticities.

Of course there are other differences in model specifications that might play a role for the results obtained. Even if the differences are of minor importance in a comparative static analysis they might not be negligible in a full equilibrium analysis based on real data. Alternative specification and differences in the underlying data are the major reason why the findings of empirical work are less unique than those of the analytical work. While the models of Goulder (1995a) and Bovenberg/Goulder (1996) trace no double dividend for the US economy, the models of Jorgenson/Wilcoxen (1994) and Ballard/Medema (1993) confirm the double dividend hypothesis for the same economy.

Finally, one should reconsider that the analysis traces economic benefits or losses only. The welfare measure chosen abstracts from any (individual) environmental benefit. As the simulation results indicate, a tax on CO_2 reduces also emissions of other (energy related) pollutants like SO_2 or NO_x. The sum of benefits from slowing environmental degradation might outweigh occuring costs easily. Integrating economic costs and environmental benefits within one assessment framework is also a topic of our future research.

4 Conclusion

National economic concerns still hinder the implementation of effective policy measures that would enable a significant reduction of greenhouse gas emissions. Considering other economic problems like unemployment, the enforcement of Maastricht criterions and highly tensed social security systems, the global warming problem seems to be of less priority for national policy makers. Nevertheless, the international political pressure forces governments to do something and show at least political will. To escape the dilemma countries choose their strategy with respect to least cost criteria; the potential environmental benefit is of less impor-

[24] see also Stern et al. (1976), Reinert et al. (1992).

tance as long as the reductions achieved relieve the political requests of the international society.

The purpose of this study was to point out clearly the assumptions underlying the model specification and their potential impact on the empirical outcomes of the analysis. There is a wide range of factors that play an important role for the double dividend issue. Apart from the labour market specification this concerns in particular the 'openness' of the economy assumed, the number of agents considered and the factor mobility imposed. While alternative assumptions with respect to these issues might explain some of the divergence of results that can be found in the literature, a major disadvantage remains: there is (at least at the current state of economic modelling) no complete reliability either to reject or to confirm the existence of a double dividend - everything is linked to the reliability of the model specification and its parameterisation.

Our simulations with a multi-country multi-sector CGE-model for the European Union indicate that a revenue neutral environmental tax reform, where CO_2 emissions are taxed and the tax receipts are used to release the employers' contribution to social security, might yield a double dividend. But the positive effect obtained on the economic side is rather small (for Germany, e.g., the positive welfare could be 0.02 percent of GDP) and it depends crucially on the institutional setting of the policy chosen (unilateral versus EU-wide coordination) as well as on the labour market regime imposed (flexible versus rigid labour market).

In a classical labour market specification Germany obtains a double dividend if a unilateral policy is imposed. The coordinated policy makes Germany worse off as a small welfare loss is obtained. There are two reasons for explanation: First, the two policies (unilateral: 10% CO_2 reduction in Germany, coordinated: 10% CO_2 reduction EU-wide) are not neutral with respect to the environmental goal obtained and second, the burden sharing between countries in the coordinated policy is not considering the 'polluter pays principle.' The allocation of a country's contribution is obtained by the (coordinated) minimum cost solution on the EU level. Taking into account the differences in potential[25] emission growth rates (which is below average for Germany) yields a reduction for Germany that is much higher in the coordinated case compared to the one by unilateral action.

Nevertheless there is an overall benefit of coordination. Therefore, more sophisticated coordination schemes that include side payments could enable the acceptance of such a coordinated policy by all parties (countries).

The outcomes are different if one specifies the labour market alternatively. Assuming fixed real wage rates makes Germany indifferent to the two policies; there is a small welfare gain in both policies. If real wage rates are adjusted according to the decrease in labour productivity - the latter is linked to every policy that reduces labour cost - no double dividend is obtained. Hence, justification or rejection of the double dividend hypothesis depends crucially on the labour market assumptions made.

---

[25] If no policy action would have been undertaken.

From an empirical point of view, obtaining an economic dividend in terms of welfare by undertaking an 'intelligent' environmental tax reform remains at least possible, but relying on it seems rather optimistic as there are numerous uncertain influences that might alter the sign of the welfare effect. Obtaining a second dividend in terms of employment seems to be more robust. At any rate, even if there might be no ‘no regrets’ policy, the economic impact of such a reform seems - from the perspective of consumer welfare - within acceptable limits.

Table 1: The impact of an unilateral environmental tax reform in Germany (DE-nc) (percent changes from baseline)

a)

	EV in % of GDP	GDP (%)	Production (%)	Priv. Consumption (%)	Investment (%)	Exports (%)	Imports (%)
Period	**1-10**	**10**	**10**	**10**	**10**	**10**	**10**
Belgium	-0.01	0.00	-0.01	-0.01	0.0	0.0	0.0
Germany	0.02	0.09	-0.20	0.13	-0.2	-0.1	-0.2
Denmark	0.01	0.00	0.00	0.01	0.0	0.0	0.0
France	-0.01	0.00	0.00	-0.01	0.0	0.0	0.0
Greece	-0.01	0.00	0.00	0.00	0.0	0.0	0.0
Ireland	0.00	0.00	0.00	0.00	0.0	0.0	0.0
Italy	-0.01	0.00	0.00	-0.01	0.0	0.0	0.0
Netherlands	-0.04	0.00	-0.01	-0.02	0.0	0.0	0.0
Portugal	0.00	0.00	0.00	0.00	0.0	0.0	0.0
Spain	0.00	0.00	0.00	0.00	0.0	0.0	0.0
Great Britain	-0.01	0.00	0.00	-0.01	0.0	0.0	0.0
EU-11	0.00	0.02	-0.05	0.03	-0.1	0.0	0.0

b)

	Employment (%)	After tax real wage rate (%)	Non-labour income (%)	CO_2-tax (ECU/ton CO_2)	CO_2-emissions[1] (in % of reference)	CO_2-emissions[1] (in % of base)	CO_2-projection[2] (%)
Period	**10**	**10**	**10**	**10**	**1-10**	**1-10**	**1-10**
Belgium	0.00	-0.01	-0.01	0.00	0.3	2.3	2.0
Germany	0.26	0.47	-0.31	8.24	-7.0	-10.0	-3.0
Denmark	0.00	0.01	0.00	0.00	0.4	17.4	17.0
France	0.00	-0.01	0.00	0.00	0.3	7.3	7.0
Greece	0.00	0.00	0.00	0.00	0.1	13.1	13.0
Ireland	0.00	0.00	0.00	0.00	0.1	13.1	13.0
Italy	0.00	-0.01	0.00	0.00	0.1	9.1	9.0
Netherlands	0.00	-0.02	-0.02	0.00	0.1	3.1	3.0
Portugal	0.00	0.00	0.00	0.00	0.1	28.1	28.0
Spain	0.00	0.00	0.00	0.00	0.0	5.0	5.0
Great Britain	0.00	-0.01	-0.01	0.00	0.1	4.1	4.0
EU-11	0.05	-	-	-	-2.0	1.7	3.7

[1] considering underlying growth (see [2])

[2] projection of business-as-usual: conventional wisdom and no emission reduction policy (i.e., rates include economic growth and efficiency improvements)

Table 2: The impact of a coordinated environmental tax reform in the EU (**EU-co**) (percent changes from baseline)

a)

	EV in % of GDP	GDP (%)	Production (%)	Priv. Consumption (%)	Investment (%)	Exports (%)	Imports (%)
Period	**1-10**	**10**	**10**	**10**	**10**	**10**	**10**
Belgium	1.87	0.54	-0.11	1.47	0.62	-0.69	-0.24
Germany	-0.22	0.13	-0.55	0.13	-0.55	-0.37	-0.80
Denmark	0.56	0.35	-0.03	0.72	0.23	-0.44	-0.42
France	0.34	0.12	-0.27	0.34	-0.11	-0.33	-0.38
Greece	-1.71	-0.28	-0.78	-0.92	-1.32	0.21	-2.05
Ireland	1.47	0.09	-0.20	1.07	-0.08	-0.57	-0.13
Italy	-0.21	0.09	-0.30	-0.04	-0.38	-0.29	-1.01
Netherlands	0.32	0.47	-0.09	0.51	-0.01	-0.29	-0.58
Portugal	0.24	0.52	-0.29	0.67	0.05	-0.65	-0.76
Spain	0.57	0.31	-0.25	0.81	0.21	-0.41	-0.09
Great Britain	-0.38	0.30	-0.44	0.06	-0.36	0.01	-0.81
EU-11	0.01	0.19	-0.37	0.23	-0.31	-0.22	-0.59

b)

	Employment (%)	After tax real wage rate (%)	Non-labour income (%)	CO_2-tax (ECU/ton CO_2)	CO_2-emissions[1] (in % of reference)	CO_2-emissions[1] (in % of base)	CO_2-projection[2] (%)
Period	**10**	**10**	**10**	**10**	**1-10**	**1-10**	**1-10**
Belgium	1.03	3.08	-0.12	22.01	-17.5	-15.5	2.0
Germany	0.58	0.85	-0.83	22.01	-15.0	-18.0	-3.0
Denmark	0.65	1.48	-0.28	22.01	-13.4	3.6	17.0
France	0.30	0.68	-0.32	22.01	-14.3	-7.3	7.0
Greece	-0.06	-1.18	-0.73	22.01	-24.1	-	13.0
Ireland	0.39	1.60	0.05	22.01	-12.2	0.8	13.0
Italy	0.26	0.20	-0.58	22.01	-12.0	-3.0	9.0
Netherlands	0.97	2.02	-0.71	22.01	-14.9	-11.9	3.0
Portugal	1.17	1.97	-0.63	22.01	-13.7	14.3	28.0
Spain	0.94	1.51	-0.26	22.01	-15.3	-10.3	5.0
Great Britain	0.63	0.48	-1.20	22.01	-10.3	-6.3	4.0
EU-11	0.54	-	-	-	-13.7	-10.0	3.7

[1] considering underlying growth (see [2])

[2] projection of business-as-usual: conventional wisdom and no emission reduction policy (i.e., rates include economic growth and efficiency improvements)

Table 3: The impact of alternative labour market regimes (percent changes from baselines)

	Classical labour market						Real wage rate fix						Real wage rate according to labour productivity					
	Germany alone			EU (coor)			Germany alone			EU (coor)			Germany alone			EU (coor)		
	EV (% of GDP)	Consumption	Employment	EV (% of GDP)	Consumption	Employment	EV (% of GDP)	Consumption	Employment	EV (% of GDP)	Consumption	Employment	EV (% of GDP)	Consumption	Employment	EV (% of GDP)	Consumption	Employment
Belgium	-0.01	-0.01	0.00	1.87	1.47	1.03	0.00	0.00	0.00	0.58	0.31	1.77	0.00	0.00	0.00	0.24	0.20	1.78
Germany	0.02	0.13	0.26	-0.22	0.13	0.58	0.08	0.04	0.51	0.08	0.05	1.11	-0.02	0.01	0.54	-0.12	-0.02	1.17
Denmark	0.01	0.01	0.00	0.56	0.72	0.65	0.02	0.01	0.02	0.53	0.34	1.44	0.03	0.02	0.02	0.38	0.28	1.50
France	-0.01	-0.01	0.00	0.34	0.34	0.30	0.00	0.00	0.00	0.20	0.10	0.63	0.00	0.00	0.00	0.06	0.05	0.65
Greece	-0.01	0.00	0.00	-1.71	-0.92	-0.06	0.01	0.00	0.01	-0.17	-0.11	0.25	0.01	0.01	0.01	-0.32	-0.15	0.24
Ireland	0.00	0.00	0.00	1.47	1.07	0.39	0.00	0.00	0.01	0.26	0.14	0.96	0.00	0.00	0.01	0.01	0.06	1.00
Italy	-0.01	-0.01	0.00	-0.21	-0.04	0.26	-0.01	-0.01	-0.01	0.00	-0.04	0.39	-0.01	-0.01	-0.01	-0.13	-0.08	0.40
Netherlands	-0.04	-0.02	0.00	0.32	0.51	0.97	-0.02	-0.01	-0.01	0.30	0.13	1.75	-0.03	-0.01	-0.01	0.05	0.05	1.76
Portugal	0.00	0.00	0.00	0.24	0.67	1.17	0.00	0.00	0.00	0.80	0.39	2.15	0.00	0.00	0.00	0.43	0.29	2.24
Spain	0.00	0.00	0.00	0.57	0.81	0.94	0.00	0.00	0.00	0.32	0.21	1.66	0.00	0.00	0.00	-0.04	0.07	1.72
Great Britain	-0.01	-0.01	0.00	-0.38	0.06	0.63	0.00	0.00	0.00	0.07	0.04	0.95	0.00	0.00	0.00	-0.14	-0.05	0.99
EU-11	0.00	0.03	0.05	0.01	0.23	0.54	0.02	0.01	0.10	0.14	0.07	0.97	-0.01	0.00	0.11	-0.05	0.00	1.00

References

Ballard, C.L. and S.G. Medema (1993): 'The Marginal Efficiency Effects of Taxes and Subsidies in the Presence of Externalities', Journal of Public Economics, Vol. 52, p. 199-216.

Bergman, L. (1990): 'Energy and Environmental Constraints on Growth: A CGE Modeling Approach', Journal of Policy Modeling 12, p. 671-691.

Bergman, L., S. Ahlroth and K. Skanberg (1995): 'A CGE Analysis of Sulphur Deposition and Sweden's "Green" Net National Product', Paper prepared for the Sixth International CGE Modeling Conference, University of Waterloo, October 26-28.

Boehringer, C., A. Pahlke and T.F. Rutherford (1996): 'The prospects for a double dividend in Germany', First Interim Report, IER University of Stuttgart, March.

Burniaux, J.M., J.P. Martin, G. Nicoletti and J. Oliveira (1991): 'GREEN -- A Multi-Region Dynamic General Equilibrium Model for Quantifying the Costs of Curbing CO_2 Emissions: A Technical Manual', Department of Economics and Statistics Working Paper 104, OECD, Paris.

Bovenberg, A.L. and R.A. de Mooij (1994a): 'Environmental Levies and Distortionary Taxation', American Economic Review, Vol. 94/4, p. 1085-1089.

Bovenberg, A.L. and R.A. de Mooij (1994b): 'Environmental Taxes and Labor-Market Distortion', European Journal of Political Economy, Vol.10, p. 655-683.

Bovenberg, A.L. and F. van der Ploeg (1994): 'Green Policies and Public Finance in a Small Open Economy', Scandinavian Journal of Economics, Vol. 96, p. 343-363.

Bovenberg, A.L. and L. H. Goulder (1996): 'Optimal Environmental Taxation in the Presence of Other Taxes: General Equilibrium Analysis', American Economic Review, Vol 86, No.4, p. 985-1000.

Boyd, R., K. Krutilla and W.K. Viscusi (1995): 'Energy Taxation as a Policy to Reduce CO_2-Emissions - A Net Benefit Analysis', Journal of Environment Economics and Management 29, p. 1-24.

Carraro C. and M. Galeotti (1996): 'Environmental Fiscal Reforms in a Federal Europe', in: S. Proost and J.B. Braden (eds.), Climate Change, Transport and Environmental Policy: Empirical Applications in a Federal System.

Conrad, K. and T.F.N. Schmidt (1996a): 'National Economic Impacts of an EU Environmental Policy - An Applied General Equilibrium Analysis', in: S. Proost and J.B. Braden (eds.), Climate Change, Transport and Environmental Policy: Empirical Applications in a Federal System.

Conrad, K. and T.F.N. Schmidt (1996b): 'Economic Impacts of a Non-Coordinated vs. a Coordinated CO_2 Policy in the EU - An Applied General Equilibrium Analysis', Economic Systems Research, Special Issue, forthcoming 1998.

Conrad, K. and M. Schröder (1991a): 'The Control of CO_2-Emissions and Its Economic Impact', Environmental and Resource Economics 1, p. 289-312.

Conrad, K. and M. Schröder (1991b): 'Demand for Durable and Non-durable Goods, Environmental Policy and Consumer Welfare', Journal of Applied Econometrics 6, p. 271-286.

Dean, J.M.(1992): 'Trade and Environment: A Survey of Literature', in: Patrick Low (ed.): International Trade and the Environment, World Bank Discussion Papers, The World Bank, Washington, D.C..

Glomsrød, S., H. Vennemo and T. Johnsen (1992): 'Stabilization of Emission of Carbon Dioxide: A CGE Assessment', Scandinavian Economic Journal 94, No.1, p. 53-69.

Goulder, L.H. (1995a): 'Effects of Carbon Taxes in an Economcy with Prior Tax Distortions: An Intertemporal General Equilibrium Analysis', in: Journal of Environmental Economics and Management, Vol 29, No.3, p. 271-297.

Goulder, L.H. (1995b): 'Environmental Taxation and the "Double Dividend:" A Reader's Guide', International Tax and Public Finance, Vol. 2, p. 157-184.

Hazilla, M. and R.J. Kopp (1990): 'Social Cost of Environmental Quality Regulations: A General Equilibrium Analysis', Journal of Political Economy 98, No. 4, p. 853-873.

Jaffe, A.B., S.R. Peterson, P.R. Portney and R.N. Stavins (1995): 'Environmental Regulation and the Competitiveness of U.S. Manufacturing: What does the Evidence Tell us?', Journal of Economic Literature, Vol. 33, p. 132-163.

Jorgenson, D.W. and P.J. Wilcoxen (1993): 'Energy, the Environmental and Economic Growth', in: A.V. Kneese and J.L. Sweeney (eds.), Handbook of Natural Resources and Energy Economics, Vol 3, Elsevier, Amsterdam, p. 1267-1349.

Jorgenson D.W. and P.J. Wilcoxen (1994): 'Reducing U.S. Carbon Emissions: An Econometric General Equilibirum Assessment', in: D. Gaskins and J. Weyant (eds.) The Cost of Controlling Greenhouse Gas Emissions, Stanford, CA, Stanford University Press.

Manne, A.S., R. Mendelsohn and R. Richels (1995): 'Merge - A model for evaluating regional and global effects of GHG reduction policies', Energy Policy, Vol.23, No.1, p. 17-34.

McKibbin, W.J. and P.J. Wilcoxen (1992): 'G-Cubed: A Multi-Sectoral General Equilibirum Model of the Global Economy (Quantifying the Costs of Curbing CO_2 Emissions', in: Brookings Discussion Papers in International Economics No. 98, The Brookings Institution.

Nordhaus, W.D. (1994): 'Managing the Global Commons - The Economics of Climate Change', MIT Press, Cambridge, Mass..

Nordhaus, W.D. and Z. Yang (1996): 'A Regional Dynamic General Equilibrium Model of Alternative Climate Change Strategies', American Economic Review, 86, 4, p. 741-765.

Perroni, C. and R.M. Wigle (1994): 'International trade and environmental quality: how important are the linkages?', Canadian Journal of Economics, 27(8), p. 551-567.

Pethig, R. (1996): 'Ecological Tax Reformand Efficiency of Taxation: A Public Good Perspective', (revised version, July 1996), Diskussionbeitrag Nr. 57-96, Universität Siegen.

Proost, S. and D. Van Regemorter (1995): 'The double dividend hypothesis, the environmental benefits and the international coordination of tax recycling', in: C. Carraro, D. Siniscalco, Environmental Fiscal reform and Unemployment, Kluwer Academic Publishers.

Reinert, K.A., D.W. Roland-Holst and W. David (1992): 'Armington Elasticities for United States Manufactoring Sectors', Journal of Policy Modeling, Vol.14, No. 5, p. 631-639.

Schmidt, T.F.N. (1997): Unpublished working paper, forthcoming in spring 1998.

Stephan, G. (1989): 'Pollution Control, Economic Adjustment and Long-Run Equilibrium - A Computable Equilibrium Approach to Environmental Economics', Springer, Berlin.

Stern, R.M., J. Francis and B. Schumacher (1976): 'Price Elasticities in International Trade', Macmillan Press, London.

Welsch, H. (1996): 'Recycling of Carbon/Energy Taxes and the Labour Market - A General Equilibrium Model for the European Community', Environmental and Resource Economics,Vol.8, p. 141-155.

PART III

POLICY OPTIONS

10 Possible Contributions of Renewable Energies to the Solution of the Global Climate Problem

Helmuth-Michael Groscurth, Sigurd Weinreich
Zentrum für Europäische Wirtschaftsforschung (ZEW)
P.O. Box 10 34 43, D-68034 Mannheim, Germany

Abstract. This paper describes results of a project aimed at developing scenarios for an European energy system in 2050 which emits 80% less carbon dioxide than in 1990 and which does not use nuclear energy. It will, thus, be based on the efficient use of energy and on renewable energies. It is assumed that the necessary transition is started by value changes within the society which will then accept extensive norms and regulations leading to a more efficient energy use, application of renewable energies, and – in limited cases – renunciation or limitation of certain forms of energy use. It can be concluded that there are in principle no technical or financial bottlenecks. However, substantial changes of the economic, legal, and institutional frameworks, in which energy systems are operated, are inevitable.
Keywords: Renewable Energies, Energy services, Energy Supply.

1 Objectives, Assumptions, Methodology

Mitigating global climate change and solving a number of other key environmental problems will require massive reliance on renewable energies together with a sharp increase in energy productivity. Considering the fact that most of the technologies necessary for such a development have already been invented and tested at least on the level of prototypes, today's most important questions are: How do we get there? How can renewable energies be integrated into the European energy system at sufficiently high diffusion levels? What are the costs and benefits of such a strategy? What are the major obstacles and bottlenecks for such a development?

These questions were tackled within the project "Long-Term Integration of Renewable Energies into the European Energy System and its Potential Environmental and Economic Impacts" – shortly denoted as the *LTI-Project* hereafter. For this purpose, a *Sustainable Scenario* of a future European Energy system was developed. Starting points of the scenario are the objectives to reduce CO_2 emissions by 80% and to phase-out nuclear energy. Even though the CO_2-reduction goal is ambitious and in line with demand from the IPCC and many national committees, it must be stressed that such a reduction will not buy a 100% insurance against negative consequences resulting from potential changes of the global climate. It is much rather the rate of emissions which will hopefully lead to limiting climate change to a level that both developed and developing countries can cope with. Nevertheless, it may turn out that further reductions of CO_2 emissions are necessary, especially when considering the need to provide acceptable living conditions not only in the industrialised part of the world, but everywhere.

It is self-evident that the purpose of such a scenario cannot be to predict or even plan future developments. It rather serves as a vision that may help society to make the necessary decisions today which will start the transition process. Its objectives and instruments will have to be evaluated and, if necessary, adjusted over and over again. The scenario also evaluates the technical and economic feasibility of a sustainable energy system. In so doing, it counters various other studies which deny the feasibility of such a system.

It is assumed that equal living standards will be reached all over the European Union (EU) by the year 2050. They will be similar to those currently experienced in Northern Europe. The rest of the world was not modelled explicitly, but considered as behaving in such a way that the development in Europe is not hampered. This would, for instance, be the case if the other countries followed the EU with a time lag of about 10 years. It is self-evident that the development in Europe should not occur at the expense of other regions of the world.

As a prerequisite for the scenario design, we derived a comprehensive database on the energy demand situation in Europe, the potentials of renewable energy sources, and the efficiency and cost developments of conventional and renewable energy technologies. Next, assumptions were made on the future development of the energy demand in different regions of the EU and for different energy services. To meet the remaining energy demand, renewable energy sources and technologies are continuously introduced into the energy market. Lacking an appropriate dynamic model to map this process, the selection was done normatively. For 2050, the technical consistency of the resulting energy system was checked. Finally, economic and environmental impacts of the restructuring of the European energy systems were estimated using an input-output model.

2 Technologies and Potentials

The renewable energy sources considered in the LTI-Project are biomass, solar heat, photovoltaics, and wind energy. Other sources like geothermal or wave and tidal energy were discussed in the project, but not used due to the fact that a joint assessment of their overall availability and their costs showed only small EU-wide potentials. This does, however, not mean that these technologies should be disregarded. They should rather be used whenever there is a promising local potential.

The following technologies for exclusive electricity production were taken into account: Wind turbines, solar-thermal power plants, photovoltaics, existing large and additional small hydro power plants. Heat may be produced by solar collectors, solid and gaseous biofuels, hydrogen, heat pumps, or electricity, and it may be stored in hot water tanks. Combined production of heat and electricity is possible via cogeneration units and fuel cells. Biomass may be turned into different gaseous fuels ranging from biogas to pure hydrogen for the use in reversible fuel cells. In addition, it may be converted into liquid fuels for transport purposes. In some cases, where efficiency measures and renewable energies are not yet sufficient to meet the demand for energy services, fossil fuels may be used. This is mainly the case in the transport sector and for some residual backup.

The technology, potential, and cost data are documented in detail in the project report [1]. As an example, the left-hand side of Fig. 1 shows the possible electricity production from wind energy which includes all areas of the European Wind Atlas with wind speeds larger than 5 m/s taken into account. To account for ecological concerns, only 30% of the original capacity derived is considered. The data on photovoltaics shown on the right-hand side of Fig. 1 assume that all sufficiently southward-oriented roof areas and facades are covered with PV cells. The average system efficiencies estimated for 2050 are 18% and 11% for roofs and facades, respectively. The added potential of wind and PV in 2050 is more than two times larger than the total electricity production in 1990.

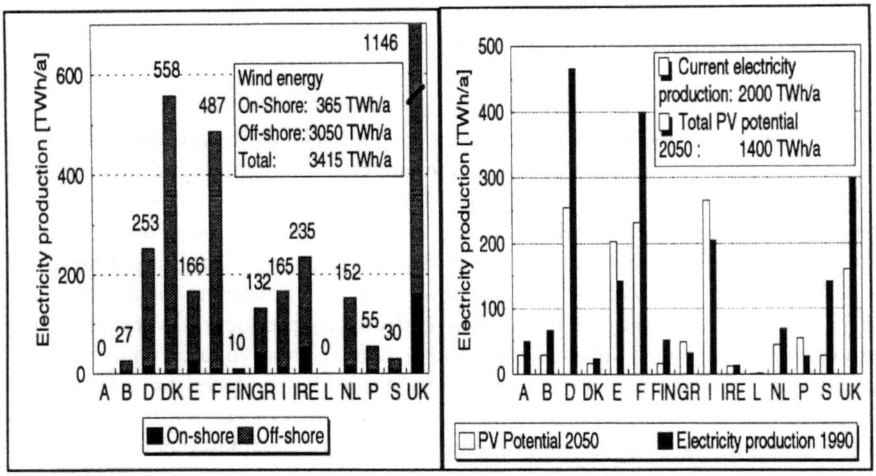

Fig. 1: Estimated potential for electricity production from wind energy (left-hand side) and photovoltaics (right-hand side) in the European Union.

3 A Sustainable Scenario of the Future European Energy System

3.1 Basic Assumptions

The *Sustainable Scenario* is based on the precautionary principle. Hence, it tries to minimise anthropogenic impacts on the ecosystem subject to the restriction that comfortable living standards have to be maintained. Consequently, its guidelines are to keep open development options of future generations and peoples outside the EU, to avoid wastes or restrict them to substances which occur naturally, and to rely on energy technologies which do not endanger political stability. In addition, some uses of energy have to be consciously renounced due to the principle of sufficiency. The world of the Sustainable Scenario is, thus, not characterised by *having more* but by *feeling better off*. This different perception of wealth includes higher valuation of immaterial goods like education, health, security, absence of violence, liberty, equal rights etc. A society following these principles will probably show a greater identification with the regions in which people live. It is there that people can become involved with most of the decisions that are important for their everyday life and for implementing a sustainable economy.

The idea of sustainability has consequences for almost all areas of society. In residential planning, for example, it will be necessary to find common spatial structures for industrial, commercial, and living areas. Only then, it will be necessary to reduce traffic to an environmentally compatible level. The diet has to change considerably, the most important factor being less consumption of meat. The latter is a prerequisite for changing land use in a way that 10% of the total area can be set aside as totally protected regions, that the use of fertilisers, pesticides and herbicides may be reduced by the factor five, and that areas, on which agricultural products exported from and imported into the EU are grown, are of similar size.

For the energy system, this means moving away from a situation in which heat is produced individually, while electricity is supplied from centralised power plants. In the future, municipal supply structures may be favourable in which heat and power generation are integrated and combined with short-distance district heating.

3.2 Energy Demand

For the industrial sector, it is assumed that the material output is reduced by a factor of 4. This can be achieved by careful product design, reducing its weight and size, by longer product lifetimes, and by changing consumption patterns. The latter may, for instance, consist in sharing products which are seldom used. Combined with an average recycling rate of 50%, the input of raw materials may thus be reduced by a factor of 8. For some industrial sectors, however, specific assumptions were made: Due to the still increasing floor space per capita and the necessity of retrofitting the stock of buildings, we assumed a constant demand for cement and a 20% increase of glass production. A sub-study of our project, which examined energy efficiency improvements in industry sector by sector, found possible savings of 20-50% for fuels and 10-50% for electricity, on the average. In some areas, using secondary materials will result in even larger savings. Secondary aluminium, for example, may be produced with 40% less fuel and 97% less electricity than the primary material.

In the residential sector, it is assumed that floor space per capita will rise to $42m^2$ until 2050, which is equal to the current top level within the EU. Household sizes will decrease to 2.24 persons, while the number of households will at the same time increase from 140 to 164 million. The availability of appliances is taken to be similar to the current situation in Denmark, Germany, and the Netherlands. Their energy efficiency can be improved by 60-85%.

The specific heating demand of existing buildings will be reduced by retrofitting from 150 $kWh/(m^2a)$ to 40 and 30 $kWh/(m^2a)$ in Northern and Central Europe, respectively. These values hold for new buildings as well.

The tertiary sector will grow by a factor of 1.5 until 2050 making up for at least some of the jobs lost in industry. This will mainly be due to additional services in the areas of recycling, repairing, and remanufacturing. Simultaneously, efficiency gains for office heating, computers etc. will limit the increase of the respective energy demand.

The most drastic assumptions are made for the transport sector, thus, changing its development from the current 7% growth per year to an annual 6% decrease during the years before 2050. For passenger air transport, it is assumed that all short and medium distance flights are abandoned. This will cover most inner European flights. Air freight is automatically reduced due to the decreasing throughput of industry. For passengers, equal shares of public and individual transport are assumed. People will travel shorter distances, especially during leisure time, according to the sufficiency principle. The number of trips, however, is not decreased by the same degree. Transport of goods will go down by 60% due to the changes in industry discussed above. Major efficiency improvements, e.g., cars which use only 2 liters of fuel per 100 km and trucks which use 34% less diesel fuel, will in addition contribute to a total decrease of energy demand for transport by almost the factor 10. However, relaxing these restrictive assumptions significantly will still not compromise the scenario objectives.

The assumed overall development of the energy demand within the European Union is shown in Fig. 2. In 2050, it amounts to 38% of the demand in 1990.

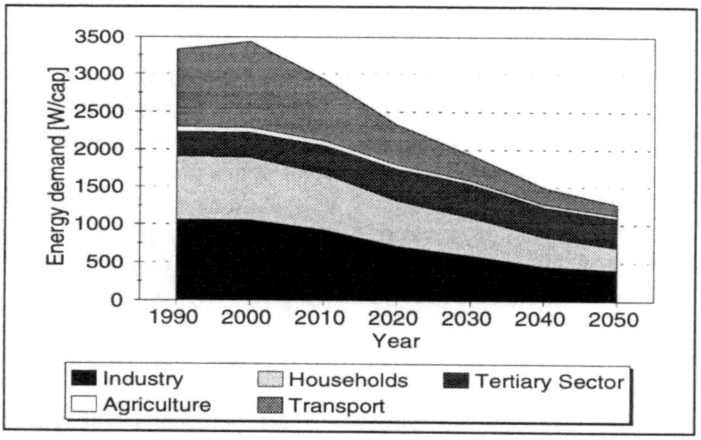

Fig. 2: Development of the energy demand in the Sustainable Scenario.

3.3 Energy Supply

Fig. 3 shows how the different renewable energies develop from 1990 to 2050. The final shares of the technologies are set normatively due to heuristic evaluations of conceivable figures. The trajectories are then determined by logistic S-curves starting in the year 2000.

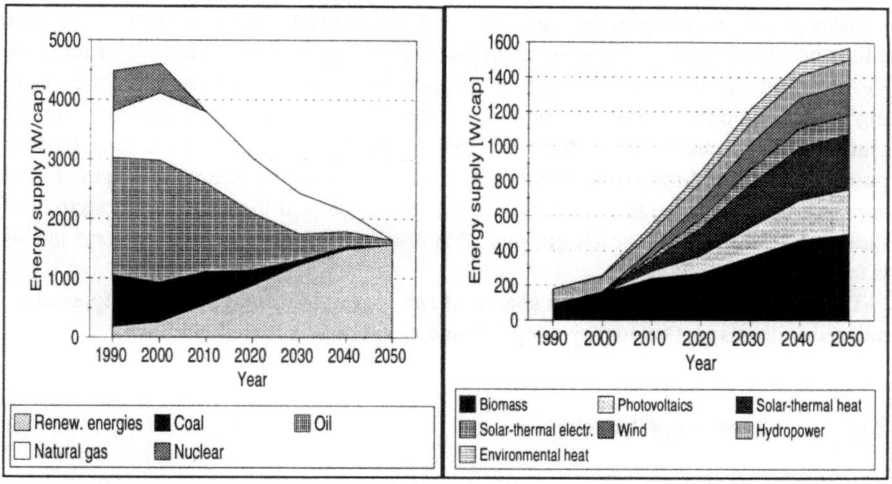

Fig. 3: Left-hand side: Phase-out of fossil and nuclear energy in the Sustainable Scenario. Right-hand side: Phase-in of different renewable energies.

Biomass for energetic uses will come from energy crops and residues. The latter will consist of waste wood and textile materials from industry and of the 50% of food which is on the average lost somewhere in the supply chain in industrial societies. It will, however, not comprise residues from forestry or agriculture since these may have to be left on or brought back to the respective areas in order to maintain their content of nutrients. Relaxing this constraint would substantially increase the biomass potential. Energy crops can be grown on excess land which will be available if the diet is changed as discussed in Section 3.1. The energetic use of specifically grown biomass will, however, have to compete with food production for exports to countries with less favourable conditions for agriculture. Biomass will partly be burnt directly and partly gasified. In both cases, fuels from biomass have the advantage that they can be stored to compensate fluctuations in the energy supply from other renewable sources. Where necessary for electricity production, biofuels will be burnt in cogeneration units.

In 2050, *photovoltaic modules* will have been installed on 50% of all suitable roofs and facades. The respective potential was estimated using statistical methods linking population densities with roof and facade areas. It was calibrated with German data and then extrapolated to the EU level. Module efficiencies will have risen from 10 to 18.4% until 2050.

The remaining 50% of the suitable roof and facade areas will be used for *solar collectors*. They will cover 80% of the hot water demand and 50% of the room heating demand in Northern Europe. In Southern Europe, they will supply all the room heat and hot water, and there will even be some space left on which additional photovoltaic modules may be installed for cooling purposes. The temporal mismatch of supply and demand requires the use of individual heat storages.

Solar power plants will be used for electrolysis producing hydrogen. This will either be distributed via pipelines and burnt directly, or it will be fed to fuel cells for electricity generation. No large *hydro-power plants* will be added to the existing ones. In contrast, the exploitation of the technical potential of small hydro plants will increase from today's 20-25% to 90%. Some of the existing hydropower plants will have to be retrofitted to serve as pumped energy storages. *Environmental heat* can be utilised via heat pumps which will then have a performance index of 4.5. Finally, a small amount of liquid *fossil fuels* will still be used in the transport sector in 2050.

Table 1 in Sec. 4.2 shows the shares of the renewable energy technologies considered with respect to energy supplied and capacities installed.

3.4 Consistency Checks

A consistency check performed with an hour-by-hour simulation model showed that the scenario has a potential problem supplying sufficient low-temperature heat in winter [2]. This is due to the higher demand and lower insulation during that time of the year and may be tackled either by introducing seasonal heat storage or, in the worst case, by supplying the lacking heat by fossil fuels. An additional model run revealed that adding 450 W/cap in fossil fuels would solve the problem. Nevertheless, this would not compromise the original objective of reducing CO_2 emissions by 80%. Alternatively, other measures than using additional fossil fuels are possible. More hydrogen and biomass-derived fuels could be produced and dwellings could be even better insulated. The photovoltaics capacity could be reduced, since there is an electricity surplus in summer. Thus, roof area for additional solar-thermal installations would be freed, which, in turn, may also require additional heat storage capacity. However, the consistency checks make it clear that building codes are one of the crucial bottlenecks for the success of any attempt to improve the European energy system.

4 Economic and Environmental Impacts

4.1 Technology Costs

Together with the efficiencies and potentials of the renewable energy technologies, their specific costs were estimated. These estimates were based on literature surveys, expert consultations and curve fitting. All figures are given in 1990 ECU. The interest rate used is 3% (real). The trajectories for specific investments and electricity production costs for the technologies considered in the LTI-Project are shown in Fig. 4. Electricity production costs are calculated by deriving annuities from the investments using the lifetime of the equipment, adding annual operation and maintenance costs, and dividing the sum by the energy produced within one year. Finally, specific fuel costs are added, if applicable.

For small hydro power plants, that is those smaller than 10 MW, we assume no cost reductions, because the technology is mature and no significant economies of scale can be expected. The technical lifetime of this technology is taken as 60 years and operation and maintenance costs come to 1% of the investment per year.

For wind turbines, which have a lifetime of 20 years, the current exponential decrease of costs is expected to continue. Off-shore installations are assumed to be 30% more expensive per kilowatt-hour than on-shore ones.

The largest cost reduction is expected for photovoltaics due to a sharp increase of module efficiencies and strong economies of scale once this technology is produced in large numbers. The lifetime is taken as 25 years, operation and maintenance requires 1% of the investment per year. The production cost depends heavily on the solar irradiation. For example, the same technology installed in Spain will produce 54% more electricity than in Germany. Thus, in Fig. 4, average costs are shown.

Solar-thermal electricity production is only considered for Southern Europe, that is Italy, Spain, Portugal, and Greece. Costs are estimated for an artificial technology representing an average of parabolic trough plants, tower plants, dish/Sterling plants, and anabatic wind power plants. The lifetime is 20 years and the energy yield is expected to increase from 950 kWh/(kW_p a) in 1990 to 1550 kWh/(kW_p a) in 2050.

Solar-thermal collectors which have a lifetime of 20 years can be used for room heating and hot water production. Fig. 4 shows an average of the specific costs in Northern and Southern Europe.

To calculate average costs of biomass utilisation is difficult because of the many different technologies involved. Nevertheless, an average technology was established using data from numerous studies. The lifetime of the equipment is 20 years, operation and maintenance costs are 8% of the investment per year. The fuel costs were estimated to be 2.7 ECU/GJ for the industrial biomass residues and 13.8 ECU/GJ for energy crops. Since the scenario assumed 6 000 full load hours per year, we use this figure for the cost calculation, but perform a sensitivity analysis using a value of 3 000 hours for comparison. Heat pumps for utilisation of environmental heat will last 20 years, operation and maintenance requires 2.5% of the investment per year.

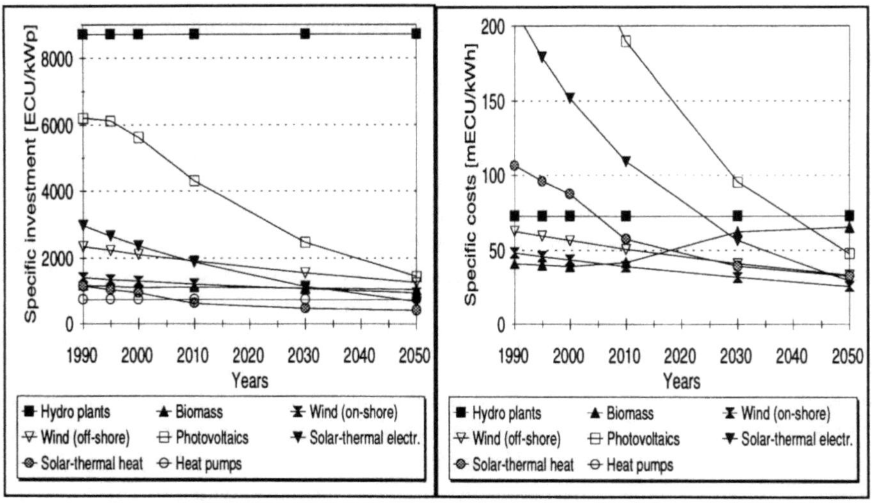

Fig. 4: Specific investments (left-hand side) and production costs (right-hand side) of the renewable energy technologies considered in the Sustainable Scenario.

For comparison, conventional energy technologies are taken into account via three aggregated technologies providing electricity, low-temperature heat, and high temperature heat. The assumed specific capital costs for coal-fired power plants are 19.0 mECU/kWh in 1990 rising to 22.3 mECU/kWh in 2050 and for gas-fired plants 9.3 and 11.9 mECU/kWh, respectively. For nuclear electricity, a constant value of 50.1 mECU/kWh is presumed until 2010 after which this technology is not used anymore. The specific capital costs of low-temperature heat equipment are 11.3 and 11.9 mECU/kWh for gas- and oil-fired installations, respectively. The corresponding values for high-temperature heat from gas and oil are 8.8 and 10.5 mECU/kWh. Investments in the transport sector are not considered as belonging to the energy system. In this case, only fuel costs matter. Fuel prices in the transport sector are assumed to increase from 40-50 mECU/kWh in 1990 to 60-80 mECU/kWh in 2050 (including consumer taxes, but without VAT). Prices of other fuels increase from 10-20 mECU/kWh to 20-40 mECU/kWh based on estimates made by the German Enquete Commission in 1995 [3].

4.2 Economic Impacts

In the LTI-Project, no data were available on the cost of efficiency measures which were detailed enough to determine the development of the full system costs of the energy system. Thus, we restrict ourselves to calculating the cost of the energy-supply equipment and its operation for conventional and renewable energy sources.

Using the cost estimates of Sec. 4.1, the annual costs for operating the existing renewable energy installations in the European Union are calculated to be about 42 billion ECU (1 billion = 10^9) in 1990. In the Sustainable Scenario, this figure will rise to 253 billion ECU/a by the year 2050, which seems rather high at a first look. However, it is only half the cost of the full energy system today (cf. Fig. 6). Furthermore, it corresponds to only 4.8% of the 1990 gross domestic product (GDP) of the EU and amounts to less than 700 ECU per citizen and year. While renewable energy costs rise by a factor of 6 during the 60 year time period considered, the installed capacity will increase from 144 GW_p to 2 300 GW_p , that is by a factor of 16. The left-hand side of Fig. 5 shows the development of the cost figures with respect to the different technologies. The right-hand side of this figure indicates the investments that have to be made in the respective years to build-up the renewable energy system, including substitution of technologies reaching the end of their lifetime.

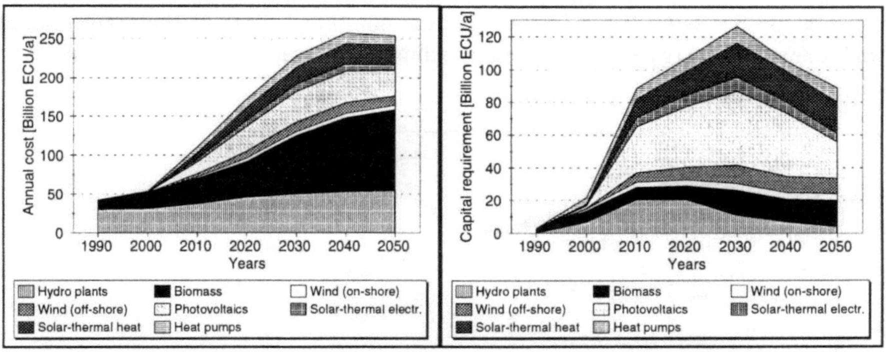

Fig. 5: Development of the total annual costs of the renewable energy technologies (left-hand side) and necessary annual investments in these technologies (right-hand side) in the Sustainable Scenario.

There is a large jump in the amount of money required in the year 2000 which is the starting point of most developments. In practice, however, this demand for investments can probably be spread over a number of years. Table 1 shows the shares of the different technologies in 2050 with respect to energy supplied, capacity installed, and annual costs.

Table 1: Shares of the different renewable energy technologies in 2050

Technology	Energy supplied	Capacity installed	Annual costs
Biomass	31.8 %	12.0 %	41.4 %
Solar-thermal heat	20.5 %	40.8 %	9.9 %
Wind energy	13.6 %	10.4 %	6.9 %
Solar-thermal electricity	11.3 %	7.1 %	2.6 %
Photovoltaics	9.8 %	18.2 %	13.3 %
Hydropower	8.8 %	7.7 %	21.2 %
Environmental heat	4.3 %	3.8 %	4.6 %

To allow for a better judgement of the renewable energy costs, they are compared with the remaining costs of conventional energies in Fig. 6. These were calculated using the cost assumptions described in Sec. 4.1 and the trajectories of the Sustainable Scenario. Fig. 6 shows that the combined costs of the conventional and renewable energy supply decrease to almost half the amount spent in 1990.

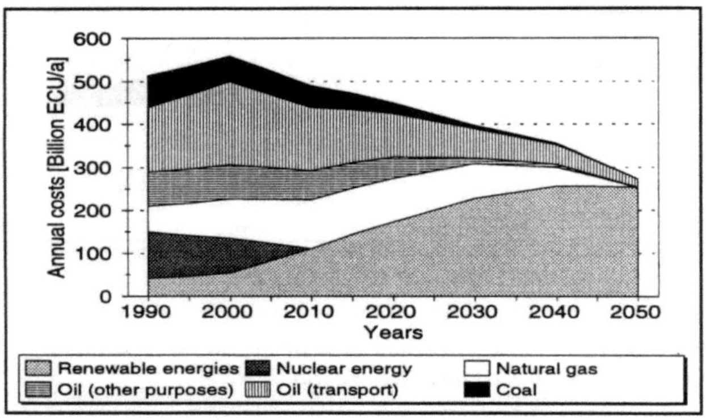

Fig. 6: Annual costs of the conventional and renewable energy parts of the energy systems in the Sustainable Scenario.

However, it has to be taken into account that these figures do not include the costs of efficiency measures which involve substantial restructuring of the economy and could thus not be calculated in this project. One may conclude that the large amount of money saved in the energy-supply sector presents a sufficient basis for financing efficiency measures. Finally, we consider a reference case for the year 2050 in which the energy demand of the Sustainable Scenario is met exclusively via the best fossil-fuel fired technologies available. The respective technologies are gas-fired combined-cycle power plants, condensing value boilers, and heating plants as well as oil-fired boilers. The total annual cost of such an energy system amounts to 201 billion ECU per year and is thus 25% cheaper than the system derived in the Sustainable Scenario consisting of the renewable part, which costs 253 billion ECU/a as discussed above, plus 19 billion ECU/a for the remaining fossil fuel in the transport sector. The cost difference of 71 billion ECU/a can be viewed as the cost of mitigating global warming and other energy related externalities. It amounts to about 19 mECU/kWh which is moderate. Taking into account that calculations on future costs bear large uncertainties, one may nevertheless conclude that even though average specific energy costs will rise by almost 50% until 2050, the total cost of the energy supply system will most probably not increase significantly.

The static input-output model EMI 2.0 was applied to calculate employment effects of the restructuring of the energy-supply system according to the Sustainable Scenario [4]. Unfortunately, the input-output tables were only available for West Germany in 1988 which was thus used as an approximation for the whole of Europe. Being a highly rationalised region with high work force productivity, this may, however, be a good reference case with respect to the assumed future convergence of living standards and economic structures in Europe. For each of the eight renewable energy technologies considered, two input vectors of the intermediate demand from other sectors were put together. The first one maps the construction costs attributed to the year in which the investment is made. The second one includes the operation and maintenance costs as well as fuel costs for biofuels over the whole lifetime of the equipment. Keeping in mind the uncertainty of the calculations, 4 million person years of employment per year (at 1988 labour productivity) are necessary to construct and operate the renewable energy based part of the energy system in 2050 (cf. Fig. 7). Of the 1.6 million person years needed for operating biomass technologies in 2050, two thirds are created in agriculture and forestry.

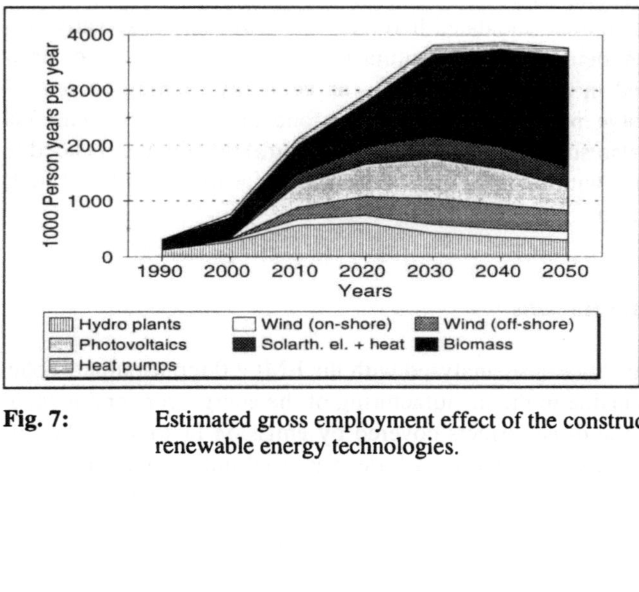

Fig. 7: Estimated gross employment effect of the construction and operation of renewable energy technologies.

For comparison, the employment effects of the reference case were calculated. In that case, building and operating the energy system (again without the part of the transport system still based on fossil fuels in 2050) requires 1.7 million person years per year in 2050. However, this is not the only relevant effect. To make the two cases comparable, we assume that the energy cost difference of 71 billion ECU per year is spent on additional private consumption in the reference case. Consequently, 1.5 million person years of employment are created by private consumption if the average 1990 German savings rate of 14% and the 1988 German economic structure are assumed. If all the additional costs are spent on consumption, which is likely during the long time horizon considered, the figure rises to 1.7 million person years. If only one national economy was considered, one would have to subtract the amount of goods imported. Since most of the trading of EU member countries occurs inside of the EU, this effect is neglected here. In total, there are 3.2 to 3.4 million person years of employment in the reference case.

Comparing the employment effects of the Sustainable Scenario with those of the respective reference case yields a net effect of 340 000 to 580 000 additional person years for the renewable energy based system. When assessing these results, one has to keep in mind that one reason for the relatively small difference is the fact that the production of goods for private consumption in 1988 is more labour intensive than the energy sector. In addition, the assumptions for the reference case are very optimistic and lead to overestimating the cost difference to the Sustainable Scenario.

The effects of the fundamental restructuring of the economy assumed in the Sustainable Scenario are almost identical for the reference case, since both have the same energy demand. Under business-as-usual assumptions, the energy demand and costs of the reference case would be substantially higher, leaving less money for additional consumption.

Consequently, the net figures calculated here can be viewed as conservative estimates. This allows for being carefully optimistic that – in spite of the many rough estimates involved in these calculations – an economy running on renewable energies will involve more employment than one driven by conventional energies. This result is also supported by the fact that domestic labour is needed to produce the capital goods within the EU which substitute for imported fossil fuels in the Sustainable Scenario.

4.3 Environmental Impacts

The emission of pollutants was also analysed with the EMI 2.0 input-output model to include indirect effects due to the manufacturing of the energy conversion technologies in addition to the direct effects covered by emission factors for the different fuels. Again, the 1988 German data base was used and scaled to the EU level proportionately to the primary energy consumption. In a test run, this approach yielded 1990 CO_2-emissions which deviated by only 1% from the statistical value given in Table 2.

To calculate the remaining emissions, the energy demand of the different economic sectors in the input matrix of the model is reduced to the values given in the scenario description for 2050. When renewable energies are used, direct emission factors are set to zero. Fossil fuels are only utilised in the transport sector. The resulting emissions are shown in Table 2 together with the respective 1990 values.

Table 2: Emissions of pollutants (in million t/a; without production and operation of renewable energy technologies).

Pollutant	1990	2050	Reduction
CO_2	3060	308	90 %
NO_x	10.8	4.01	63 %
SO_2	19.8	0.175	99 %
CO	38.3	11.6	70 %
VOC	6.52	1.76	73 %
Dust	2.32	0.215	91 %

The CO_2-emissions are reduced from 3 100 million tons in 1990 to 310 million tons in 2050, that is by 90%. Thus, the objective of the project is more than fulfilled. Table 2 does not include the emissions released by renewable energy technologies in 2050. An upper limit for these emissions can again be determined using the 1988 economic structure, that is assuming that these technologies are still produced with fossil fuels. The CO_2-emissions from construction and operation of renewable energy technologies are below 65 million tons in 2050. Using this figure would increase the emissions given in Table 2 by 20%, but would still meet the project objective. As a crude estimate, one could further assume that the emissions from the production of renewable energy technologies are reduced by the factor calculated for the whole economy. In this case, only 6.5 million tons or 2% would have to be added for the construction and operation of renewable energy technologies in Table 2.

5 Conclusions

It has been demonstrated that the European energy system can be converted towards a more sustainable use of energy within the next 60 years. Massive application of energy efficiency measures and substantial use of renewable energies will allow for a total phase-out of nuclear energy and a dramatic reduction of fossil fuel use at the same time. In the medium- and long term, the revised energy system will most probably not be more expensive than the current one and will create more jobs than are lost in the conventional energy system. Additional investments will be necessary in an early phase to trigger off this development. However, there are no general technical or financial bottlenecks hampering the desired development. Nevertheless, the economic, legal, and institutional conditions of the energy-supply systems have to be changed substantially and rather soon.

In general, one could either increase the prices of conventional energy by an average factor of two or pass extensive regulations and norms on the use of energy to facilitate the transition. In practice, one would probably rely on a mixture of instruments. In addition to an economically acceptable, but steady and significant increase of conventional energy prices, renewable energy technologies near economic efficiency should be made cost-efficient by subsidies based on the amount of energy delivered. Those technologies that are rather expensive at the moment should be continuously supported to maintain the knowledge base and to trigger further cost degression by technology improvements. Housing codes have proven to be one of the crucial issues which must be tackled soon. Any house newly built or renovated today without sufficient improvement of its energy efficiency will contribute to a severe bottleneck for the next 20-30 years. For grid-bound energy forms, it is essential to ensure the access of individual energy producers to the grid at economically attractive conditions. Finally, the support of the population for the necessary restructuring of the energy-supply system has to be ensured. As a prerequisite, the knowledge about energy efficiency and renewable energies needs to be improved on all levels of the society, be it schools, public debate, or decision makers' committees. Also, decision processes have to be transparent and the people affected should be involved as early as possible. In addition to technical requirements, this is another argument calling for much more regional responsibility for the energy supply.

Simultaneously with removing these external bottlenecks, the internal coordination of the energy efficiency and renewable energies industry has to be improved. This involves better integration of the different technologies as well as offering local energy-services derived from local sources. It will, thus, circumvent direct investments by private households and small consumers and will at the same time ensure professional management and maintenance of the installations.

Above all, the desired development towards a sustainable energy supply will only take place if a broad consensus is found and actively pursued by the European governments. A start should be made as soon as possible since each year passing without decisive action will create additional bottlenecks due to additional money invested in long-living conventional equipment.

Acknowledgments

This paper is based on the executive summary of the LTI-Project which was carried out by an international team of researchers: A. Andrew, J. Bougard (Faculte Polytechnique, Mons, Belgium); P. Audinet, Ch. de Gouvello, P. Matarasso, M. Poppe (CIRED, Paris, France); W. Bräuer, H.-M. Groscurth, O. Hohmeyer, I. Kühn, S. Weinreich (ZEW, Mannheim, Germany); H. Connor-Lajambe, S. Krüger-Nielsen, B. Kuemmel, B. Sørensen (University of Roskilde, Denmark); B. Drees, H. Lehmann, T. Reetz, K.O. Schallaböck (Wuppertal Institute, Germany).

References

[1] A. Andrew, P. Audinet, J. Bougard, W. Bräuer, H. Connor-Lajambe, B. Drees, H.-M. Groscurth, Ch. de Gouvello, O. Hohmeyer, S. Krüger-Nielsen, I. Kühn, B. Kuemmel, H. Lehmann, P. Matarasso, M. Poppe, T. Reetz, K.O. Schallaböck, B. Sørensen, S. Weinreich. *Long-Term Integration of Renewable Energy Sources into the European Energy System and Its Potential Economic and Environmental Impacts.* Report on the EU-APAS Project RENA-CT95-0041, Zentrum für Europäische Wirtschaftsforschung (ZEW), Mannheim 1996.

[2] B. Sørensen. *Consistency Checks of the Scenarios.* In: [1].

[3] Enquete-Kommission "Schutz der Erdatmosphäre" des Deutschen Bundestags. *Mehr Zukunft für die Erde.* Economica, Bonn 1995.

[4] O. Hohmeyer, J. Kirsch and Stefan Vögele. *A disaggregated Model Linking Economic Activities and Emissions.* In: S. Bringezu, M. Fischer-Kowalski, R. Kleijn, V. Palm (eds.), Regional and National Material Flow Accounting: From Paradigm to Practice of Sustainability, Proceedings of the ConAccount workshop 21-23.1.1997 in Leiden, Wuppertal 1997, p. 204-210.

11 What Hope for a Sustainable Energy System?

Nick Eyre
Eyre Energy Environment
7 Abbotsgate, Kirkby Lonsdale, Cumbria, GB-LA6 2JS

Abstract. Emissions reductions to achieve climate stabilisation require fundamental changes to the energy system - high energy efficiency and large scale use of renewable energy sources. It has been shown that these are technically possible and would not impose huge costs over a 60 year period. In evaluating the likelihood of this transition, technical progress and economics are far from the whole story. There is a long standing failure of all economies to achieve the levels of energy efficiency which are known to be economic. This is caused by non-economic barriers. Similar constraints apply to the use of renewable energy. The problems lie in the nature of the energy system developed in industrialised countries, with structures biased against investment in decentralised supplies and end use energy efficiency. There are signs that this is changing, but the transformation required is so substantial it calls for a fundamental change in social attitudes to energy. The climate problem could be a key element in producing such a change.

Keywords: Renewable energy, Energy efficiency, Energy policy, Sustainable development.

1 Introduction - What Are the Key Findings of the Sustainable Scenario?

Helmuth-Michael Groscurth's paper on a Sustainable Scenario for energy in Europe is in the tradition of energy analyses started by Amory Lovins in *Soft Energy Paths* (Lovins, 1977). Of course, in the late 70's the main concern was that the world had insufficient fossil fuel resources to meet future needs. Nowadays, the concern is almost the opposite, that we have so much fossil fuel that burning even a fraction of it might seriously alter our climate. However, for the energy system the challenge is much the same - to meet the needs of developed and developing countries without dependence on the unsustainable sources which have fuelled industrial societies for the last 200 years.

The results of the Sustainable Scenario show that energy needs can be met whilst achieving climate stabilisation by a combination of high levels of energy efficiency and the use of renewable sources. The broad conclusion is not in itself new. A demonstration for the global energy system by 2100 was developed some years ago (Lazarus, 1993). However, the Sustainable Scenario makes two important new contributions to analyses in this field:

- It shows that a sustainable energy system is achievable in Europe within two generations. And, if it is possible in Europe with its high population density, high living standards, and relatively limited solar resources, it is reasonable to infer that it should also be possible elsewhere. This is important because it implies that sustainability and energy security are compatible. There is no need for developments like huge Saharan solar plants with hydrogen piped to Europe.
- It also shows that the incremental costs over 60 years are small, so that the economics of the transition are practicable. Indeed the employment benefits projected look very attractive in the current European economic context.

2 How Should We Interpret the Sustainable Scenario?

It is important to be clear what the Sustainable Scenario represents. It certainly should not be taken as a firm prediction for Europe's energy system in 2050. Like any good scenario it is only an existence proof - a demonstration of what is possible not what will necessarily happen. It is no more guaranteed than any other development path.

The Sustainable Scenario is refreshingly honest about the subjective assumptions which underpin it. It is only consistent with a Europe which is concerned about social justice, good public services, decentralisation of power, and environmental protection. Of course the Sustainable Scenario is not unique in having implicit ideological assumptions. All scenarios do - it is just not that all authors state their assumptions or even recognise that they have made them.

One of the strengths of the energy system presented in the Sustainable Scenario is its technological diversity. Diversity is important in any system because it tends to provide resilience (Stirling, 1994). In the scenario no single source provides more than 20% of the energy in 2050 (assuming there is a mix of biofuels). Similarly, the large improvements in energy efficiency require many independent technologies. The broad outline of the scenario will not collapse if a single technology fails to develop as expected.

However, resilience once established is not the only concern in assessing the plausibility of the scenario. The likelihood is also dependent upon the difficulties in organising a transition from the current energy system. Large scale deployment of renewable energy requires an energy system which is very different from the one we have today. In particular, it will require massive changes in land use and it will virtually end centralised energy production. Similarly, major improvements in energy efficiency require a redistribution of investment away from energy supply towards the point of energy use. These changes will have ramifications throughout the economic and social systems. It is these issues, as much as the political willingness to accept incremental costs, which affect the plausibility of the transition to a sustainable energy system.

3 What Are the Non-Economic Barriers to Sustainable Energy?

The existence of non-economic barriers to energy efficiency is now well established. A typical taxonomy of the problem is as follows:

- imperfect information to energy consumers,
- separation of costs and benefits (e.g., the landlord/tenant barrier),
- high costs of capital,
- tariffs below marginal costs,
- non-inclusion in prices of externalities, and
- bounded rationality (i.e., behaviour which does not profit/utility maximise).

The barriers are usually thought of as market imperfections - that is as reasons why economic efficiency is not optimal in real energy markets (see, e.g., Sanstad and Howarth, 1994). They explain the common observation that "top-down" energy models generally produce higher projections of future energy use than more technologically detailed "bottom-up" models (Wilson and Swisher, 1993). The former tend to assume existing energy use is optimal and therefore underestimate the scope for "no regrets" improvements. The latter show that existing energy efficiency is sub-optimal, but tend to assume that future use can be optimised. Needless to say neither has a monopoly of wisdom, but the difference in results shows that optimal efficiency is a bad assumption.

A similar analysis can be developed for renewable energy sources, although because the economic case has traditionally been less good, the analysis of non-economic barriers to renewables is less well known. However, it is clear they exist (see, e.g., Grubb, 1995). Some barriers such as the omission of external costs in pricing and the higher costs of capital to smaller projects are shared with energy efficiency. Others are more particular to renewable energy. These include:

- the neglected benefits of embedded generation (reducing system losses etc.),
- poor information on resource availability,

- subsidies to existing land uses and energy sources, and
- more arduous land use planning constraints than for conventional projects.

Analyses of the barriers facing both the energy efficiency and renewables industries have some clear common themes. They are not perceived as "serious energy options" by the more traditional forces in industry and government, or by many citizens. They come in small projects, not the multi-MW chunks familiar to energy institutions. They require far greater involvement with customers and local communities. And they involve interaction with life outside the energy industry in activities as diverse as construction, consumer goods, land use planning, agriculture, forestry, and waste management.

It is important to understand that the barriers are not just an unfortunate accident. They are an inevitable product of the way modern energy systems have developed. To support mass production and the development of a consumer society, it was critical that energy systems provided reliable and low cost energy. The solution was large scale systems (Hughes, 1983) and the features of the industry that developed to provide such systems are that:

- they are highly centralised and distinct from other sectors of the economy,
- there is a clear distinction between consumers and producers,
- energy is sold as a commodity with little attention to the services it provides, and
- government energy institutions have close links with major producers.

These characteristics underpin the barriers to energy efficiency (Eyre, 1997). They are also applicable to the barriers to renewable energy sources.

In the context of the Sustainable Scenario, the question is "Can the barriers be lowered significantly?" so that sustainable technologies can fulfil their potential. It would be wrong to think that all that is required is the commitment of the political class to develop a sustainable energy system. Changes are required in much of society and its attitudes; most importantly, fundamental changes are needed in the energy system. The transition to a sustainable system therefore requires political commitment, but also fundamental social, industrial, and institutional change.

It is therefore possible to characterise what will be required to make the energy system sustainable as a paradigm shift. Such shifts have occurred in the past in the energy sector, notably as the steam engine replaced human and horse power as the dominant energy source and a century later as Edison and his contemporaries developed the electric power network. But they are relatively rare and they only occur when the existing system fails to meet society's needs. In looking at the plausibility of the transition to a sustainable energy system, the relevant questions are:

- Will current energy systems fail to deliver what society expects of them? And,
- What role could responses to the climate problem play?

4 Will the Sustainable Scenario Happen?

The Sustainable Scenario requires massive investment in home insulation, local CHP plant, photovoltaic roofing, wind farms, and new energy crops. If this happens, energy technology will have to be part of the landscape in both urban and rural areas, and interaction with it will play a part in the lives of far more people. There is no reason in principle why this cannot happen, but it involves a reversal of the trends towards remote production and long range distribution observed for much of this century.

There are signs that energy system developments are changing:

- The growth in unit size has stopped. Combined cycle gas turbine technology and new renewables have produced new power stations smaller than those they replace for the first time in the history of the electricity industry. And as the larger gas and oil fields are depleted, the same is true in the fossil fuel extraction sector.
- Market liberalisation is undermining the monopolies which have dominated the energy sector. The process is uneven across the world, depending on political factors, but the trend is everywhere the same - towards competition except in the natural monopolies of grid distribution and tighter regulation of those residual monopolies.
- There is a shift away from "constructed" towards "manufactured" energy (Flavin and Lennsen, 1994). It is ironic that the energy sector, whose development underpinned Fordist production and the consumer age, has been one of the last sectors to be penetrated by mass production. But technologies requiring huge construction projects are in retreat in the face of technologies which have greater flexibility and more scope for cost reduction.
- Consumer and environmental awareness are at odds with the remote and unaccountable structures of the traditional energy supplier. Increasingly consumers want a more complex package of services not just a commodity. And large sections of society have lost faith in the techno-centrism which characterised the mid-20th century.

None of this makes a sustainable energy system inevitable. The "traditional" energy industries will not just disappear. Job losses in traditional sectors have already provoked widespread concern in many countries. The nuclear industry has sought to use the climate change problem to its advantage. And public service commitments are being used to defend traditional monopoly practice. The changes noted above could be either just an adjustment of the existing system to new pressures or the beginnings of a paradigm shift.

If the sustainable technologies are to make a major breakthrough, there will have to be a change in the perception that the energy systems which have developed over the last century are in some way natural and immutable. For most people in industrialised countries fuel is something that they expect to come out of an electric socket or a gas pipe. Maximum convenience is part of marketing strategies, and minimum participation of customers and communities in decision-making is the norm. The common expectation is of little or no interaction with the process by which energy comes into the home, office, or factory. An energy system built around high efficiency and renewables requires these social attitudes to change. Whilst a reversal of history is an unrealistic expectation, some re-engagement between the population and its energy supply is necessary if that supply is to be sustainable.

The prospects for change depend critically on culture. This is already clear from comparisons of behaviour both within and between countries. Energy efficient behaviour is observed most strongly amongst social groups which are most knowledgeable and concerned about environmental issues (e.g., Strang, 1995). And the countries which have emphasised social cohesion and environmental protection in a market economy have generally achieved higher levels of energy efficiency than either command economies or more free-market oriented countries.

Change in the energy sector will depend on the way information-rich, post-modern societies develop. If economic activity is dominated by trans-national corporations, with most energy users simply seeking the lowest cost commodities, there is little hope of fundamental structural change. But if a post-materialist society develops, with increasing emphasis on community control, local production, and useful employment, then the old energy system will be untenable, and a new sustainable system can emerge.

The Sustainable Scenario is more likely to be achieved in a culture where environmental concern plays an important role. The climate problem therefore probably has a role to play at the level of both the political elite and wider civil society.

The global nature of the climate change debate helps raise the political profile of energy related emissions. Fifteen years ago a British Prime Minister remarked (of a minor war for ownership of some small islands in the South Atlantic) that it was nice to be dealing with a real crisis rather than "humdrum issues like the environment". Within 7 years, the same person was insisting all her ministers attended briefings on climate change and the related issues for energy policy (see Currie, 1989). The recognition that it was a real international political issue had changed energy's position on the domestic political agenda - a trend which still continues.

For many years, low energy scenarios were widely considered as an environmentalist dream, marginalised from establishment debates about energy. But now it is common to hear calls for the sort of emissions reductions in the Sustainable Scenario from leading figures in the political and scientific communities. What remains unclear is the extent to which those calls will lead to the necessary changes to energy policy when faced with the inevitable opposition of entrenched vested interests.

The EU has taken a leading role in promoting emissions reduction targets within the Framework Convention for Climate Change. It seems likely that this will continue up to and beyond the Conference of Parties in Kyoto where it is intended to negotiate binding commitments. However, the policy measures taken to reduce emissions within the EU and the results achieved have been less impressive. It remains an open question whether this mismatch will be resolved by a weakening of political commitment on the climate problem or a more radical energy policy. The Sustainable Scenario provides ammunition to those arguing for the latter.

5 Conclusions

The Sustainable Scenario shows that Europe could respond positively to the threat of climate change by restructuring its energy system. Whether we actually will respond this way is not such an easy question to answer. Although the incremental costs of the change are well within the range which developed economies could bear, the social, institutional, and political changes required should not be underestimated.

The key distinction between energy efficiency and renewable energy technologies and the "conventional" energy system lies in the technical and institutional structure. Conventional energy systems are large and centralised, they distinguish clearly between production and consumption, deliver units of fuel to consumers with little thought about how it is used, and have minimal interaction with most other sectors of the economy. If we are to have a sustainable energy system all that has to change.

There are promising signs. The energy systems in many countries are already changing to smaller scale, more flexible systems - technically and organisationally. Climate change is part of the process through which there is increasing recognition that change is necessary. But the scale of social and industrial change needed has not been fully recognised even by most people involved in the sustainable energy industries. Only when this realisation is widespread in society will it be possible to judge the probability that the Sustainable Scenario will actually happen.

In 60 years time the energy paradigm may have changed. Our grandchildren may look back and laugh at our extraordinary behaviour. Why, they might ask, did we:

- try to build fusion reactors on Earth instead of using energy from one we already had?
- tunnel underground to extract ancient biomass instead of simply growing it?
- burn fuel in power stations, then waste most of the energy? and
- use resistance heaters as lights half a century after the development of the fluorescent lamp?

I hope they will, but it is too early to tell.

References

Currie, W.M. (1989) *Options for Mitigating the Greenhouse Effect.* A presentation to the Prime Minister's Seminar on Climate Change of 26 April 1989. ETSU R-54. HMSO, London.

Eyre, N.J. (1997) Barriers to Energy Efficiency - More than Just Market Failure. *Energy and Environment* 8(1) 27-45.

Flavin, C. and Lenssen, N. (1994) Reshaping the Electric Power Industry. *Energy Policy* 22 (12) 1029-1044.

Grubb, M.J. (1995) *Renewable Energy Strategies for Europe.* Earthscan/RIIA, London.

Hughes, T.P. (1983) *Networks of Power.* John Hopkins University Press, Baltimore.

Lazarus, M. (1993) *Towards a Fossil Free Energy Future. The Next Energy Transition. A Technical Analysis for Greenpeace International.* Stockholm Environment Centre, Boston.

Lovins, A.B. (1977) *Soft Energy Paths: Towards a Durable Peace.* Pelican Books, London.

Sanstad, A.H. and Howarth, R.B (1994) "Normal" Markets, Market Imperfections and Energy Efficiency. *Energy Policy* 22 (10) 811-818.

Stirling, A. (1994) Diversity and Ignorance in Electricity Supply Investment. Addressing the Solution rather than the Problem. *Energy Policy* 22 (3) 195-216.

Strang, V. (1995) "Cultural and Behavioural Issues", in DECADE - Domestic Equipment and Carbon Dioxide Emissions. Second year report. (Ed. Boardman, B.) Environmental Change Unit, Oxford.

Wilson, D. and Swisher, J. (1993) Top Down versus Bottom Up Analyses of the Cost of Mitigating Global Warming. *Energy Policy* 21 249-256.

12 Comment on: Potentials of Joint Implementation Measures

Henning Rentz
RWE Energie AG
Kruppstr. 12, 45128 Essen

When taken seriously, global warming prevention is a daunting challenge. It cannot be mastered solely by government action, but requires far-reaching entrepreneurial initiative. A large-scale transfer of capital and know-how is required in order to enable other countries to benefit from the technical standard achieved in the developed countries in terms of energy efficiency and environmental protection.

If the state of the art currently applied in Germany was put into practice in industrial production world-wide, emissions of CO_2 could be cut by 3.8 billion tons. If the latest state of the art in electricity generation and measures taken in the transport sector are included, the achievable reduction potential is in excess of 7 billion tons of CO_2: this is equivalent to close to 40% of the current level of emissions.

Several fundamental requirements have to be met if capital is to be mobilised on an appreciable scale:

1) Projects must have the support of the governments involved.
2) Investment risks must not be disproportionately high.
3) Capital must earn adequate return.
4) Earnings must be convertible and exportable.

Joint Implementation is an often quoted instrument supposed to induce the necessary capital flows. This possibility to carry out climate protection measures jointly with partners in other countries has already been mentioned in the Framework Convention on Climate Change.

No decisions on Joint Implementation were taken at the first Conference of the Parties under the Framework Convention on Climate Change in Berlin in March/April 1995. Instead, it was agreed to set up a pilot phase of "Activities Implemented Jointly" and to check the Joint Implementation mechanism for its practicability. The following criteria apply to the pilot phase to be evaluated at the fifth Conference of the Parties in 1999:

1) All activities implemented jointly under the pilot phase require prior acceptance, approval, or endorsement by the governments of the parties participating in these activities.
2) Activities implemented jointly should bring about real, measurable, and long-term environmental benefits related to the mitigation of climate change.
3) The financing of activities implemented jointly shall be additional to the financial obligations of the developed countries within the framework of the financial mechanism of the Framework Convention on Climate Change as well as to current official development assistance flows.
4) No credits shall accrue to any party as a result of greenhouse gas emissions reduced or sequestered during the pilot phase from activities implemented jointly.

The third point seems to be particularly critical. There are no criteria yet by means of which it could be determined which investment is made in addition. This will shortly have to be the subject of international negotiations in order to find clear and workable regimes.

Several requirements have to be met from industry's point of view for Joint Implementation to be translated successfully into practice.

The present number of about 50 pilot projects carried out under activities implemented jointly has produced a large number of different results. They do have, however, one thing in common: The transaction costs are high. This aspect therefore has to be given particular attention when it comes to the concrete design of Joint Implementation.

An internationally accepted, reliable Joint Implementation regime has to be established which is binding for all participants. This is the only way to create the security required for investments.

Joint Implementation must be easy to handle, have a simple structure and require only little bureaucracy. It is with a feeling of unease that I have been observing current trends in authorities and research institutions which are already picturing an omniscient Joint Implementation bureaucracy spanning the world. Hosts of specialists travelling around the world to get an idea of proper erection and operation of plants would inevitably frustrate this instrument because the transaction costs would clearly outweigh the cost benefits of controlling emissions. The E7 projects have taught us anyway that transaction costs are high. We are confident, however, that costs will decline significantly in the long run.

The emission reductions achieved must be measurable and verifiable. Once they are certified they must be tradable to ensure the degree of flexibility required by the market.

As far as "additionality" is concerned, nothing has been finalised yet. I believe, however, that the scale of the task before us does not allow us to subsidise economically non-viable investments and to send wrong signals for the allocation of the scarce production factor capital. I am keenly looking forward to our forthcoming discussions on this subject.

We need a broad approach in order to cover as many aspects of Joint Implementation as possible. The most varied projects and project types for CO_2 reduction must be eligible. Moreover, the reduction of other greenhouse gases such as methane and nitrous oxide will also have to be given consideration.

Incentives for governments and companies alike are the precondition for Joint Implementation to become feasible in the first place. Binding and quantified reduction objectives (QELROs) for the Annex 1 countries are required in particular. At the companies' level various crediting options are available. In Germany, we have taken the approach of voluntary agreements which can be a good basis if they specify emission quantities. However, each country will have to establish its own crediting mechanisms.

I pointed expressly to the voluntary commitment of German industry and consider this to be a practicable, if not the only feasible way forward. Here, too, we are anxiously looking forward to what the future international negotiations but also the creativity of our federal government have in store for us.

I should like to conclude on a word of warning. Joint Implementation can always only make a limited contribution to protecting our climate. Most of the work required to reduce the emission of greenhouse gases must be done in the industrial countries currently accounting for three quarters of world-wide carbon dioxide emissions. Nobody can seriously expect the population in the newly industrialising and developing countries to be prepared to cut emissions on a per capita basis and in absolute terms. Quite on the contrary: These countries need additional energy services and will meet this demand. Even a functioning Joint Implementation mechanism will not obviate the need for us in the developed countries to undertake great efforts for climate protection at home.

Joint Implementation and potential certification systems, which may be following, foster innovation and industry's own initiative to a greater extent than regulatory measures taken by the government. Joint Implementation and emissions trading regimes are both aimed at quantity reduction and can hence be ideally combined with each other.

For global warming prevention to be effective it is necessary to deploy the world-wide capital, which is in limited supply, combined with sophisticated technology and know-how wherever its benefits are greatest both in terms of global environmental protection and for all participants.

Governments, industrial companies, and consumers alike are called upon to rise to the global ecological challenge of global warming prevention with economically appropriate strategies on the basis of partnership and across national borders.

13 Assessment of Voluntary Agreements in Environmental Policy- A Neoliberal Perspective

Klaus Rennings, Karl Ludwig Brockmann, Heidi Bergmann
Centre for European Economic Research (ZEW)
P.O.Box 10 34 43, D-68034 Mannheim, Germany

Abstract. A trend towards "softer" regulation, especially in the form of negotiated environmental agreements, is observable in national and international environmental policies. Such agreements are controversial, because there are fears that government will relinquish its responsibility for environmental protection. This paper analyses recent experiences with voluntary agreements in Germany. Topical German examples that have prompted public debates include the takeback agreement for cars, the voluntary agreement made by a number of industries on a CO_2 reduction by the year 2005 and the voluntary agreement made by the automobile industry on the development of energy-efficient cars.

Proponents of voluntary agreements argue that this instrument provides incentives to the business sector for the development of efficient, innovative, and environmentally-friendly solutions. Analysing the examples mentioned above, we conclude that it is hard to detect solutions deserving such attributes. These agreements are unlikely to produce results that go beyond what industry would have done in any case and they avoid using economic incentives. The agreements are non-binding and unenforceable, with the negotiating process leading to a watering-down of the environmental goals government had originally aimed at. A preference for negotiated solutions on principle, as currently espoused by the Federal Government in Germany, seems to be "counterproductive." If the government clearly signals its willingness to refrain from using regulatory or economic instruments in favour of industry agreements, it weakens its negotiating position. The government also limits its options should the implementation of the agreement prove unsatisfactory. Government needs to be "in control" in order to leave its choice of policy instruments open and to be flexible. In a last step, we derive some general conclusions concerning reasonable strategies and applications of voluntary agreements within the European Union.
Keywords: negotiated agreements, climate protection, circular economy, economic instruments of environmental policy.

1 Introduction

By reverting to common command and control measures preventive environmental protection can entail a considerable extension of government intervention. The German Federal Government believes that this should be increasingly counteracted through cooperation between government and the business community, amongst other things through voluntary environmental protection measures. However, these "soft" instruments involve the risk of government relinquishing its responsibility for the environment. Negotiated agreements made by the business community, like the pledges to develop so-called 3-litre cars, to take back old cars, or to reduce carbon dioxide emissions, are therefore controversial from an ecological and economic point of view.

The economic perspective is specified here from a neo-liberal perspective. In a liberal approach, markets should coordinate the allocation of goods as far as no market failure can be observed. But when external effects, information deficiencies, or inflexibilities exist, the state is responsible for correcting these kinds of market failure. However, the state has to choose the option which minimises market distortions. In other words: Market-based instruments like taxes or tradeable emission permits are generally preferable. Nevertheless, other types of instruments like voluntary agreements can be used alternatively or additionally when they have advantages with regard to certain criteria (e.g., efficiency, institutional controllability, minimisation of unwanted side-effects).

Especially the German Ordo-Liberal-School in the tradition of Eucken argues in favour of a government correcting market failure, and against an interventionist state. Principles and criteria of the ordo-liberal school will be specified in chapter 3 where we in general develop an analytical framework for the valuation of environmental policy measures with special interest in the application to voluntary agreements. Since ordoliberalism is mainly a German economic school, it should be mentioned that similar ideas have been established in other countries. For example, the so-called Oxford-Liberals have created the guideline "as much competition as possible, as much planning as necessary" (GROSSEKETTLER 1991:106).[1]

---

1 However, the ordoliberal school focuses more on the responsibility of the state to establish basic rules for a regulatory framework within long-term oriented economic policy, while the Oxford-Liberals give higher weight to policy measures in the short run.

2 Attributes of Voluntary Agreements in Germany

2.1 Negotiated Agreements: Soft, Non-voluntary, and out of Keeping with a Market Economy

Negotiated agreements made by the business community are generally called voluntary. However, what government does is more like showing the instruments of torture to the victim as a first step of torture. Comparable to what, in former times, torturers would do with their instruments, nowadays the Minister for the Environment presents a draft for an ordinance so as to achieve "voluntary" concessions (MURSWIEK 1988: 985). Hence, in principle such cooperative solutions can be interpreted as barter transactions in which the business community imposes an obligation on itself to act in a certain manner, and government in return refrains from enforcing the desired conduct.

In the debate on environmental policy, voluntary agreements are sometimes pictured as "a free-market instrument." This description is only justified in the case of instruments that use the price mechanism. There are, however, only a few exceptional voluntary agreements which fall back on this fundamental free-market principle. These include the "Grüne Punkt" ("Green Dot"), which in economic terms can be characterised as a levy helping to charge the costs relating to the disposal of packaging in accordance with the polluter pays principle. Yet, as a rule negotiated agreements shy away from such "tough" economic instruments and do not touch the structure of relative prices; after all, negotiated agreements do not spring from the market system, but rather from political negotiations between government and trade associations. However, an approach that is essentially based on negotiated solutions should not be characterised as market-based, but as a corporatist approach (HOLZHEY/TEGNER 1996:426-427). The main difference is that consumers have commonly no possibility to participate in the negotiation process, although the consumer surplus is highly influenced by the result of the negotiations (improvement of environmental quality).

2.2 Sovereign versus Corporatist Approach in Environmental Policy

Voluntary agreements can contain all kinds of instruments. Goals or technical instructions for a specific industry, but also scales of charges and information systems are conceivable. That is why the agreements do not have any working mechanism in their own right, rather the latter depends on the instruments provided in each specific case. Having said this, there are typical characteristics of negotiated agreements concerning

- political decision-making and
- enforcement of agreements.

2.2.1 Political Decision-Making

Voluntary environmental protection measures differ markedly from the classic model of a rule-issuing government trying to achieve its environmental policy goals through commands and prohibitions. The government refrains from using any instruments of formal power, i.e., from enacting legal norms, and instead enters into negotiations on the realisation of the environmental goals with the groups affected. In doing so, the government uses the "threat" of enacting restrictive legal norms as a starting point for negotiating a voluntary agreement and thus exerts a guiding influence on the business community's conduct (HOFFMANN-RIEM 1990: 400, 426; BROHM 1992: 1025, 1027). In contrast to traditional sovereign actions, government seeks a solution via consensus-building. However, this bargaining process holds dangers. Unlike any cooperation in the form of the right to be heard, the right of participation or of involvement, agreements are based on a system of service-and-service-in-return. For the government side the price for an agreement often consists in reducing the announced level of environmental protection. In order to attain a consensus, government accepts "a downward correction of goals". This "lowering of the regulatory level" can turn out to be more or less marked. However, as the case studies examined in connection with this study bear out, a "decrease in the stringency of regulations" is always observable (BAUER 1987: 241, 254). With regard to the demands made in terms of environmental policy, the upshot is that there is a tendency for agreements to fall short of the opportunities arising from the implementation in the form of laws and ordinances (MÜGGENBORG 1990: 909, 915). The reasons for this danger of relativising the level of environmental protection in connection with cooperative solutions are complex. The mode of action "agreement" by itself suggests a shift away from maximum demands, since a joint solution can only be found in the intersection of possible regulations that is deemed acceptable to both sides. Government should realise that the readiness to negotiate expressed by the business community inevitably only goes as far as the latter's economic self-interest. Protecting the environment is not a value in its own right to the companies concerned. Instead it has to coincide with their own (economic) interests. Finally, what may also be relevant is the fact that it is not individual businesses that negotiate with government, instead as a rule they are represented by their associations, which, experience shows, tend to take a more rigid stand.

It is also significant that voluntary agreements do not only affect the interests of the two negotiating parties, but also to a variable extent the positions of third-parties. Agreements are always associated with the danger of consensus-building at the expense of non-involved parties, which have little opportunity to integrate their interests into the agreements being negotiated. Furthermore, consideration has to be given to the powers of participation that would have to be taken into account in connection with a legislative procedure or the issuing of an ordinance, but that never features in informal negotiations on voluntary agreements. The mere right to be heard should not be completely excluded from the negotiation procedure.

2.2.2 Enforcing the Agreement

Concerning the enforcement of an agreement, what has to be stressed is that a mere promise still fails to guarantee its actual implementation. Companies not complying with the pledges (given by their association) do not risk any fines, penalties or other coercive measures (cf. HOFFMANN-RIEM 1990: 400, 438; MURSWIEK 1988: 985, 988). There is no claim to performance in the case of a voluntary agreement; indeed, what characterises agreements is the lack of any binding force and of enforceability. In contrast to a contractual relationship, the party pledging voluntary measures never enters into any legal obligations. Contracts and agreements are similar in that both have an exchange of services in common. The relationship "do ut des", in other words a mutual give-and-take, is typical of contractual relations and agreements alike. However, the binding force of a contract - "pacta sunt servanda" - is completely foreign to negotiated agreements. Agreements are more likely to be categorised as a gentlemen's agreement, which does not entail any legal consequences. Even if in connection with an agreement the government side was granted the right to monitor compliance of the agreement, this does not constitute a claim to performance. In fact, the point is that the businesses underpinning the agreement do not want to be bound the way they are in contracts. The companies (or rather their associations) make a voluntary agreement without the so-called intention of legal consequences. The agreement solely consists of a declaration of the companies' willingness to act in a certain way, tolerate something or refrain from doing something; the government side is not granted any claim to performance. Consequently, any legal action founded on the agreement brought against the declarant would be doomed to failure. Whether or not compliance with the agreement is monitored makes no difference from a legal point of view, since violation of the agreement does not entail any legal consequences (cf. BEYER 1986: 277ff.; 282; HENNEKE 1991: 267, 271ff.; BROHM 1994: 1025, 1034).

3 Analytical Raster for Assessing Voluntary Agreements

The Freiburg ordoliberal school holds that the absence of markets, and any functional defects of existing markets, will result in corrective requirements which cannot be covered solely by an evolutionary competition of institutions - a view which will be pursued here below. This school regards as inadequate the process of natural selection emerging from international competition between institutions evolving as a quasi-random process, but not specified by the state, and the resultant elimination of inefficient institutional bodies. On the contrary, the Freiburg neo-liberals' concept of a free-enterprise economy presupposes that the state must selectively create institutions for countering existing defects and challenges (GROSSEKETTLER 1991:104-106). The efficiency of such institutions must be measured in terms of how far the principles of free enterprise are respected.

Seen from the perspective of ordoliberalism voluntary agreements are regarded with scepticism. Government no longer - as postulated in the ordoliberal model - stakes out the regulatory framework within which entrepreneurs can dedicate themselves to profit-making. Instead government delegates this responsibility to the businesses themselves. However, in the opinion of ordoliberals a laissez-faire policy with a market economy left to its own devices tends to destroy itself. Failure to protect and promote competition may lead to closed-off markets, cartels, price fixing and ultimately to a distorted price structure. A lack of competition and the wrong price signals in turn may entail a spate of government interventions, e.g. measures to monitor prices and regulatory requirements concerning the application of given environmental technologies (due to a lack of price incentives). That is why those economists who consider government to be a powerful custodian of the system in the tradition of Eucken's ordoliberal school view the trend in environmental policy towards voluntary agreements with great concern (MAIER-RIGAUD 1995).

However, the following assessment will test these hypotheses by applying the ordoliberal principles to concrete case studies of voluntary agreements. The generalised analytical raster in table 1 comprises a catalogue of check criteria applying to the selection of instruments for eliminating deficits in a free-enterprise economy (cf. RENNINGS et al., 1996 for a detailed explanation of the criteria involved).

Table 1: Systematised check criteria for evaluating economic policy measures in terms of regulatory efficacy

Step 1: Goal formulation and operationalisation

- Formulating the targeted goal system
- Indicators
- Assignment of goals, means and implementing agencies

Step 2: Legitimisation of the action's goal in terms of contract theory:

- Hypothetical justification (Rawls)
- Reference to concludent action

Step 3: Selection of the decision-making level/process:

- Subsidiarity principle
- Congruence principle:
  - Equivalence: the user group must coincide with the payer group for a collective good
  - Democratic monitoring: the group of the decision-impacted must coincide with the group of entitled monitorers

Step 4: Economic legitimisation of actions formulated:

- **Choice of instruments:** selection of conceivable instruments for goal implementation
- **Effectiveness (goal-conformity)**:
  - Degree of goal attainment (direction and dosage)
  - Speed of goal attainment
  - Invariance against changes in the macro-economic boundary conditions
- **Necessity (system-conformity)**:
  - Market-conformity:
    - Instrumental subsidiarity: designing measures with minimised impact on individuals' powers of decision-making (centralised/decentralised)
    - Creation of fully functional markets (free price formation, fully functional competition)
    - Minimising intervention into the functioning of existing markets
    - Priority of regulatory before process policy: formulation of a long-term orientation framework, and avoidance of stop-and-go measures
  - Minimisation of detectable unwanted side-effects:
    - Stability-policy goals (economic compatibility)
    - Distribution-policy goals (social compatibility)
- **Economic efficiency**:
  - Static economic efficiency (cost-efficiency):
    - Purpose/avoidance costs
    - Transaction costs
  - Dynamic economic efficiency (innovation efficiency)
- **Institutional controllability**: implementability in the political process and allowance for the possibilities for abuse in the political/administrative apparatus

Source: In broad conformity with GROSSEKETTLER (1991: p. 114).

With regard to voluntary agreements, these criteria can be interpreted and applied by the following questions:

Step 1 and 2: Goal operationalisation and justification

- What environmental goal does the measure examined in a given case study refer to?
- How can the targets be measured?
- Can this goal be justified hypothetically (Rawls theory of justice) or empirically (democratic decisions, concludent action)?
- Was the goal originally pursued watered down as early as during the negotiation process or were there any delays?

Step 3: Selection of the decision-making level/process

- Are decisions taken at an appropriate level with regard to the principles of subsidiarity and congruence?
- Is any environmental responsibility delegated and, if so, how is the delegation to be assessed?

Step 4: Economic legitimisation of actions formulated

- Choice of instruments: what are the relevant instruments agreements have to be compared with?
- Goal conformity: Is it possible to achieve the specific underlying environmental policy goal with the help of the given agreement?
- System conformity: Is the measure in keeping with the system of a social market economy? Is the system strengthened, does it remain unchanged or is it weakened? Are detectable unwanted side-effects minimised?
- Economic efficiency: Are the costs higher or lower than when using other instruments? What incentives are provided to achieve technological progress?
- Institutional controllability: How immune are voluntary agreements to political influences that can dilute their desired effect? How do the associations deal with the problem of free riders?

4 Choice of Case Studies

The assessment scheme to evaluate environmental policy measures is applied to topical examples taken from the fields of climate protection and circular economy. The problem area of the promotion of clean technology is cross-sectional in nature because it is relevant for both fields of environmental policy.

4.1 Climate Protection

The Chlorofluorcarbon (CFC)-phase-out and the reduction in carbon dioxide emissions will serve as case studies from the area of climate protection. The CFC-case study derives its relevance from the fact that the progress made in phasing out CFCs is often cited as a case for the economic and ecological advantageousness of using voluntary agreements. Voluntary agreements, the argument goes, result in the ecological goal being attained more quickly, trigger off an innovation drive in industry and safeguard industry's export capability. The carbon dioxide case study is the obvious choice, not only because of its particularly prominent role in the global warming problem. Accounting for 60% of the damage, anthropogenic carbon dioxide emissions are the largest cause of the greenhouse effect. Moreover, with regard to the carbon dioxide problem the German business community presented an updated voluntary agreement in March 1996. This German negotiated agreement has come to be considered a model for the European Union's climate policy.

4.2 Circular Economy

Particularly in the field of German waste disposal policy, voluntary agreements dominate other measures. The significance of negotiated agreements in the area of waste disposal policy is explicitly stressed by the Federal Government. In order to place greater emphasis on waste avoidance, Germany created the Kreislaufwirtschafts- und Abfallgesetz (Act on a Circular Economy and Waste), which has come into force since October 1996. Producers of, say, electronic goods and batteries can pre-empt industry-specific ordinances through negotiated agreements and can voluntarily comply with the legal obligations. The voluntary agreement on the disposal of old cars made by the automobile industry, which is being used as a case study, also has to be seen against this background. In addition, an analysis of the Duales System Deutschland (DSD, Dual System Germany) will be carried out.

4.3 Clean Technology

A further case study to be examined is the so-called 3-litre car, which serves as an example of clean technologies. Although there is overlap between that case study and others - for clean technology is used in climate protection as well as in strategies of material flow management aiming at establishing a circular economy - this area is to be dealt with separately. It seems likely that in the future clean technologies will be of great importance on the road to a sustainable economy, and the question arises how such a development can be encouraged without repeating the mistakes made in the past (regulatory enactment of best available technologies) and what part voluntary environmental protection measures can play in this.

5 Goal Operationalisation and Legitimation

5.1 Climate Protection

To start off with, one has to examine whether in individual environmental policy areas operational goals have been defined which could serve as a yardstick for the evaluation of voluntary agreements. As for the case study taken from climate protection, there are specific national and international reduction goals for greenhouse gases and ozone-depleting substances to which one can revert without any reservations. The goals are operational and can be justified by theoretical reasoning and concludent actions.

5.2 Circular Economy

It is much harder to answer the question how goals relating to the establishment of a circular economy can be described in concrete terms and in a manner being compatible with a liberal market system. On the whole the disposal crisis in the German waste disposal industry has eased, thus no bottlenecks concerning dump or incineration capacities are foreseen for the future. Regional crises can primarily be put down to compulsory (regional) disposal self-sufficiency. Nationally solid waste production has stabilised or is even falling. Recycling capacities are being built, numerous plants are in the planning stage and there are reserve capacities (cf. BRENCK et al. 1996:2-4).

Although the situation has been defused, a goal has to be formulated for a future-oriented waste disposal policy which is oriented towards controlling solid waste production and towards an adjustment of the latter to existing dump capacities. For the case studies chosen from the sphere of waste disposal policy, goals appropriate to the problem have to be specified. The case studies were chosen to illustrate two different aspects of the waste disposal problem. Whereas packaging waste primarily constitutes a quantitative problem, the problem relating to the disposal of old cars is primarily qualitative in nature.

The disposal of old cars creates a need for environmental policy measures, especially because of the large volume of shredder waste that has accumulated after the disposal of some 2.6 million old cars each year. This waste is a mixture of plastics, glass, textiles, and wood fibre substances, which cannot be recycled and can only be put on a dump. Its volume amounts to about 1.5% of the domestic waste volume in Germany and is therefore fairly small, but owing to its explosive make-up, this waste is classified as hazardous waste.

Packaging waste, on the other hand, primarily poses a quantitative problem that in terms of quality is not explosive in nature. With regard to weight, packaging waste accounts for roughly 30 per cent of domestic waste, twenty times the annual shredder volume from the disposal of old cars.

The voluntary agreement on environmentally-sound recycling of old cars made by the automotive industry aims at improving the recycling of old cars as well as of old car components, and especially at reducing the amount of shredder waste, processing old cars in keeping with the needs of the environment and at going easy on dump capacities and raw materials.

The catalogue of measures included in the voluntary agreement distinguishes between the processing of old cars that were registered before the voluntary agreement came into force ("existing car fleet") and the processing of old cars registered after the coming into force of the voluntary agreement ("future car fleet"). This distinction makes sense, since the producers' scope for reducing resource consumption as well as solid waste production differs. With respect to the future car fleet, additional options present themselves due to changes in design, construction, materials chosen etc.

The most important measures mentioned in the voluntary agreement to attain the goals that were set are:

- Establishment of a nation-wide take-back system of certified businesses for the taking back and recycling of old cars within two years after the regulation takes effect.
- Old cars from the existing fleet of cars are taken back at generally accepted market terms.
- Old cars from the future car fleet that meet certain criteria (for instance, the initial registration must not date back more than twelve years) are taken back free of charge.

5.3 Clean Technology

Finally, regarding the last case study the question has to be answered whether supporting clean technology should be a goal of environmental policy and how such a goal can be formulated in operational terms. Although supporting clean technology is a sensible environmental policy goal provided that the increased application of this technology promises improvements in solving environmental problems, from an economic point of view environmental goals should not be formulated in the form of technological standards, because this restricts the available freedom for adjustment unnecessarily. Instead there should be an orientation towards long-term environmental quality goals that leaves the decision on the technology applied open for the time being. Such an orientation towards long-term environmental goals which ideally should be formulated in such a way that they cover various pollutants also constitutes an important stimulus for the application of clean technology.

Against this background an environmental policy goal like the "development of a 3-litre car", to which representatives of the German automotive industry pledged themselves at a meeting with the premiers of their respective German states, has to be viewed critically, as it largely determines the choice of technology. Similar in nature is the voluntary pledge given by the Verband der Automobilindustrie (VDA, Association of the Automotive Industry) of March 1995 of a 25 per cent reduction, based on the 1990 figure, by the year 2005 in the average fuel consumption of the cars/estate cars it produces and sells in Germany (ZIMMERMEYER 1995:2; VDA 1995). It is planned to upgrade the pledge by the year 2000 to over 30 per cent.

These pledges are part of the efforts made in the past few years to reduce the fleet consumption of motor vehicles (BMV 1995; SRU 1994).[2]

It has to be stressed that goals aiming at supporting clean technology - for example, fuel-efficient vehicles - are only justifiable if they form part of a broader overall plan. The latter should be oriented towards the attainment of certain environmental quality standards (e.g., climate stability). If such an overall plan exists, promoting the development of fuel-efficient motor vehicles, for example, can certainly be one of several "action-oriented goals" to pursue a higher environmental quality goal.[3]

6 Choosing the Decision-Making Level and Process

6.1 International Free-riders Push Negotiated Agreements in Climate Policy

According to the principle of congruence included in the theory of public goods, the club deciding on the provision and funding of a good should ideally be identical to the club benefiting from the good. A stable global climate benefits the entire world population, consequently, a global institution would have to be founded to decide on the provision and funding of the good "climate protection" in an economically optimal manner. However, as long as there is neither an "Environment

2 The European Commission is also backing the goal of reducing fuel consumption. An essential instrument of this strategy is the reduction in the average consumption to 5 litres per 100 kilometres for new vehicles with a petrol engine and 4.5 litres per 100 kilometres for new diesel vehicles. In this context the Commission regards the deadline for implementation that the European Parliament has in mind, the year 2005, as very ambitious (ÖKOLOGISCHE BRIEFE 1996 (5)). On account of the different carbon dioxide content in fuels like petrol, diesel, and natural gas, a standard limit of 120 milligramme of carbon dioxide per kilometre is to be fixed (SZ of 6 March 1996).

3 According to a definition given by the ENQUETE-KOMMISSION „SCHUTZ DES MENSCHEN UND DER UMWELT“ („Protection of Humans and of the Environment“ Commission of Inquiry) (1995:4), environmental action-oriented goals „indicate the steps necessary to achieve the characteristics of the environment or the states described in environmental quality goals.“

Security Council" nor a comparable institution, the road via international agreements that has been used so far, has to be followed further. In that connection one has to keep in mind that any solution below the global level entails serious additional problems of free rider-behaviour.[4] Amongst other things, the implications of this free rider-behaviour have so far been seen in the fact that none of the leading industrialised nations is willing to play a pioneering role in introducing climate taxes (MULLER 1995). This national wait-and-see attitude in the follow-up process to Rio has led to an environmental policy standstill in the field of global climate protection. Precisely in view of this standstill in international climate policies, voluntary agreements seem to be an instrument politicians and business representatives are taking up readily, because it allows a certain degree of national activities without businesses having to accept serious cost disadvantages in international competition. Voluntary agreements in the field of climate protection have become common all over the world.

6.2 Negotiated Agreements in Climate Policy Arise from a "No Regrets" Approach

Originally, the policy of voluntary climate protection measures was pushed by the US administration in particular, which had committed itself at a very early stage to a "no regrets" strategy concerning climate protection (KRAUSE/KOOMEY/ OLIVIER 1994; RENNINGS 1994:83-86). Unsure about possible climatic damage, the administration concluded that to be on the safe side, it would only order measures to reduce greenhouse gases that even from a managerial point of view were at least cost covering. To date the "no regrets" strategy has contrasted with the "insurance buying" policy of some Western European states like Germany. According to the latter policy, the climate protection standards that are fixed are arrived at on the basis of a global warming deemed just about acceptable and on the basis of reduction goals derived from the latter. In order to reduce the risk of climatic damage, the costs involved in achieving the standards are accepted as a quasi-insurance premium. While most industrial countries have so far merely formulated the stabilisation of climate-relevant emissions as a climate protection goal, the Federal Government is pursuing a relatively ambitious goal with an absolute reduction of 25 % by the year 2005 compared with the 1990 level.

Now the question is whether this national reduction goal can be attained with "no regrets" measures - and one can hardly expect more on the basis of voluntary agreements. An optimistic answer to this question is fed by estimates in the latest report submitted by the Intergovernmental Panel on Climate Change (IPCC 1995b:20), according to which "no regrets" measures have a reduction potential of

4 which, admittedly, a global club does not necessarily abolish either, if, say, decisions relating to the funding of climate protection programmes and to the allocation of these funds to the beneficiaries are taken independently of one another.

10 to 30 per cent in the next 20 to 30 years. The enormous potential of "no regrets" measures, which microeconomic studies have also identified in the OECD-countries including Germany, is attributed to the fact that profitable investments in improved energy efficiency have so far never materialised due to substantive market failure and to shortcomings in the coordination between institutions. Examples of these impediments are a lack of information and of economic incentives to conserve energy in the case of public utilities. The possibilities to overcome the impediments that are mentioned include deregulation measures, provision of advice and training and upgraded financial support programmes. In addition, cooperative solutions, for instance in the form of voluntary environmental protection measures, are seen as a way of overcoming institutional impediments. Seen against this background, voluntary agreements on a reduction in carbon dioxide emissions definitely have a potential.

In contrast, initial experience gained in the US with the Climate Change Action Plan (CCAP) of 1993, which essentially contains voluntary measures on the part of public utilities and companies, is not promising. It seems as if current measures are not sufficient to attain the US stabilisation goal by the year 2000. Only individual states - such as New York, which is able to replace decommissioned oil- and coal-fired power plants with gas-fired ones in a cost-effective way - are expected to achieve the stabilisation goal. Yet, wherever costly adjustments would be necessary that go beyond "no regrets" measures, the voluntary agreements do not appear to be working (SANGHI 1995).

Once the goal in climate policy is in danger of being missed, the real Achilles heel of voluntary agreements is revealed. For now it becomes evident whether the tough stance in environmental policy that was threatened in this case will really be taken up. The Federal Government knows full well that the success of negotiated agreements greatly depends on how credible the threat looming in the background is. That is why it explicitly emphasised (BMU 1995:3) it would "remain in control" and not hesitate to "use regulatory and fiscal instruments once it emerges that the pledge given by the business community amounts to little more than 'business as usual' or fails to be complied with."

6.3 Takeback of Cars: Watering Down and Postponement of Goals

The voluntary agreement on the disposal of old cars resulted from years of negotiations between the German Ministry for the Environment and associations of the automotive industry. The desired goal, reducing the production of solid waste from the disposal of old cars, is meant to help to provide a public good, namely scarce dump capacities. The beneficiaries of this public good are the business sector and the consumers. The business sector along with the buyers of new cars and the last owners of old cars also has to meet the costs arising from the creation of the network of take-back businesses as well as from the development of new types of construction, new materials and new forms of recycling. Only the business sector, i.e., the automotive industry, was involved in the negotiating process.

Since there was no parliamentary debate either, an essential group of beneficiaries and payers of the public good, the car buyers, was excluded even from ways of exerting indirect influence on the decision-making process. This group does not have any supervisory powers either. Thus, the principle of congruence has only partly been met.

The corporatist approach in environmental policy is being criticised, amongst other things, because of the danger of environmental policy goals being watered down. For example, the willingness to make more far-reaching concessions to the recycling of old cars is greater on the part of individual car producers than on the part of the industry's associations.

The negotiating process concerning the disposal of old cars reveals that as early as 1992 there was a complete draft for an ordinance including all essential elements of the voluntary agreement made in 1996. However, due to the cooperative principle, on which the waste disposal law is based and to which the Federal Government attaches great importance in waste disposal policy, a negotiating process that lasted several years followed between the government and the business sector. Therefore, there was no noticeable acceleration of the decision-making process, on the contrary, a delay occurred.

One comes to realise that once the government commits itself to a corporatist style of environmental policy, the other negotiating partner is granted a potential to delay and water down goals that should not be underestimated. The example shows that one has to urge politicians to keep their options open. A binding commitment giving priority to cooperative solutions can deprive the instrument of voluntary agreements of the basis for effective environmental policy improvements. For example, there is some noticeable watering down of goals in the negotiated agreement on the disposal of old cars with regard to the quotas fixed and the period scheduled during which cars are taken back free of charge.

6.4 Clean Technology: Negotiated Agreements as an Accompanying Instrument

At what level should the decisions on the application and promotion of clean technology be taken? According to the principle of subsidiarity, one has to check whether the decision on the application of the technology can also be taken at the lowest level, i.e., in the markets. In principle, there are no objections to this when it is a case of applying clean technology. Having said this, a decisive prerequisite for a coordination via the price mechanism is that:

- first, there is an ecological framework that does not in general discriminate against the application of environmental technology.[5]

---

5 Empirical studies on inducing environmental innovations (GREEN et al. 1994:1051; HEMMELSKAMP et al. 1995:19) show that vast numbers of environmental innovations are caused through the existence and anticipation of environmentally-relevant re-

- second, this ecological framework is designed in such a way that it does not especially impede the application of end-of-pipe technologies to the detriment of clean technology. Disadvantages clean technologies suffer from compared with end-of-pipe solutions that can be put down to existing regulations, e.g., to an orientation towards individual environmental media and pollutants, should be reduced (HOHMEYER/KOSCHEL 1995).

Under the aforementioned conditions, markets offer a suitable coordination mechanism for decisions on technology application. On this basis voluntary agreements can fulfill an accompanying function and be used wherever rules remain incomplete and have to be supplemented.

7 Evaluating the Instrument

7.1 Goal Conformity

7.1.1 CFC-Negotiated Agreement: In Certain Applications in Conformity with the Goals

So far voluntary agreements to reduce ozone-depleting substances have definitely been successful in ecological terms. The goals were exceeded, although there were some critics who argued the phase-out could have been achieved even faster (KOHLHAAS/PRAETORIUS 1994:89). As a rule the monitoring of the negotiated agreements is done by someone who is neutral. Negotiated agreements governing the reduction of CFCs enjoyed special advantages (compared with the case study on the reduction of carbon dioxide), since substitutes had already been discovered which when applied did not lead to costs soaring and since additional pressure was exerted by the demand side (slump in sales of sprays containing CFCs). Thus, in an entrepreneur's cost-benefit calculations, opting for a reduction in CFCs involved little risk. On the other hand, continuing to manufacture products containing CFCs would have been much riskier from a managerial point of view. This is also borne out by the fact that although CFCs were speedily replaced in products, CFC-substitution in production processes was slow to materialise, because the switch was costlier and there was less pressure by the demand side.

---

gulations. Apart from that, costs and revenue aspects play a role, with product innovations primarily aiming at increasing sales and process innovations, on the other hand, at cutting costs.

7.1.2 No Impetus for Absolute CO_2-Emission Reductions

The business community's voluntary agreements on a reduction in carbon dioxide of 1991 and 1995 came in for a lot of criticism, because they did not contain any noticeable initiatives that clearly went beyond "business as usual." Criticism was levelled in particular at the fact that it was hard to check whether the goal was being achieved. The following minimum requirements for information (listed by the German Federal Environmental Protection Agency) that should be included in a negotiated agreement on carbon dioxide reduction (ÖKOLOGISCHE BRIEFE 1996 (2)) were not met:

- reference and target year as well as a reduction path in the form of a timetable with detailed information on partial goals,
- exact fixing of emission or energy conservation goals,
- absolute energy consumption listed according to fuel,
- development of primary consumption,
- development of specific consumption per technical unit,
- reductions achieved,
- in-depth comment on and analysis of the figures provided (e.g., information on whether reductions are attributable to additional climate protection activities, an economic slowdown or to modernisation investments that would have been made anyhow), as well as
- detailed list of the "special efforts" promised.

The updated version of the negotiated agreement of March 1996 meets these minimum requirements. The plan on carbon dioxide monitoring presented by the BDI (Federation of German Industries) provides for associations to record their reductions in a total of eight tables (BDI 1996). Total fossil fuel input, net power supplied externally, energy input as well as specific carbon dioxide emissions calculated from this and specific energy input are to be stated for the base year, the previous year and the year under review. The demanded comment on and analysis of the figures and the list of special efforts are also included. The reports are checked by a neutral expert.

What is particularly striking is that no fewer than 12 out of 19 associations pledge to reduce absolute carbon dioxide emissions. According to the Federal Government (BUNDESREGIERUNG 1996a:3), the pledges correspond to a 20 per cent reduction in emissions in these sectors. However, when one takes a closer look it is precisely the achievement of absolute reduction goals that turns out to be the agreement's real Achilles heel, something the statement made by the Vereinigung Deutscher Elektrizitätswerke (VDEW, Association of German Electric Power Stations) (VDEW 1996) illustrates.

According to the VDEW, the absolute reduction potential in the electricity industry until the year 2015 amounts to 25 per cent compared with the base year of 1987. However, compared with the base year of 1990, which the Federal Government is now taking as a basis, estimates merely put the potential at 12 per cent. The figures are even more off target if the target year of the Federal Government is taken as a basis: The potential for the period up until the year 2005 is, compared with the reference year of 1990, down to 8 to 10 per cent. Compliance with these pledges is even conditional on ambitious prerequisites such as:

- a consensus in society on the exploitation of nuclear energy on the basis of existing law,
- an increase in the service life and capacity of existing nuclear power stations,
- the Mülheim-Kärlich nuclear power station going into operation,
- the undisturbed operation of existing nuclear power stations and
- unrestricted choice of fuels for the power stations on the part of the companies.

On the whole with regard to goal conformity, one can stress that shortcomings of the instrument of voluntary agreements have been identified and in part abolished. Still, when a comparison with the Federal Government's climate protection goals is made, one has to question the goal conformity of this instrument, in particular if all other measures are discontinued due to the agreement that was made. To date the development of absolute carbon dioxide emissions is by no means following a path that makes this goal seem feasible (KOHLHAAS/PRAETORIUS 1995:278). Nor does the updated declaration issued by the German business sector indicate a new trend concerning this path. Additional need for action can be deduced, in particular for the German states that made up the FRG prior to unification. If the VDEW puts the absolute carbon dioxide reduction potential of German electric power stations, which are responsible for about a third of all German carbon dioxide emissions, at a mere 8 to 10 per cent compared with the target year of 2005 (VDEW 1996:5), the question arises as to who is to contribute the above-average reductions needed to offset the expected increases in carbon dioxide emissions in areas like transport.

7.1.3 Takeback of Cars: Business as Usual

The assessment of goal conformity in the case of voluntary agreements on the disposal of old cars is oriented towards the question whether the instruments chosen are suited to achieve the top goal, the reduction in solid waste production and, in particular, in shredder waste. First of all, one has to examine what the substantive contribution is in the automobile industry's agreement compared with the way in which old cars have been disposed of so far. If some substance can be established, the next question is whether the measures planned constitute an effective contribution towards the solution to the problem.

Seen against the background of the conditions in the existing car-disposing industry, the measures have to be considered out of keeping with the goals, since they do not represent any change vis-à-vis the status quo. There already is a nation-wide take-back network made up of independent businesses. Besides, even today the last owner is able to have his or her old car disposed of at generally accepted market terms. These terms can mean that the last owner still manages to get a positive market price for the old car.

The pledge to take back old cars from the future car fleet free of charge, provided they are no older than 12 years, does not constitute a substantive measure either. Firstly, most of the cars needing to be disposed of, do not come under this category. According to the ADAC (the main association of motorists in Germany), the mean age of a car whose registration has been cancelled is 13.2 years. If one takes data provided by the manufacturers as a basis, no fewer than 8 out of 23 producers say that the average life of the vehicles exceeds 15 years. Secondly, the pledge of taking cars back free of charge is obsolete, because according to the Arbeitsgemeinschaft Deutscher Autorecyclingbetriebe GmbH (ada, Association of German Car Recycling Businesses Ltd.) all old cars fitting the definition still fetch a positive market price (AUTORECYCLING, issue 1/96).

So far the bottom line of the study is that essential components of the take-back-agreement are "business as usual" measures. More specifically, they do not constitute new measures, as they will not lead to a substantive change in the way things have been regulated so far. It is alarming that the negotiated agreement encourages manufacturers and importers to assume control of the car-disposing industry. Here there is a danger of the paradox that due to the voluntary agreement not only will polluters - i.e., the producers with whom product responsibility lies - not be charged costs in line with the polluter pays principle, but the polluters are even provided with a framework for the opening-up of a novel and lucrative business segment.

On the whole the voluntary agreement on the recycling of old cars made by the German automotive industry and decisive components supplying industries, the car component distribution sector and the car disposal industry, can be characterised as a "business as usual" measure. The degree to which the environmental policy goal is met is small.

7.1.4 Clean Technology: Policy Mix Including Economic Instruments Makes Sense

Efficient measures to promote clean technology should take into account a company's investment cycles and adjustment deadlines. In principle, voluntary agreements can support this process, but by themselves they are often unsatisfactory with regard to desired targets.

Agreements presupposing a commitment to refrain from introducing accompanying economic instruments in return, have to be considered inappropriate to the goal. A case in point seems to be the meeting of three German premiers with representatives from the automotive industry of March 1995 at which moderation was pledged concerning additional government measures such as an increase in mineral oil tax. However, it is precisely the pledge to refrain from adopting these accompanying measures that undermines the prospects of also marketing an energy-efficient car successfully.[6]

With respect to the standards of vehicle fleets, even at a company level it is not sure whether the goals will be achieved, since compliance with the standards depends on the demand for individual models, which is something the company concerned finds hard to anticipate. Moreover, so far all fuel-conserving models have been vehicles with diesel engines and even modern diesel engines fitted with soot filters emit three times as many carcinogenic substances as petrol engines. Consequently, the German Council of Environmental Advisers warns of a one-sided policy on the reduction of greenhouse gases and demands a greater reduction in carcinogenic substances instead.

7.2 System Conformity

7.2.1 Non Binding Agreements: No Sanctions Against Unfair Players

Voluntary agreements can constitute solutions in keeping with the system, if they establish binding standards for the parties involved and if free-rider behaviour can be prevented. However, since associations normally do not possess any effective mechanisms to punish their members, and hence no binding rules of the game can be agreed, the approach to problem-solving shifts to the moves of the game, that is to say to the wrong level in the system. As KREUZBERG (1993:308) writes, voluntary agreements hinge on "a disproportionately high participation of 'honest companies' in environmental protection activities and therefore lead to a redistribution of burdens to the detriment of 'honest' players."

7.2.2 CFC-Phase Out: "Soft" Instruments Appropriate for Preventive Strategies

Prohibitions are definitely measures to avert concrete and acute damage to the environment that are in keeping with the system. In such situations the liberal principle of maximising freedom cannot be used as a yardstick for policy design on constitutional grounds (BROCKMANN et al. 1995:70). That means that instruments with low intervention intensity and a high degree of freedom, such as negotiated agreements, can only take precedence as long as it is safeguarded that acute environmental damage will be averted. If this is not the case, "tough" in-

---

6 The European Commission, which wants to see negotiated agreements on a reduction in the consumption of the motor vehicle fleet, also seems to have realized this. The Commission's estimation is that such an agreement would entail the risk of the newly-developed vehicle models failing in the market. That is why the Commission believes such a measure has to be combined with fiscal incentives for the consumers.

struments will be applied. This is also evident in the CFC-example: The realistic alternative to a voluntary reduction in the production of CFCs and ozone-depleting substitutes would have been a quick ban on these substances. In order to bring about a complete phase-out of CFC-application in products, finally the complementary CFC-halon-prohibition ordinance was issued in Germany. The aerosol industry did not agree to a voluntary phase-out desired by the Ministry for the Environment (KOHLHAAS/PRAETORIUS 1994:89).[7]

7.2.3 Climate Policy: Long-Term Danger of Intervention Spiral

A characteristic of voluntary environmental protection measures in the field of climate protection frequently cited as being particularly in keeping with the system, is that these measures, compared with energy and carbon taxes, cause fewer side-effects on goals relating to stability and distribution policies. The "double dividend" of an ecological tax reform is often overrated, the argument goes, i.e., the simultaneous achievement of ecological (reduction in greenhouse gases) and economic goals (e.g., job creation) due to the revenue-neutral compensation of ecological taxes (KOSCHEL/WEINREICH: 1995).

However, if emitters are charged the adjustment costs caused by a climatically-sound restructuring of capital assets, this has to be regarded as being in keeping with the market, irrespective of the compensation question. A "policy of little steps", which the German Council of Environmental Advisers advocates, would primarily send the necessary price signals to consumers for a more economical use of energy. This would be the crucial advantage over voluntary agreements. Secondly, a policy of little steps would minimise the side-effects on goals in the fields of stability and distribution policies. For voluntary agreements are primarily not in keeping with the market and the system, because such price signals usually are not included in the instruments they are provided with. A climate policy based on voluntary agreements is thus always associated with the danger that even a high "no regrets" potential that could be siphoned off is overcompensated by increasing energy consumption - not only in the sphere of transport.

In the long run failure to attain the goals leads to new government intervention in the form of limits, technical instruction, and rules of conduct. As long as there are no permanent and effective incentives to buy and use energy-efficient technologies or to practise energy-conserving behaviour, failure to attain goals has to be abolished with the help of such new regulations. Regulations on driving behaviour, the purchase of energy-efficient motor vehicles, or on using state-of-the-art technologies for heat insulation purposes are examples of such a catalogue of

[7] The way to protect the ozone layer that is more in keeping with the system than anything else would be to apply „tough" instruments presenting economic incentives, as was done in the US. There, politicians successfully opted for the application of a mix of instruments comprising levies and permits (COOK 1996:4).

measures. Unless in the long run plans for voluntary climate protection measures are accompanied by price instruments which are in keeping with the market, there is a risk of setting a spiral of intervention in motion that should have been stopped through this instrument in the first place.

7.2.4 Circular Economy: Effects of Negotiated Agreement on Competition

Cooperation between individual trade associations and government can lead to informal marketing agreements, to markets being shielded and to barriers to market entry for third parties being erected. That these fears cannot be dismissed is shown by the case studies taken from the problem area circular economy. From the point of view of competition policy, buzz words in environmental policy like "product responsibility from cradle to grave" involve the danger of encouraging trends towards concentration and vertical integration.

As for the impacts negotiated agreements on the disposal of old cars have on the structure of the disposal industry, it is the effects on the number of receiving points and on the number of businesses doing the recycling that are decisive. In general, trends towards concentration in the industry are expected on account of more stringent disposal standards.

In principle all specialised businesses meeting the criteria for recognition or certification enjoy unrestricted access to the system. However, there are dangers involved in the formulation of the criteria licensed businesses have to meet. Here unnecessary barriers to access could be erected favouring financially-strong companies, for example, because large investments prove necessary. The Association of German Car Recycling Businesses (ada) is talking of an existential threat to the small and mid-size structure of its industry (AUTORECYCLING, issue 1/96). Besides, if vehicle manufacturers largely control the disposal industry, one also has to fear that the option of recycling becomes less attractive to disposers, for manufacturers will probably prefer to sell new parts.

Competition in the recycling industry is indeed in danger of being harmed to the detriment of small to medium-sized businesses owing to cooperation between large manufacturers and suppliers. This can happen at the expense of the hitherto dominating small to medium-sized businesses; a pattern already familiar from the development of the Dual System Germany.

Irrespective of the question of licensing criteria for businesses, a network could, once it has been set up, be subject to the abuse of power. In analogy to the relationship between large motor vehicle manufacturers and suppliers, ominous structures can develop between manufacturers and recyclers (BENZLER/LÖBBE 1995:157). The take-back network can also lead to a monopolisation of the spare parts market and of shredder activities. For example, the Association of the Automotive Industry (VDA) had originally planned to couple the free take-back with brand commitment concerning spare parts, but this point was dropped in the course of the negotiations (HOMBURGER 1996).

7.2.5 Enhancing Cleaner Production: Fair Rules Instead of Voluntariness

It is obvious that end-of-pipe technologies will not automatically start to be superseded by cleaner production (SRU 1994:Tz 262). According to the German Council of Environmental Advisors, the prospects for clean technology greatly depend on progress being made in environmental policy with respect to the correction of the structure of relative prices, which still encourage an overexploitation of the natural life-sustaining systems.

On the other hand, voluntary agreements, by virtue of their approach, are not up to the job if instead of accompanying a correction of relative prices, they are used as a substitute thereof and made without giving due consideration to the market.[8] Even if, for example, fuel consumption in motor vehicles is successfully cut, unless additional price incentives are provided to use energy economically and to buy these vehicles, these innovations will turn into part of the decoration at motor shows, without large numbers of them being sold to the public.

As long as the wrong rules apply, it seems unlikely that a player is voluntarily going to behave in the ecological manner desired. However, whenever a desired form of behaviour fails to materialise voluntarily, government can respond in two ways. Either it puts up with failures in environmental policy and thereby shows that it leaves the design of the political framework to influential pressure groups, or sooner or later it helps things along with coercive measures. In this way voluntary measures, despite their own low intervention intensity, can in the long term entail a spiral of intervention with unacceptably high intervention intensity.

7.2.6 Green Dot: Market-Oriented Approach with Some Market Imperfections

As for cooperative solutions, it is in principle also possible to develop environmental policy instruments that take free-market principles into account. Due to the problems of charging social costs relating to the dumping of waste with the help of dump charges, the ordinance governing packaging, for instance, and the Dual System it spawned represent two variants of instruments of control that make sense in economic terms and are in keeping with the market. Although juridically

---

8 Economically speaking, for companies the application of clean technology is in the short term linked with increasing costs and financial risks. Disadvantages of integrated technologies from the point of view of an entrepreneurs' investment calculations, are higher access and information costs, adjustment and changeover costs, funding bottlenecks, long decision-making horizons, and greater economic risk. Here negotiated agreements may be able to provide an additional impetus for adjustment and changeover, but an adjustment to ecological requirements in a way that minimises the costs for the overall economy is not to be expected with voluntary agreements in view of the restriction to certain technologies and the expected free rider-behaviour on the part of the companies.

speaking take-back duties are commands, these commands are less questionable from an ordoliberal point of view, since they provide the standard addressee with scope for individual adjustment (SPIES 1994:309).

In the field of packaging, too, reservations from the point of view of competition policy have to be expressed. German disposal markets are highly cartelised and oligopolised. The Dual System is a demand cartel legitimated by the ordinance on packaging. There is always just one disposer per territorial division who is given the tasks of collecting and sorting.[9]

However, the Dual System also shows that non-uniform action in different countries can lead to distorted competition in foreign trade. Negative impacts of the Dual System on the European disposal industry become apparent in the case of waste paper, for example. German recyclers get their waste paper free of charge, whereas other European recyclers have to bear the collecting and sorting costs themselves. This cost advantage not only de facto represents an import impediment, but also leads to increasing German exports of waste paper. For example, in France this has the ecologically-ridiculous consequence that growing amounts of waste paper have to be incinerated or put on dumps for lack of demand (WUPPERMAN 1993:453).

7.3 Economic Efficiency

7.3.1 CO_2-Agreement is Not Cost-Efficient

Solutions to the problem of reducing emissions are economically cost-effective if each emitter fixes his contribution to the reduction in such a manner that the overall economic avoidance costs are minimised. Individual emitters neither know the avoidance costs nor are they interested in including them in their decision-making, hence, these signals have to be sent through environmental policy. Levies and permits in particular are classic textbook instruments meeting the economic efficiency criteria, because they send out these signals in the form of a correction of relative prices. On the other hand, a reduction in emissions with minimal costs for the overall economy is unlikely to be achieved via a voluntary agreement on carbon dioxide or eco-efficient cars, for instance, due to the restriction to certain sectors and the free-rider behaviour to be expected on the part of the members of the associations.

---

[9] The Federal Ministry for the Environment justifies the formation of cartels saying that new environmental technologies will be developed. In a diametrically opposed view the Federal Cartel Authority in Berlin regards functioning competition as the essential driving force for the development of new environmental technologies (WELT of April 15th 1996). The amendment of the ordinance on packaging, which is to take effect soon, is to promote competition by stipulating that disposal activities (collecting, sorting, recycling) linked to dual systems have to be put out to tender (UMWELT, 4/1996).

Moreover, voluntary agreements are not convincing when it comes to their dynamic efficiency, i.e., their effect on technological progress. Here it is possible to compare voluntary agreements to regulations. Once the goal has been achieved, there are no further incentives to reduce emissions. Compared with regulations, there does not even inevitably have to be an incentive to maintain the standard once it has been attained (KREUZBERG 1993:309).

7.3.2 Cost-Efficient CFC Phase-Out

When assessing the economic efficiency of German agreements on reducing ozone-damaging substances, consideration has to be given to three unusual features in particular that accelerated the substitution on managerial grounds alone:

- the ban on CFCs as propellants for most aerosol products in the US as early as 1978,
- the declining demand for products containing CFCs (e.g., aerosols), and
- the availability of low-cost substitutes.

The economic cost efficiency of negotiated agreements on CFCs has to be regarded as positive. Each agreement covered the main polluters, which were enabled to implement the reduction at the lowest possible cost.

Concerning dynamic efficiency, here, too, one has to repeat: Once the goal has been achieved, there are no further incentives for progress in terms of environmental technology. However, with respect to the development of environmentally-friendly CFC substitutes, the bottom line is that owing to the aforementioned unusual features there had already been sufficient incentives for technological progress.[10]

7.3.3 Fundamental Distortions of Waste Markets

Although the distribution of scarce environmental resources takes place in an allocation-efficient fashion via a change in relative prices, waste disposal in Germany is confronted with fundamental market distortions. The capacity problem regarding dumps would theoretically best be solved by charging dump users in

---

10 What has turned out to be a particularly efficient instrument in the US to protect the ozone layer is the simultaneous application of permits and a tax for the protection of the ozone layer. The American environmental agency EPA estimates that the administrative costs amounted to only 10 per cent of the administrative work a regulation would have entailed. Furthermore, it was possible to quickly adjust the licenses issued to the modifications of the Montreal Protocol. Whereas in 1988 the cost of halving CFC consumption was put at $ 3.50 per kilogramm, only two years later it was possible to lower these estimates to $ 2.20 per kilogramm (COOK 1996:4f.).

line with the polluter pays principle all internal and social costs that arise, that is to say, by increasing dump charges. Among the costs not included fully today, the costs relating to dump-scarcity, dump operation, and dump emissions feature prominently. So far costs relating to the environment have been externalised as much as possible. However, on account of an inflexible scale of charges, dump operators are not able to charge users all their operational costs (BRENCK et al. 1996:1). Inefficient forms of organisation in turn lead to price increases; on the whole serious distortions in the cost structures ensue.[11]

In view of the increase pledged in recycling quotas, the establishment of the disposal system for old cars involves considerable costs. As the establishment and maintenance of the disposal system represent sunk costs, there is a growing incentive for businesses to partially recycle the old cars that were taken back, or to recycle the raw materials they contain. Yet, it still depends on the relative price for newly purchased raw materials or for newly produced car parts whether there will in fact be a drop in the volume of waste to be dumped or in net resource input.

The fact that cars are taken back free of charge does justice to the polluter pays principle, since the producer bearing product responsibility is charged the disposal costs. Thus, the solution is dynamically efficient, for the producer is provided with incentives to look for technological innovations in competition with other producers (SACKSOFSKY 1996:103). However, the question arises whether the disposal costs are high enough in relation to other production costs to initiate recycling-oriented research and development.

7.3.4 Green Dot Economically Efficient

In the packaging sector, founded on the ordinance on packaging, an economically-efficient system was created on a voluntary basis by introducing the Green Dot system. A graduated system of royalties paid by packing material producers and distributors to the Dual System Germany Ltd. shows a tendency of charging costs in accordance with the polluter pays principle and thus results in the marginal avoidance costs being compensated for.

However, reservations result from the changed conditions in the disposal and recycling of packaging following the establishment of the Dual System. Strong concentration effects in the disposal and recycling markets make monopoly and oligopoly profits possible, and sometimes transaction costs have risen considerably compared with the status quo ante. The financial crisis experienced by the

[11] When shredder residues were classified as hazardous waste in 1990, this was a measure that aimed at containing shredder residues from the disposal of old cars and that had a great impact on prices. Dumping costs rose from around DM 70 per ton to between DM 500 and DM 1,000 per ton (1992) (WEILAND 1995:58). However, it is still too early to comment on the measure's directive effect. According to theAssociation of German Car Recycling Business (ada) the share of waste from the disposal of old cars - in particular plastics - is still increasing.

DSD in 1993 halfway through the year illustrates the level of transaction costs associated with this system. On the expenditure side, collecting and sorting costs had soared on account of higher-than-expected volumes. As for the disposal of the remnants, the problem that emerged was that disposal businesses owned by private persons or local authorities are able to charge monopoly prices. With respect to the revenue side, the Green Dot was frequently used without making the corresponding royalty payments to the DSD. By September 1993 a deficit amounting to DM 710 million had accumulated (WUPPERMAN 1993:450).

7.4 Institutional Controllability

7.4.1 Enforceability Depends on Induced Costs and External Pressure

In principle the enforceability of voluntary agreements in associations depends on three factors:

- the costs of the agreement made,
- the costs of the government ordinance looming in the case of non-performance of the agreement, and
- the effectiveness of possible punishment meted out by the associations themselves (e.g., expulsion from the association).

Whereas "no regrets" measures are unlikely to meet with much opposition in an association, pledges going beyond them involve the danger of an internal allocation struggle, which can let a withdrawal from the commitment seem worthwhile. In such cases certain passages in the declaration providing scope for interpretation, or preconditions for the agreement that were never met can provide a welcome opportunity to terminate cooperation.

Problems regarding enforceability are not expected to emerge in the case of the automotive industry's fairly undemanding voluntary agreement on reducing fuel consumption. According to the German Council of Environmental Advisers latest annual report (1996:Tz. 166, translated by the authors) "it is widely believed that the goal of cutting fuel consumption by 25 per cent over 15 years, based on a relatively high average consumption, is in line with the expected technological development already. The pledge to launch car models by the year 2000 with diesel engines consuming three to four litres per 100 km cannot claim to be a special feat of engineering, either."

On the other hand, problems relating to the enforcement of voluntary agreements on carbon dioxide reduction in associations have been revealed, e.g., in talks conducted by the Federal Environmental Protection Agency (UBA) with representatives from the associations in connection with the climate protection initiative launched by the German business community. For example, UBA's annual report reads (1993:167, translated by the authors): "The talks revealed fundamental difficulties whenever negotiated agreements made by associations are to contain binding requirements that are usually laid down in a statutory basis. Many individual businesses refuse to recognise declarations made by their associations as binding. It is virtually impossible to impose sanctions following non-compliance of the agreements". Thus, the instrument of negotiated agreements reaches limits which result from its voluntary nature.

7.4.2 Redistribution at the Expense of Third Parties

Since the instrument of voluntary agreements per se does not send out any price signals that lead to automatic adjustments on the part of the players, decisions have to be made on a case-by-case basis on how the overall reduction targeted is to be divided up between individual groups. This, some people say, involves the danger that groups that are inferior in terms of the way they are organised, like households, ultimately have to bear the largest adjustment burden. This has to be qualified by saying that this argument can also be used against economic instruments. In real life the political representation of interests also plays an important part with economic instruments, for instance when exceptional areas are stipulated that will be exempted from energy taxes. Even Denmark, a country usually regarded as exemplary when it comes to introducing ecological taxes, has special arrangements for particularly energy-intensive businesses, and the burden of levies is mainly borne by private households (MEZ 1995:109 and 126). As the example shows, the uniform price signals economic instruments want to bring about are frequently watered down in the political process.

7.4.3 Relinquishment of Political Scope

The updated declaration presented by the German business community on the prevention of damage to the climate once again clearly indicates the limits of voluntary agreements. If the environmental goal obviously runs counter to the individual economic interests of the associations, people - like some representatives of the VDEW recently - suddenly talk of the "fetish year 2005" in connection with the Federal Government's timetable. Although, for example, the VDEW's agreement already is a scaled-down version (providing for a reduction of merely 8 to 10 per cent by the year 2005), undesired political measures (e.g., in nuclear energy policy) may prompt termination of the agreement. In any case, a government has to expect being asked whether it is possible for a pledge with little substance to be worth so much that the government in return, for decades to come, puts up with being deprived of a great number of potential courses of action relating to climate and energy policies.

8 General Conclusions and Future Perspectives

8.1 General Conclusions

The examples analysed here have offered weaknesses of voluntary agreements with regard to goal-conformity, system-conformity, cost-efficiency, and institutional controllability. However, it can be argued that hardly any instrument will meet all these criteria. If it does theoretically, it may be watered down in the political process (or perhaps will not be implemented at all). Thus, it should be mentioned that there may be reasonable applications for voluntary agreements, but the instrument should be used very carefully. Against this background, we want to draw some general lessons from the German experience and give some recommendations for the international discussion of using voluntary agreements within the European Union.

Opportunities for the application of voluntary agreements that are welcome from an ordoliberal point of view especially exist if they are used within a mix of policy-instruments. If economic incentives are introduced on a voluntary basis (example: Green Dot) or if agreements are used to accompany economic instruments (example: combination of carbon taxes and voluntary agreements), agreements would really deserve the attribute of being market-based. If economic instruments cannot be employed, for instance when substances have been completely banned, negotiated agreements as "soft" instruments of environmental policy can certainly serve to accompany "tough" regulatory measures or - provided it is not a matter of warding off acute dangers - to replace them (example: CFC-phase out). Except for these areas, however, from an ordoliberal point other instruments are preferable.

By combining negotiated agreements and economic instruments, it is possible to avoid a lack of incentives, which is one of the worst flaws of most voluntary agreements in terms of design. To date a lack of incentives has led to voluntary agreements either being undemanding with regard to the contents and being phrased accordingly, or, in the case of more ambitious pledges, to them being associated with a great amount of time and effort needed for enforcement and monitoring. It is also due to a lack of incentives that as a rule, only specific and no absolute reductions are achieved. Without any signals in favour of a way of using environmental resources that is generally less harmful, even with demanding specific reduction goals, there is the risk that an overall increasing consumption of environmental resources will overcompensate for these goals.

The increasing significance of voluntary agreements in current environmental policies may have to do with them being labelled "market based instruments" and the fact that their increased use is associated with hopes of thus strengthening the market system. However, a more in-depth analysis reveals that although in principle it is possible to design voluntary agreements in a manner that is in keeping with a market economy, in most cases people avoid doing it. As the examples that were studied showed, such a free market-oriented design does not come about voluntarily and spontaneously, rather it requires standards and a framework set by government. However, once such a framework is set, agreements may be used as a tool for implementation within an environmental policy-mix.

For this reason one has to strongly advise the government against making a commitment to the effect that in return for voluntary declarations on environmental protection, the government will not make use of any other instruments. This imposes disproportionately severe restrictions on the politicians' latitude in how they act in the future and thus on their capacity for problem-solving. Such a policy tends to neglect other solutions that may be more appropriate to the problem, such as the application of a mixture of environmental policy instruments.

If a decision to give preference to voluntary solutions in general is made or if a decision in favour of such solutions is taken at an early stage, this too is counter-productive, because the substance of negotiated solutions - the governmental "potential for threats" - is weakened and delays in the form of a stamina contest are provoked. The examples that were examined confirmed that without considerable governmental pressure, voluntary agreements do not yield any pledges that go beyond "business as usual" or "no regrets" measures. In order to make it absolutely clear that the government really "is in control", the scope for design in environmental policy has to be kept unrestricted and flexible. On the other hand, giving priority to voluntary solutions on principle imposes disproportionately severe restrictions on government when it comes to quickly reverting to "tough" environmental policy instruments following unsatisfactory negotiation results or delayed implementation.

8.2 European Perspectives

Although our case studies represent only German experiences, some lessons can be drawn for the use of voluntary agreements on a European level. As it seems, the European Commission is aware of several potential shortcomings of agreements being stated here.

In November 1996 the European Commission issued a communication examining the use of environmental agreements as an instrument of EU environment policy (Com (96) 561). The (non-binding) paper deals with the instrument in general, presents guidelines on "environmental agreements" and provides a survey over the use of voluntary (environmental) agreements in the Member States.

The guidelines given by the Commission point out seven topics that should be taken into account when public authorities design environmental agreements on national or local level:

- consultation,
- contractual form,
- quantified objectives,
- monitoring of results,

- public information and transparency,
- independent verification of results, and
- additional guarantees.

Before an environmental agreement is concluded, the paper says, all interested parties (companies, business associations, environment groups, and public authorities) should have the opportunity to comment on the draft. Their opinions should be taken into account.

The legal status of an agreement plays an important role for the success of an agreement. According to the Commission, binding agreements provide in general better safeguards in terms of achieving environmental objectives. Contracts (binding on both parties) offer a well-defined framework that may include sanctions for non-compliance and is enforceable through the courts.

The weakness and bad reputation of certain past agreements partly derive from the lack of quantified objectives, leaving room for the perception that agreements were used to avoid or delay effective action, the Commission states. Objectives have to be quantified in figures as opposed to "best efforts" clauses. Also intermediate objectives should be set to show the effectiveness of an agreement.

According to the Commission's findings, results of agreements have to be monitored, the agreement itself should define "how."

The failure of agreements, the paper goes on, is often connected with a lack of public information and transparency. Transparency is crucial to assure third parties that non-regulatory obligations are kept. Agreements should be published in the national Official Journal or an equallly public document. Even a public register of agreements should be considered.

In some cases it may be appropriate to set up a committee or independent body to collect, evaluate, or verify results. This is particularly important in cases where the measuring methods differ or where the disclosure of business secrets has to be avoided.

As an additional guarantee that the agreement will be fulfilled, dissuasive sanctions such as fines and penalties could be foreseen for case of non-compliance, the Commission concludes.

Measuring the German way of using environmental agreements at the standard set by the Commission's guidelines, one will find that the national agreements fall behind in nearly every respect. Apart from the fact, that all the German agreements are non-binding, the procedure while negotiating an agreement and the structure of the final commitment show a whole lot of deficiencies.

Thus, fundamental flaws of the negotiations may be overcome when the guideline will be followed. From our point of view it is important to mention that, according to the communication of the Commission:

1) A cost-effective use of agreements will be made as a part of policy mix together with, for example, regulatory or economic instruments and

2) agreements may be under certain circumstances an efficient tool for implementing environmental policy, but they are not appropriate for environmental target setting. General targets should be set through legislation.

These basic conclusions go along with the findings of our study. The politician's "art of the possible" may now be to look carefully for reasonable applications of agreements without falling back to traditional command-and-control measures. As experiences in Belgium (Seyad et al.:1996) and the United Kingdom (Eden:1996) show, neither a strategy of strong regulation of voluntary agreements (which makes the agreements obsolete) nor a strategy of far-reaching deregulation (or self-regulation) of environmental target-setting seems to be promising. If not driven to one of these extremes, there may be several applications for more flexible ways of environmental policy.

Acknowledgements
We would like to thank our colleague Henrike Koschel for valuable discussions on former versions of this paper. Thanks also to two anonymous referees for helpful comments. This work was supported with a grant from the German Federal Ministry of Economics.

References

Autorecycling 1996 (1): Herausgegeben von der Arbeitsgemeinschaft Deutscher Autorecyclingbetriebe GmbH. Köln.

Bauer, Hartmut (1987): Informelles Verwaltungshandeln im öffentlichen Wirtschaftsrecht, VerwArch 1987, S. 241 - 268.

BDI - Bundesverband der Deutschen Industrie e.V. (1996): Aktualisierte Erklärung der deutschen Wirtschaft zur Klimavorsorge. Köln.

Benzler, Guido, Klaus Löbbe (1995): Rücknahme von Altautos - Eine kritische Würdigung des Konzepte. In: RWI-Mitteilungen, Jg. 45, 1995-2, S. 141-162.

Beyer, Wolfgang (1986): Der öffentlich-rechtliche Vertrag, informales Handeln der Behörden und Selbstverpflichtungen Privater als Instrument des Umweltschutzes, Dissertation Universität Köln.

BGW - Bundesverband der deutschen Gas- und Wasserwirtschaft (1996): 1. Zwischenbericht zur Klimaschutzerklärung der deutschen Gaswirtschaft vom März 1995 bis zum Jahr 2005. In: BDI - Bundesverband der Deutschen Industrie e.V.: Aktualisierte Erklärung der deutschen Wirtschaft zur Klimavorsorge. Köln.

BMU - Bundesministerium für Umwelt, Naturschutz und Reaktorsicherheit (1995): Staatssekretär Jauck fordert mehr Kooperation von Staat und Wirtschaft im Umweltschutz. BMU-Pressemitteilung 117/95, Bonn.

BMV - Bundesminister für Verkehr (1995): Statement des Bundesministers für Verkehr anläßlich der Präsentation des Opel Corsa Eco 3 am 7.9.1996 in Bonn. Fs. Nr.: 0340 070995 1430.sch

Brenck, A. et al. (1996): Zwischen Entsorgungsnotstand und Überkapazitäten - Wege zur effizienten Organisation einer umweltverträglichen Abfallwirtschaft. Schlußbericht zur Studie "Optimierung des Verbrauchs knapper Deponiekapazitäten" für den Bundesminister für Wirtschaft, Zusammenfassung der Ergebnisse. Berlin.

Brockmann, Karl Ludwig, Henrike Koschel, Isabel Kühn, Klaus Rennings (1995): Ordnungspolitische Grundfragen einer Politik für eine nachhaltige, zukunftsverträgliche Entwicklung. Expertise im Auftrag des Bundesministeriums für Wirtschaft. Zentrum für Europäische Wirtschaftsforschung, Mannheim.

Brohm, Winfried (1992): Rechtsgrundsätze für normersetzende Absprachen - Zur Situation von Rechtsverordnungen, Satzungen und Gesetzen durch kooperatives Verwaltungshandeln, DÖV, S. 1025 - 1035.

Brohm, Winfried (1994): Rechtsstaatliche Vorgaben für informelles Verwaltungshandeln, DVBl., S. 133 - 139.

Bundesregierung (1996a): Erklärung der deutschen Wirtschaft zur Klimavorsorge. Pressemitteilung Nr. 118/96, Bonn.

Cook, Elizabeth (1996): Marking a Milestone in Ozone Protection: Learning from the CFC Phase-out. In: World Resources Institute. Issues and Ideas, January 1996. Washington.

DIW (1996): Energieverbrauch und CO_2-Emissionen in Deutschland in der ersten Hälfte der neunziger Jahre. In: DIW-Wochenbericht 4/96, S. 73 - 79.

Eden, Sally (1996): Problems of Environmental Consensus: Packaging Regulation and Self-Regulation in the UK. Paper presented on the Sixth Annual Greening of Industry Network Conference, Heidelberg, November 1996.

Enquete-Kommission "Schutz des Menschen und der Umwelt" des Deutschen Bundestages (Hrsg.) (1994): Die Industriegesellschaft gestalten - Perspektiven für einen nachhaltigen Umgang mit Stoff- und Materialströmen. Bonn.

Environment WATCH (1995): Commission Plans EU Codes for Green Taxes, Voluntary Agreements. 27.10.1995, S. 13.

European Commission (1997): Communication to the Council and European Parliament on Environmental Agreements. In: Supplement to Europe Environment Nr. 492 - January 28, 1997.

Green, Kenneth, Andrew McMeekin, Alan Irwin (1994): Technological Trajectories and R&D for Environmental Innovation in UK Firms. In: Futures 1994, 26 (10), S. 1047 - 1059.

Grossekettler, Heinz (1991): Zur theoretischen Integration der Finanz- und Wettbewerbspolitik in die Konzeption des ökonomischen Liberalismus. In: Erik Boettcher et al. (Hrsg.): Jahrbuch für Neue Politische Ökonomie, Band 10, Tübingen, S. 103 - 144.

Hemmelskamp, Jens, Georg Licht, Egbert Oldenboom, Jan Willem Velthuijsen (1995): The Impact of Parameters provided by Environmental Policy on the Innovative Behaviour of Companies in selected European Countries. Final Report of a Study commissioned by the European Commission DG XII, Contract Nr. EV5V-CT94-o376, Mannheim, Amsterdam.

Henneke, Hans-Günter (1991): Informelles Verwaltungshandeln im Wirtschaftsverwaltungsrecht und Umweltrecht, NuR, S. 267 - 275.

Hoffmann-Riem, Wolfgang (1990): Reform des allgemeinen Verwaltungsrechts als Aufgabe - Ansätze am Beispiel des Umweltschutzes, AöR, S. 400 - 447.

Hohmeyer, Olav, Henrike Koschel (1995): Umweltpolitische Instrumente zur Förderung des Einsatzes integrierter Umwelttechnik. Gutachten im Auftrag des Büros für Technikfolgenabschätzung beim Deutschen Bundestag, Mannheim.

Holzhey, Michael, Henning Tegner (1996): Selbstverpflichtungen - ein Ausweg aus der umweltpolitischen Sackgasse? In: Wirtschaftsdienst 1996/VIII, S. 425 - 430.

Homburger, B. (1996): Altauto-Recycling endlich auf dem Weg. Erklärung der umweltpolitischen Sprecherin der F.D.P.-Bundestagsfraktion vom 21.2.1996.

IPCC - Intergovernmental Panel on Climate Change (1995a): Second Assessment Synthesis of Scientific-technical Information Relevant to Interpreting Article 2 of the UN Framework Convention on Climate Change. Genf.

Kohlhaas, Michael, Barbara Praetorius (1994): Selbstverpflichtungen der Industrie zur CO_2-Reduktion. Berlin.

Krause, Florentin, Jonathan Koomey, David Olivier (1994): Incorporating global warming externalities through environmental least cost plannning: a case study of Western Europe. In: O. Hohmeyer, R. L. Ottinger: Social Costs of Energy. Berlin, Heidelberg, New York, S. 287 - 312.

Kreuzberg, Peter (1993): Zur ökonomischen Rationalität "freiwilliger Kooperationslösungen" für das Klimaproblem. In: Zeitschrift für Energiewirtschaft (ZfE) 4/93, S. 304 - 309.

Maier-Rigaud, Gerhard (1995): Für eine ökologische Wirtschaftsordnung. In: Jahrbuch Ökologie 1996. München.

Mez, Lutz (1995): Erfahrungen mit der ökologischen Steuerreform in Dänemark. In: Olav Hohmeyer (Hrsg.): Ökologische Steuerreform. ZEW-Wirtschaftsanalysen Band 1, Baden-Baden, S. 109 - 128.

Müggenborg, Hans-Jürgen (1990): Formen des Kooperationsprinzips im Umweltrecht der Bundesrepublik Deutschland, NVwZ, S. 909 - 917.

Muller, Frank (1996): Energy Taxes, the Climate Change Convention and Economic Competitiveness. In: Olav Hohmeyer, Richard L. Ottinger, Klaus Rennings (eds.), Social Costs and Sustainability, Valuation and Implementation in the Energy and Transport Sector. Berlin, Heidelberg, New York, S.

Murswiek, Dietrich (1988): Freiheit und Freiwilligkeit im Umweltrecht. In: Juristen Zeitung, 43. Jg., H. 21, S. 985 - 993.

Ökologische Briefe (1996) (2): "Denkhilfe" für die Wirtschaft aus dem Umweltbundesamt.

Ökologische Briefe (1996) (5): Europäische Kommission will das Fünfliterauto.

Ökologische Briefe (1996) (8): Klimaschutzverhandlungen - Genf wird kaum Ergebnisse bringen.

Rennings, Klaus (1994): Indikatoren für eine dauerhaft-umweltgerechte Entwicklung. Stuttgart.

Rennings, Klaus, Karl-Ludwig Brockmann, Henrike Koschel, Heidi Bergmann, Isabel Kühn (1996): Nachhaltigkeit, Ordnungspolitik und freiwillige Selbstverpflichtung. Heidelberg, 1996.

Sacksofsky, E. (1996): Anmerkungen zu verschiedenen Konzepten einer Neuregelung der Altautoentsorgung. In: Zeitschrift für Umweltpolitik & Umweltrecht, Jg. 19, Heft 1, S. 99-108.

Sanghi, Ajay K. (1995): Climate for Climate Change Actions in the U.S.: The New York Experience. Social Costs and Sustainability, Valuation and Implementation in the Energy and Transport Sector. Berlin, Heidelberg, New York, S.

Schmitt, Dieter, Helmut Düngen (1992): Klimapolitik in der Sackgasse? Einsatzmöglichkeiten für Kompensationen. In: Wirtschaftsdienst, 1992/V, S. 271 -276.

Seyad, Akim, Marc de Clerq, Filip Senesael (1996): The Use of Voluntary Agreements as Instruments of Environmental Policy in the Energy Sector - the Belgian Electricity Voluntary Agreement Case. Paper presented on the Sixth Annual Greening of Industry Network Conference, Heidelberg, November 1996.

Spies, R. (1994): Der "Grüne Punkt" als ökonomisches Instrument in der Abfallwirtschaftspolitik. In: Zeitschrift für angewandte Umweltforschung, Jg. 7, Heft 3, S. 309-321.

SRU - Rat von Sachverständigen für Umweltfragen (1994): Umweltgutachten 1994 - Für eine dauerhaft-umweltgerechte Entwicklung. Wiesbaden.

SRU - Rat von Sachverständigen für Umweltfragen (1996): Umweltgutachten 1996 - Zur Umsetzung einer dauerhaft-umweltgerechten Entwicklung. Wiesbaden.

SZ - Süddeutsche Zeitung vom 6.3.1996: Fünf-Liter-Auto gemeinsames Ziel.

Umwelt (1995a): Workshop "Freiwillige Selbstverpflichtungen zur Klimavorsorge." In: Umwelt Nr. 12/1995, S. 438.

Umwelt (1996): Novelle der Verpackungsverordnung. In: Umwelt Nr. 4, S. 158.

UBA - Umweltbundesamt (1993): Jahresbericht 1993. Berlin.

VDA - Verband der Automobilindustrie e.V. (1995): Freiwillige Zusage zur Kraftstoffverbrauchsminderung. Frankfurt/Main, 22.3.1995.

VDEW - Verband Deutscher Elektrizitätswerke (1996): Bericht 1996 zur "Erklärung der VDEW zum Klimaschutz." In: BDI - Bundesverband der Deutschen Industrie e.V.: Aktualisierte Erklärung der deutschen Wirtschaft zur Klimavorsorge. Köln.

Weiland, R. (1995): Rücknahme- und Entsorgungspflichten in der Abfallwirtschaft - Eine institutionenökonomische Analyse der Automobilbranche. Wiesbaden.

Welt vom 15.4.1996: Merkel liebäugelt mit Kartellen.

Wupperman, B. (1993): Zwei Jahre Verpackungsverordnung: Weitere Machtkonzentration statt Müllvermeidung. In: Zeitschrift für angewandte Umweltforschung, Jg. 6, Heft 4, S. 448-455.

Zimmermeyer, Gunter (1995): Voluntary Fuel Economy Targets for Motor Vehicles in Germany. Presentation for the IEA/OECD International Workshop on Voluntary Approaches for Mitigating CO_2-Emissions. Bonn.

14 The Political Context for Emissions Trading in the Climate Change Negotiations[1]

Michael Grubb
Energy and Environmental Programme
The Royal Institute of International Affairs (RIIA)
London, Great Britain

0 Introduction

This paper explores the relationship of potential CO_2 trading and joint implementation to the evolution of international and national law and politics, as mediated through the climate convention processes. I will be looking first, briefly, at why emissions trading is attracting this kind of interest and at its relationship to the comprehensive approach, using multiple gases. Then I consider the question of which sectors we are talking about controlling in connection with tradeable permits, and what that implies; the politics of the process, particularly relating to the European Union rather than the United States; and, more broadly, the international perspective beyond the OECD in the relationship with joint implementation. Finally, I touch upon the role of the Kyoto protocol.

1 Why Emissions Trading?

Why are we finding ourselves in this debate? There is the classic economic argument that tradeable permits are an efficient way of implementing emission reductions, but then in principle, if the market is working well, so too are other economic instruments such as a tax. However, carbon taxes have proved highly controversial and extremely difficult to implement, as indicated by the failure of both the EU carbon tax and the US BTU tax proposals. So the first question is why

---

[1] This paper is adapted from M.Grubb, 'The political context under the Climate Change negotiations', in Proc. RIIA conference on *Controlling carbon and sulphur: joint implementation and trading initiatives*, RIIA, 1997.

there is such a big political difference between taxes and tradeable permits - instruments that many economists say are very similar.

There are some economic differences between caps and taxes noted in the economics literature, relating to uncertainty and whether one is trying to control the product or the cost incurred. In fact this argument suggests that taxation might in fact be the more appropriate instrument for CO_2 control.

But the politics point the other way, for key reasons which have not appeared much in the literature. A trading instrument creates an asset. Even if the permits are auctioned and industries have to buy permits to cover emissions, so the financial transfer is the same, the company now has an asset on its books, so from the industrial perspective it is different. If permits are granted or given out initially then the company has acquired an asset, and it looks totally different in terms of the industrial politics for that reason. Also within countries governments have a negotiating flexibility over the relationship between auctioning permits – selling them – and giving some out. There is a lot of political flexibility for the governments, crudely, to buy off the most powerful parties.

Finally, there is an institutional appropriateness in the sense that energy or carbon tax instruments are perceived as a tax instrument. This means that by and large the finance ministries and treasuries, whose primary objective is not limiting CO_2 emissions, find it on their turf. Frequently, indeed, we find a big institutional mess in governments about who is controlling the policy. Permits are more clearly an instrument that falls in the legitimate terrain of environment ministries. These are very big political differences which explain why we are now looking more at the option of emission permits.

There are of course drawbacks and limitations to tradeable permits, particularly if we are still talking mainly about exploiting rather cheap emissions limitations. In a situation where some people are saying this is a problem that will cost us the earth, and others do not even expect the Kyoto protocol to result in any measures that will cost anything, tradeable permits would at least help to reveal who is right. They would reveal prices, and potentially put limits on them.

2 Comprehensive Approach

The Simple Comprehensive Approach

The approach of including all the different greenhouse gases and sources as a basket is intellectually extremely appealing. It increases the flexibility and efficiency of the process overall. Before the Convention in 1992 the United States gave this considerable political support. The Europeans were very hesitant, however, and this resulted in a compromise in the Convention. The phrase used is 'CO_2 and other greenhouse gases' – one of the many creative ambiguities in the Convention as to whether the CO_2 should be regarded separately.

Interestingly enough, more than four years later we are now finding that in submissions for the protocol – the AGBM negotiations in December 1996 – the US focused on CO_2 trading, whereas the European Union has dropped the words 'CO_2 and other' and refers only to 'greenhouse gases'. The UK submission states that 'There is no longer any reason not to adopt a comprehensive approach'.

Despite its theoretical appeal, there is a big problem with the comprehensive approach, particularly in the context of legally binding targets and permits. If you intend to make something legally binding, still more to attach some kind of financial value to it, as would happen with trading, you need to be very sure about the monitoring capabilities – the costs of monitoring, accuracy, verifiability and the basis on which you are comparing different gases.

There is a whole range of gases and sources, starting with carbon dioxide from fossil fuels and running through the different kinds of CO_2 sources, the halocarbons, methane from a whole range of sources. These differ enormously in terms of the ability with which they can be monitored. Among the methane sources we have bovine flatulence, which in greenhouse terms I guess is the biological equivalent of car exhaust, in the form of cow exhaust. Clearly no-one can monitor methane from bovine flatulence to within the few percent accuracy that could be required for a tradeable permit system.

Two Lists Approach

We can think of a more feasible approach, if it is worth the complexity. A protocol could establish a system which starts out with what we can do in terms of the monitored and verifiable sources in a first list, and which would also specify how the different sources were to be compared. It could then establish a secondary list of sources which we could not be controlled in that way but which would still be addressed in other ways. This is important. There needs to be pressure on the other sources and sinks, for example with non-legally binding targets as appropriate – because you cannot monitor a binding target – and perhaps tackled by specified policies and measures. If – and that is an important *if* – it is considered important enough, again the protocol could establish a process by which one would seek to move over time different sources from list two to list one, as and when it was institutionally feasible.[2]

---

2 For a further elaboration of this, see M. Grubb, D. Victo, and C. Hope, 'Pragmatics in the Greenhouse', *Nature*, Vol. 354, 5 December 1991, pp. 348–50.

3 US Sulphur Model

Experience

I want now to look at the US model extrapolated from the sulphur experience and applied to CO_2. Relatively little has been said about the fact that this model is limited to certain sectors – in the US, power plants above a certain size; in the EU one could conceive that a very similar sort of system would be confined to boilers that fall under the purview of the Large Combustion Plant Directive for Sulphur, namely above 50 megawatts thermal capacity. Of course, this is not relevant to the use of 'trading' for exchanging national emission quotas, but it is important for the use of tradeable emission permits as an instrument of domestic implementation.

In terms of climate change, this immediately raises the question of the non-participant sectors. Would such a system not introduce distortions, of scale if the small boilers profit, or of sectors – is it not unfair, the power generators would say, that other sectors are not properly controlled? That will lead to political repercussions. Some thinking, therefore, about the non-participant sectors is required. They may be addressed through sub-sectoral targets – what governments are expecting each sector to contribute – or specific policies and measures.

There is a perception that the US is pushing emission targets and trading, while the European Union is pushing policies and measures. The chances are that one needs to look at and combine both, because you need different kinds of instruments for different sectors, at least if one is looking primarily at carbonising the US sulphur model.

Obstacles to Internationalising the US Model

We need to be very careful about what we are getting into, particularly concerning US promotion of carbon trading based on their SO_2 trading experience. There are countries with very different energy economic structures, for example Switzerland, Sweden, and France have essentially very close to zero carbon in their power sector. In Switzerland only 13 percent of carbon dioxide emissions comes from power generation or from heavy industry, so a carbon version of the US model is almost irrelevant. The Swiss do not gain any serious flexibility from it, and their concern is with other sectors.

Similarly there may be other kinds of problems or differences in energy economic structures – a few countries may have large electricity imports from non-participants in the scheme. These need some attention.

Perhaps more important than any of these issues, countries have very different kinds of cultures and institutions related to public policy. Clearly France and Japan have very different assumptions underlying the relationship between government and industry from, say, the United States. Indeed, much of continental Europe has seen a preference for taxation and negotiated agreements to give tax exemption for some industrial sectors. Likewise countries differ in their monitoring and enforcement capabilities.

Personally I am very much in favour of emissions trading, but I believe it will take different forms – maybe very different forms – in different countries. What the Kyoto protocol will not do is create a coherent international trading system; but it could set a framework of national commitments upon which different national trading systems can be developed and made internationally compatible.

Alternatives to the US Model

There are a number of different options. Countries might have slightly different sectoral coverage of trading systems. They can try to apply the permits at different points in the production chain – for example, to the carbon content of the fuels coming out of refineries. I have heard suggestions by both Canada and the European Union that maybe they want to focus on the traders, the companies that deal in fossil fuels, rather than the companies that burn them. I have serious reservations about the feasibility of that approach, but I may be wrong and I know it is being considered.

Likewise geographical coverage; I mentioned the Australian federal problem, and I was most interested to learn that New South Wales is considering trading within the state. Governments themselves could hold and trade emission quotas, as the basis for enforcing national emission commitments. It is an idea which sends shivers down the spine of some US industrialists I know, but not all European countries or publics or industries have quite the same aversion to governments being involved in that way.

4 CO_2 Trading in Europe

The European Situation

The situation in Europe is very tricky and may illuminate some of the issues. Why should Europe potentially be interested in trading?

The most obvious reason is the complete and rather abject failure of most EU-level policies to date, of which the most obvious is the carbon tax, but the remark applies to some extent to other policy proposals. They have tended to fall on the altar of EU politics and subsidiarity. I cannot resist saying that about three years ago we had a meeting which involved a discussion between Ernst von Weizsäcker in Germany and myself on the question of policy instruments. People had been looking forward to a healthy debate between the proponent of taxes and the proponent of permits, and they were very disappointed when we both agreed that at international level quota trading made more sense, because you did not want to interfere in the national sovereignty of instruments that countries used. At the national level, certainly in some sectors, we had no trouble agreeing that taxation may well make more sense.

Having apparently found some consensus, Ernst and I decided to write something about this together. I wrote the first draft, on European policy, and faxed it off to Ernst. A horrified reply came back; we had not realised that I had been thinking of the European Union as essentially a collection of states, whereas Ernst had been thinking of it as a unified state. That may tell you something about the nature of the European problem! Nevertheless the upshot of that dialogue was a report which eventually emerged on emissions trading involving a number of European institutes.[3]

I think it is unlikely the European Union will magically acquire powers for issues such as taxation or energy policy, which did not even appear on the agenda of the Inter-Governmental Conference. And yet if we look at the national programmes, there are a lot of no-cost or low-cost measures, but there is a process of gradual exhaustion which will lead to more and more pressure to look at more potentially costly measures, in economies which are increasingly closely entwined. That trend is most dramatic in the electricity sector where the June 1996 Liberalisation Directive establishes a process towards opening up all the member states' electricity sectors to international competition in Europe. That will set limits upon what member states can do without having instruments coordinated at European level.

Another reason why the EU needs to find a feasible approach to the coordination of instruments, I would suggest, is the international politics of the protocol negotiations. Clearly a number of countries outside Europe are watching very keenly to see the credibility of European commitments in this area. Indeed I know one government represented here has said it will block a Union signatory of a protocol unless it is very clear how the Union is itself going to deliver what it is calling upon other countries to do. That requires some credibility in the instruments that will be used at a pan-European level.

Finally I have already made reference to the legal and institutional basis of policy instruments. The European Union does not have legal competence in the fields of taxation or energy policy overall, but a tradeable permits system is arguably an emissions control/environmental policy on which there is legal basis of competence, and which in its institutional design may not even require unanimity in the voting procedures. There are thus institutional reasons why Europe might profitably take a closer look at emissions trading.

All this may sound fine in theory, but will it happen? I would agree the prognosis is not very good. There are a few individuals in a few European governments who recognise essentially that the European Union does not yet have clear and credible proposals for implementing the commitments it is calling upon the rest of the industrialised world to agree to. The Commission itself is discussing this but is

---

3 M. Grubb et al., *Implementing the European CO_2 Commitment: a joint policy proposal*, RIIA, London, 2nd edition, February 1997.

very much divided on questions of permits trading, and anyway could not drive such a big change. Politically the only way one could see it emerging would be if one country started to champion the idea.

The European Trading System

On the Kyoto time-scale the need is not to design and create a full tradeable permit system, but to establish the conceptual and political basis that would enable the Union credibly to sign legally binding commitments and over subsequent years to start to implement them, including but not exclusively using trading programmes. If that were to occur then it would probably start with the concept of intergovernmental quotas – exchange of national targets. This might become mixed in with permits which are diffused down into some sectors, power generation being perhaps the most obvious, following the process of electricity liberalisation.

My guess is that from an early stage border relationships would be established, bringing Norway into the European emissions bubble. There is a huge mutual interest in enabling that to happen, given Norway's domestic predicament combined with its role as a key gas – and increasingly electricity – exporter to the Union.

Politically then, having established some credibility in the process, the key is how Europe would address the implementation issues after the protocol, after signing on to a collective commitment. Interestingly, the time scale is quite good if that occurs. A European CO_2 trading system will not be developed by a group of ministers sitting round a table and negotiating a fair allocation. It will involve messy, bloody questions of the politics of allocation within Europe, but at the end of 1997 Europe also enters a whole new cycle of budgetary negotiations – revision of the structural funds, and associated negotiations on enlargement.

If Europe does seek to implement its keys to commitment with intergovernmental quota trading, allocation will become part and parcel of that horse trading. It could become intimately linked with proposed changes in structural expenditures in the cohesion countries, for example. It seems to me that within that rather messy political process this is in practice how any allocation negotiations will proceed.

Beyond that basic process one would then have to start seeing the development of an implementation architecture, presumably involving European-level institutions for monitoring and carrying out the exchange, sectoral definition of exactly which sectors were involved, and ultimately leading to a carbon currency. Although the prospects on both fronts are problematic, to judge from the recent furore over EMU, we might conceivably see a common carbon currency before we get to the common European currency!

5 Joint Implementation

Let me come to the question of the relationship to joint implementation among Annex I countries. I will not say anything about activities implemented jointly outside Annex I, other than that attempts to reintroduce that question in terms of official intergovernmental trading or offsets for the Kyoto protocol may risk undermining the hard-won political basis of the Berlin Mandate and the decisions on the activities implemented jointly. JI is currently an issue within Annex I, and one should proceed first on that basis.

In central and eastern Europe JI projects are beginning to happen. This has political legitimacy, and will be tied in some sense to the generation of credits, as was discussed earlier, which might become formalised into sectoral allocations leading on to permit trading. I do not see this as being a very neat or uniform process; the dissemination of these instruments will occur at varying rates.

This development will also be rather intimately related in Europe to energy trade, particularly gas and electricity trade. Norway will obviously be a big player, with Russian gas and the electricity trade being closely intertwined. To an extent it will follow the progressive liberalisation of electricity and gas networks. Ultimately liberalisation and emissions trading could become part of the *acquis communautaire* for the projected eastern enlargement of the European Union.

Over time all these processes will in some way become related to one another, and one should perhaps be thinking about whether this has any implications for the negotiation of the second phase of the Energy Charter Treaty, which is the phase designed to deal with the terms of new investment, particularly in central and east European energy.

6 Conclusions

Finally, what is the role of Kyoto in all this? The Kyoto agreement cannot, need not and should not define internal national trading implementation systems. That does not matter, providing there is some international compatibility in the systems established, and an integrity verified for the national systems. In my view, the Kyoto agreement needs to define commitments that allow for the possibility of international and perhaps inter-temporal trading and banking within fairly closely prescribed limits.

The Kyoto protocol needs to define the role of joint implementation and crediting within Annex I. As a final leg, it needs to establish a body which can do the jobs that are outlined here, namely which can certify that national systems will lead to compliance, and the compatibility of national permit systems and their integrity, as well as relationships to joint implementation where countries choose to use those instruments.

Just to remind us where we started, there are political attractions to emissions trading related to the negotiating structure of the emissions cap, related to the industrial politics of creating an asset and buying off certain quarters, and its institutional basis as an environmental measure rather than something which falls with the tax or other ministries.

Inevitably one will start with carbon dioxide. There are possibilities for expanding the regime over time, and the protocol negotiations need to think fairly carefully about whether and how they want to create a 'two list' or other system that could move the structure towards gradually encompassing a more comprehensive approach.

Emissions trading along the US sulphur model would cover only some sectors – and for some countries, only minor sectors in terms of total CO_2 emissions. Other sectors could be addressed through sectoral targets, agreement on specific policy measures, and variants of trading systems applied higher up the energy stream, for example to refineries and fuel trading companies.

The details of trading systems are likely to vary considerably between different countries. The situation in Europe is of particular interest, given the failure of previous efforts and the current predicament, but it would require some strong and imaginative political leadership to overcome the current aversion to emissions trading. Central Europe is likely to be a key area for development of joint implementation and its possible evolution towards permit trading.

Thus we will see varied national systems in which the key question, if we are to attain the benefits identified, will be to make sure that international trading between those systems can evolve. Likewise we will see the national systems evolving. In that context the Kyoto protocol will not try to establish the global or Annex I system, but it will create incentives through the binding commitments, the enabling mechanisms that I touched on, and the verification institutions, to ensure that emission trading happens in a credible way.

15 The Kyoto Protocol and the Emergence of *"Contraction and Convergence"* as a Framework for an International Political Solution to Greenhouse Gas Emissions Abatement

Aubrey Meyer
Global Commons Institute
London, UK

The Kyoto Protocol, completed in the early hours of December 11th 1997, at present is no more than a potential breakthrough in the development of effective global policy for the control of atmospheric concentrations of greenhouse gases and the mitigation of human-induced global climate changes. The core issue of the negotiations has been deferred until CoP4 in November 1998. The industrial countries have negotiated a compromise that subject to ratification will legally bind them to commitments beyond those in the UNFCCC. But the ratification of the Protocol by the US still remains contingent on achieving the *"meaningful participation"* of *"key"* developing countries in the abatement regime and the multilateral acceptance of international emissions trading. This is a struggle to define property rights. These key developing countries include India and China and they have made it clear that their acceptance of trading is contingent on the achievement of *"equitable allocations"* of emissions entitlements based on achieving equal per capita entitlements globally. COP issued instructions to the technical bodies attached to the UNFCCC to *"define the relevant principles, modalities, rules, and guidelines for emissions trading"* in time for CoP4 in November 1998 in Buenos Aires.

GCI argues that *"Contraction and Convergence"* is the approach that can break through this deadlock and welcomes the fact that major parties and interest groups in this dispute have already acknowledged that they take this approach seriously and that it has growing support throughout the world. As a leading economics commentator Peter Jay has noted,

> *... unless there is some recognition that eventually no one group of human being can expect to have an internationally recognised right to consume more of the world's limited capacity to absorb greenhouse gas emissions than any other group, it is hard to see how a globally enforceable policy can be built by consent.*

And in the words of the President of GLOBE International, *"Contraction and Convergence is not simply the right way to solve the problem, it is the only way to solve the problem."*

CHALLENGE FROM IPCC CLIMATE SCIENCE

The First Assessment Report (FAR - 1990) of the Intergovernmental Panel climate Change (IPCC) noted that atmospheric concentrations of CO_2 were 25% higher (350 ppmv) than pre-industrial (280 ppmv) and rising faster and higher than anytime in the previous 160,000 years. An extremely strong correlation between rising CO_2 concentrations and human CO_2 emissions (mainly from fossil fuel burning) was observed from around 1800 forward. The IPCC also observed circumstantial links to rising global mean temperature and stated that immediate minimum 60% to 80% cuts in human CO_2[1]emissions were necessary if atmospheric concentrations of CO_2 were to be stabilised just at 1990 levels (see chart 1). Since then, IPCC has stated that the balance of evidence suggests that there is a discernible human influence on the climate system. They also suggested the damages consequent on no abatement and further global temperature increase as being between serious and potentially catastrophic, regionally and even globally. Since 1990 there has been much investigation into what constitutes atmospheric greenhouse gas concentrations levels that do not dangerously affect the climate system.

According to carbon cycle and global climate modelers, the time frame foreseen for achieving at least 60% cuts in emissions is between 50 and 200 years, depending on the ultimate atmospheric CO_2 concentration goal. There is a real concern however, that even going to 450 ppmv (60% cuts in emissions over roughly one hundred years) may result in serious ecological and consequent socio-economic damages. The IPCC has published data derived from climate models that attempt to demonstrate the quantitative links between greenhouse gas emissions and accumulated atmospheric concentrations. (These are summarised in Appendix A).

---

1 GCI recognises the relevance of other ghgs, but also that industrial CO_2 emissions account net for over 70% of the human influence on climate change. Moreover we are primarily advocating the policy concept *"Contraction and Convergence"* for negotiating the shared ownership of greenhouse gas emissions entitlements globally. This depends on having reasonable global datasets and so far these only exist for industrial CO2 emissions.

GLOBAL SOLUTIONS ARE CONSTITUTIONAL

At the very least, sensible contingent planning requires that if the global community is to demonstrate both technically and politically that the worst of the potential damages from human-induced global climate change can be avoided, it will have to demonstrate that the cuts in emissions can be achieved. There will be a twin compromise. The rate of cutting emissions must be rapid enough to halt the rise of atmospheric greenhouse gas concentrations below levels that dangerously affect the climate system. However, it must also be gradual enough to give time for non-fossil alternative energy sources and energy saving measures to be introduced so as not to precipitate an economic crisis. Moreover, the whole operation will have to be planned for inclusivity under continuous political and scientific review. Not only is it appropriate to insist that *"a global solution is required for a global problem"* as the US has repeatedly done, it is also necessary to actually come up with a global solution. This unavoidably means having to recognise and acknowledge that a global solution is by definition, constitutional. It will be the result of having first had to determine the principles upon which the rules for globally sharing finite resources will be founded and applied. In the case of climate change, this means determining the principles and rules for sharing a future global carbon budget that is also consistent with the twin compromise above. The political struggles at Kyoto have brought the need for this unprecedented imperative into focus more sharply than before.

Deliberate limitation of CO_2 emissions from industrial activity is certainly contentious. Dollar GDP from the formal economy has so far been very closely correlated with CO_2 from fossil fuel burning (see Chart 1) and anything less than the positive growth of dollar GDP is regarded as a primary signal of macroeconomic failure. Consequently now achieving a delinking of GDP and CO_2 emissions must be a primary feature of any future economic planning. Moreover, CO_2 emissions and GDP are both historically and currently, very unevenly generated and distributed throughout the global economy. In simple language, it is those who have made the money who have also made the mess in the atmosphere. This is now increasing instability in global politics as well as in the global climate system, where the increasing risks of environmental adversity are increasing the risks of attendant social and political conflict. The issue of how to determine the "differentiated responsibilities" in any global programme required to achieve the necessary levels of emissions abatement is thorny and has confounded the UNFCCC negotiations all the way from INC1 in Washington in 1991 to COP3 in Kyoto in December 1997. The argument is fundamental, but is quite novel for being truly "global". Everybody regardless of levels of wealth and development is implicated, some as alleged perpetrators but all as probable victims.

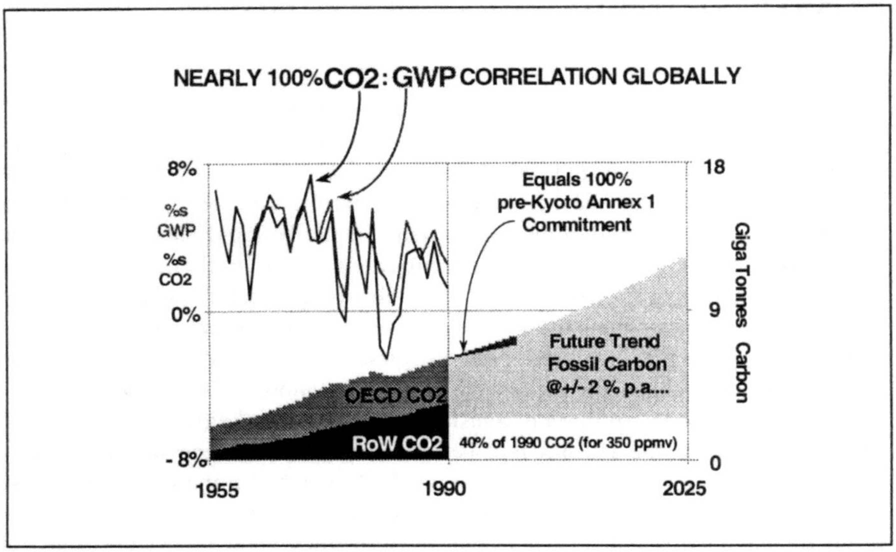

Chart 1: Global CO_2: GWP correlation

NO PROBLEM, NO REGRETS, NO SOLUTIONS WITHOUT DEVELOPING COUNTRIES

Since 1990 we have been through three periods of argument about whether human-induced climate changes were occurring. The initial period of *"no problem"* gave way to a period of *"no regrets"* (perhaps there's a problem so do what makes sense for reasons of economic efficiency). In reality this all seems to have reflected a wish to postpone any genuine engagement with the real issues. For example John Knaess, the head of the US delegation to the Second World Climate Conference, insisted that *"simple sophomore physics reveal that the problem is real"* (greenhouse gases trap heat, so more greenhouse gases trap more heat) and that the only questions were *"how much"* and *"how soon"*. It bears some reflection as to why it took seven years until June 1996 at CoP2 for the US government to attempt to get behind emissions abatement policy consistent with the acceptance of human-induced climate change as a reality. It seems probable that the real argument has always been about how to compute and then most particularly *share* the future sustainable "Global Carbon Budget." Post-Kyoto we are all now openly being called to account on this point. So, since CoP2 we have been in the third period called *"no solutions without developing countries"*. For the last eighteen months the call from across the board in the US has been that there has to be, *"meaningful participation by developing countries"* because *"global problems require global solutions."* This was in spite of the Berlin Mandate with its focus on developed country commitments only.

A STRUGGLE TO ASCERTAIN PRINCIPLES GOVERNING GLOBAL DISTRIBUTION

The period of tactical denial could well have related to the battle between two competing socio-economic arguments that were advanced at the outset for determining the international distribution of future CO_2 emissions entitlements in the carbon budget. These were flat-rate emissions cuts globally with budget shares, or emissions entitlements, proportional either to GDP or to population. Application of each argument leads to very different distributions of entitlements globally. For example the USA in 1990 had approximately 4% of global population but emitted 25% of CO_2 emissions with a GDP share to match. In the same year India had approximately 15% of global population but emitted 3% of CO_2 emissions with a GDP share to match. It is no surprise therefore to find the US favoured flat-rate cuts applied to shares proportional to initial GDP figures, while India favoured shares proportional to population. In the World Bank's Development Report of 1992 the arguments were applied, the consequences were analysed and the inverse distributional results were compared. They noted distribution from shares equal on a per capita basis accumulated between 1950 and 1990, gave an overall negative share to developed countries stating simplistically that for that reason the alternative seemed the more feasible approach.

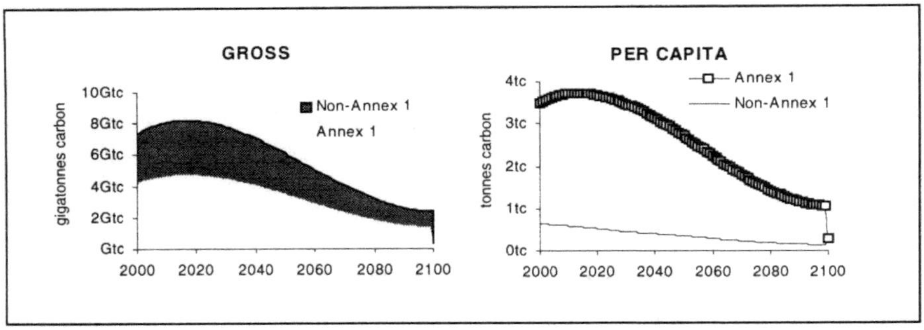

Chart 2: Entitlements reduced always proportional to initial GDP

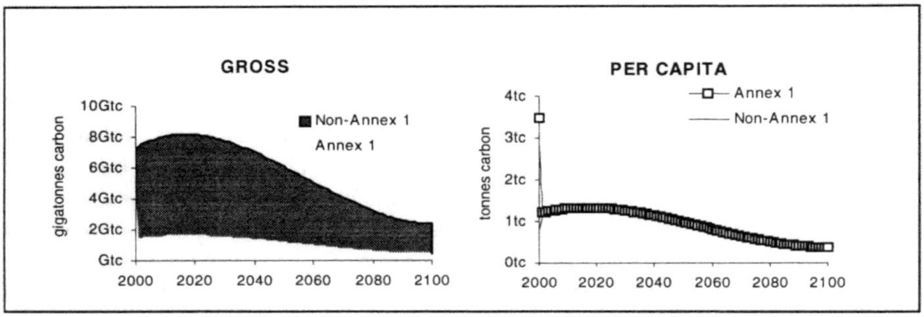

Chart 3: Entitlements reduced always proportional to initial population

However, what the arguments have in common is that they are "pro rata" arguments where entitlements would be the result of applying a central organising principle to distribution out of necessity, simply to enable the collective contraction to be computed. The point here is that you have to make some assumption other than business-as-usual about distribution if you want to deliver the purpose of the UNFCCC, pre-eminently to deliver overall contraction of greenhouse gas emissions. Over the full period of achieving the 60% cuts in global emissions, this is unavoidably a negative-sum-game on emissions for everyone. And it is self-evident that without the application of a central organising principle to the determination of future allocations of emissions entitlements, the negative-sum-game of global carbon contraction will be unachievable. Business-as-usual and the now globally competitive character of market forces tend to engender deregulation and an erosion of democratic politics. If this culture is assumed for the management of greenhouse gas emissions, the future global carbon budget and its distribution will simply be the "invisible hand" in its malignant guise. It will be the ad hoc result of each party to the negotiations continuing the attempt to maximise its shares of the budget at the expense of all its competitors. Consequently, the future global carbon budget will expand indefinitely as the global aggregate of reluctance to clean up causing dangerously raised levels of atmospheric greenhouse gas concentration. In other words, it will be the ever more visible and aggressive hand of climate change and the painful damages it will bring. This will continue unless and until there is a multilateral willingness to accept the application of a central organising principle to the determination of future international shares of what globally is a deliberately conceived and managed budget for carbon *contraction*. Put another way, more and more people now seem to accept that two hundred countries times two hundred arguments will never achieve the level of international agreement required to secure 60% cuts in emissions globally. What should also be self-evident is that the defining character of such arrangements will be based on some universally recognisable principle of equity simply to secure the necessary multilateral acceptance of Quantified Emissions Limitations Reductions Objectives (QELROS). If so it would then be recognised that any efficiencies that are achievable will be derived from - rather than give rise to - the primacy of the equity based arrangements. In other words principle and practice are inseparably linked and the old adage is true: *"principle without practice is useless and practice without principle is dangerous."*

So far, the most frequent argument of many Western economists has been that sustainable future CO_2 emissions entitlements should be distributed between countries proportional to GDP precisely because of the close correlation between CO_2 and GDP. Contrarily, most developing countries have argued that sustainable future CO_2 emissions entitlements should be distributed between countries proportional to population because the global atmosphere and climate system are a "global commons", the sustainable use of which should be the equal responsibility of all members of the global community. If the commons belongs to anybody, it belongs to everybody. Both themes are embedded in the language of the United Nations Framework Convention on Climate Change (UNFCCC). Moreover, while the "basis of equity" is recognised in the UNFCCC in the context of international per capita emissions paths which are historically disparate, "cost-effectiveness" of global emissions abatement measures is also called for, which is portrayed as pursuing the "global benefit" (of avoided emissions) at least possible cost.

CHART 4 shows that these conflicting positions can be resolved through *"Contraction and Convergence".*

We start proportional to GDP and deliberately move to being proportional with population (here by 2030).

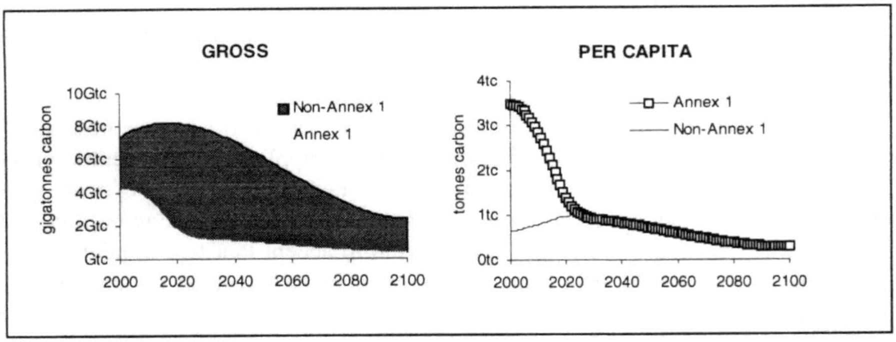

Chart 4: An example of gross and per capita *"Contraction and Convergence"*

But so far they have been polarised to an intractable seeming deadlock. Those who see entitlements as being proportional to GDP not only get the bulk of entitlements, they have also seen themselves as the main providers of entitlements in any market that may emerge where entitlements are tradable internationally. In other words if there was emissions trading and we were following the World Bank's *"most feasible"* case (see above), India would purchase entitlements from the USA. The US Mineworkers Association put out material making this case in the run up to CoP3 in Kyoto. Contrarily, because at least 50% of any year's emissions continue to reside in the atmosphere over very long time frames, countries with low per capita consumption have argued that they are innocent of causing climate change historically, currently and even into the medium term future. The Brazilian proposals to CoP3 for example attempted to define responsibility for observed global temperature rises by computing national proportions of blame as a function of nationally accumulated historical emissions. The result of this analysis shows that countries such as Brazil remain "blameless" well into the future. Bach, Koomey, and Krause did a related exercise in a report for the Dutch Government ("Energy Policy in the Greenhouse" – Earthscan 1989) where "blame" was linked to nationally accumulated historical per capita emissions. The result of this particular analysis demonstrated that if equal emissions entitlements were the currency for resolving the problems of global abatement, the industrial countries are already indebted to developing countries to the extent of total bankruptcy. There have been many other attempts at straightening out distributional methodology for the allocations of future emissions entitlements. They usually involve more or less complex combinations of weighted indicators such as carbon intensity of emissions, per capita GDP and per capita emissions (proposed by the Norwegians) and emissions per unit of export (favoured by coal-rich Australia). This has usually been applied to time frames well short of stabilising greenhouse gas concentrations and often to only sub-global country groupings such as the Annex One of the UNFCCC for example. Moreover, special exceptions have been achieved as well where for example the US has successfully negotiated that emissions associated with UN-backed military operations should not be added to the accounts of the countries undertaking the operations.

Faced with this degree of intractable complexity and quarrelling, some commentators have pooh-poohed the need for a central organising principle altogether. Some are non-contrarians suffering from real anxieties about climate change and a process which seems fundamentally unwilling to really engage in the politics beyond denial, self-exemption, and exclusivity. Often they have expectations based on accepting the continued operations of unfettered markets and link these to programs conceived in the mould of "Joint Implementation". Others suggest a general disposition towards "adaptation to" rather than "mitigation of" climate changes is a more realistic way of "submitting to" the future. This occasionally tends – probably not co-incidentally – to accompany the at least residually contrarian views of human-induced climate change, where the need for carbon contraction per se is still questioned. The adversities of climate change are seen as being on a scale from exaggerated to non-existent and the extent of human-causation of climate changes either as over-emphasised or invented and overwhelmingly irrelevant. There is also a line of reasoning which says that with or without climate change, the ability of humans to devise and operate to anything political that has a component of central planning is "ideological" and therefore undesirable and probably ineffective or unachievable in practice anyway.

"FAIRNESS" – AMBIGUITIES, STEREOTYPES AND CONFLICTS

What this reveals is that without a clearly perceived need for carbon contraction, the issue of "fairness" remains as ambiguous as ever it was. "Haves" defend unequal distribution as "fair" because rewards should be proportional to factors such as competence, initiative and sustained effort. "Have-nots" usually defend redistribution towards equal distribution as "fair" pointing to structural disadvantage usually in the context of the traditional arguments between capital and labour. In essence the "North/South" argument is no different from this albeit at a global level.

Distributional fairness, in circumstances where the increasing gap between "haves" and "have-nots" is structurally resisted, is probably easiest identified with the application of welfare economics as at least a necessary buffer against the social distortions of unmitigated market-forces. Distributional fairness, in circumstances where the increasing gap between "haves" and "have-nots" is seen as requiring reversal and even closure, probably embraces everything between the tendency to philanthropy on the one hand and the tendency to say *"when's it my turn?"* on the other. Some people (usually characterised as coming from the left and often but not always in the constituency of "have-nots") appear to be for distributional fairness. And some people (usually characterised as coming from the right and often but not always in the constituency of "haves") appear not to be. These inverse policy attitudes about "fairness" tend to centre on the distribution of socially created and privately partitioned wealth measured as GDP or the "benefit" of income alone. Seen this way arguments concerning "distributional fairness" would seem to have no altered prospects of relevance or realisation one way or the other now than in the past. Although globally the have-nots consistently have been and remain the majority, the "haves" dominate the political decision-taking related to distributional fairness. The resultant status quo embeds a trade-off between these left/right tendencies, stabilised by a well-established legal framework for continuity in property rights with much attendant academic and theoretical work explaining and justifying the "political economy" of this status quo.

In global terms, this framework has not been subjected to any pressures that seriously challenge its sustainability and therefore its legitimacy until the present and the advent of human-induced global climate changes. But in the economics of the global commons everything changes. The rules which developed for the distribution and protection of the socially "created wealth" cannot simply be transposed to encompass the "received wealth" of the commons. The commons are the nearest thing we could identify with providence itself. This is wealth we did not create nor could we. As Tim Wirth of the US State Department said on the subject last year in his lecture at Kew Gardens, *"the economy is a wholly owned subsidiary of the environment."* With this understanding of subsidiarity and the very global scale of the problem, it is not foolish for us to look for new principles of wealth protection and security and distribution related to establishing property rights in the global commons.

BEWARE THE NUMERAIRE AND THE ECONOMICS OF RELUCTANCE

An eminent contributor to the climate change debate – Professor William Nordhaus of Yale University –introduced ideas for "The Economic Management of the Global Commons". However his have relied mainly on neo-classical assumptions in favour of using "Global Cost-Benefit Analysis" for this task. He and many other economists contributing to the IPCC's Second Assessment Report (SAR) maintained that the whole question of what to do about climate change is answerable through recourse to analysis of this kind. Their early results were collated in the SAR published in 1996. The results tended to portray the costs of damages from climate changes as being less than the costs of the actions for emissions abatement necessary to avoid these damages. Bluntly, it was cost effective to go along with climate change, not to resist it. The whole exercise seemed to have the character of a self-fulfilling prophecy in favour of business-as-usual. In effect it was polluters contentiously tending to put a high price on abatement and a low price on damage. The clearest example of the latter being the cash evaluation of global mortality at CO_2 doubling where crudely fifteen dead Chinese equalled one dead American, despite one living Chinese emitting about one tenth of the industrial CO_2 of one American. It was also despite the even more skewed history of the emissions of industrial CO_2 and the fact that at least half the emissions in any year accumulate there with a residence time of about 100 years. (The US alone with an average 3% of global population over the last century and a half remains responsible for just under 35% of accumulated industrial CO_2 emissions to date). In fact all damage evaluation resorted to the snap-shot convention of expressing units of damage as cash values proportional to the average levels of local income. The predictable result was that damages of all kinds in developing countries were devalued relative to the equivalent unit of damage in a developed country. So in spite of the prediction that there would be roughly five times as much damage in developing countries as in developed countries, the overall cash value assigned to the damages in developing countries was about half that of the value of the damages predicted to occur in developed countries. And all this was contributing to a global cost/benefit comparison that broadly suggested that it was cheaper to adapt to the damages from climate changes than to prevent them. Unsurprisingly it was characterised as the economics of genocide in the Indian press.

These analytical results attracted much criticism in the fora of the United Nations such as the United Nations Framework Convention on Climate Change (UNFCCC). These fora had been created especially to mediate and resolve the international policy conflicts of human-induced climate change, not to exacerbate them. The attempt to mediate some of the dispute that followed in the IPCC itself is recorded in the Summary for Policy Makers of Chapter 6 of the Working Group Three contribution to the IPCC SAR. (It is reproduced as Appendix B). The period seems in retrospect to have been one of "stressful learning". Perhaps a more relaxed and robust attitude is now possible with regard to the need to test a whole array of relevant assumptions. Some at least of these will inevitably underpin the next round of analysis and its assessment in the IPCC's Third Assessment Report (TAR) due for completion some time after the year 2000. It seems crucial in the preparations for the TAR that the economic assumptions related to the assignment of property rights in the global commons are reassessed. This is relevant because "meaningful participation" in the UNFCCC of developing countries is one of the conditions the US attaches to its ratification of the Kyoto Protocol.

It is fair to point out that much of the economic analysis reflected in the SAR was conceived during the earlier period of "no regrets". During this period, climate change as a human-induced problem was generally downplayed, alongside the continuing efforts of climate contrarians who were attempting to demonstrate that there was actually "no problem" at all. Their efforts continue at this time in an increasingly implausible way. However, at that time uncertainties to do with the climate changes were clearly more about whether the problem really existed than with concerns about actually *under*-reading the dangers of climate change and potential catastrophes. For example the most of the potential for biogeochemical feedbacks was omitted from the climate models because of their complex non-linearity. Much of this potential would contribute to the upward forcing of global temperature if the mechanisms become active. Large releases of the greenhouse gas methane from beneath tundra and icecaps as they melt will increase warming. Little understood but globally crucial CO_2 sinks in the boreal forests and elsewhere could easily switch off as the temperature rises. Increased water vapour in the atmosphere as a result of ocean warming will compound the warming effect. The range of increased temperature predicted by the models (e.g. with a best-guess 2.5 degrees Celsius rise at CO_2 doubling) are inadequate because the models could only in effect assign a zero value to these factors. The quantitative results acquired have nonetheless achieved the status of "received wisdom" simply by virtue of the frequent reference made to them. When uncertainty is cited in this context it has been of the "even-handed" kind, which positions the results as being between either too high or too low. The quantitative absence of feedbacks in the models results clearly give results that err on the side of caution however.

Faced with these dimensions of complexity, it is not surprising that economics has been struggling to redefine itself in the face of global ecological imperatives. Even Professor Nordhaus by 1997 had loosened his neo-classical belt a little. In a paper for the IPIECA conference last year he commented as follows,

Once we open the door to consider catastrophic changes, a whole new debate is engaged. If we do not know how human activities will affect the thin layer of life-supporting activities that gave birth to and nurture human civilisation and if we cannot reliably judge how potential geophysical changes will affect civilisation or the world around us, can we use the plain vanilla cost-benefit analysis (or even the premium variety in dynamic optimisation models). Should we not be ultraconservative and tilt towards preserving the natural world at the expense of economic growth and development? Do we dare put human betterment before the preservation of natural systems and trust that human ingenuity will bail us out should Nature deal us a nasty hand?

Having asked the questions he asserts a preference for the reasoned judgement of natural and social scientists over the judgement of philosophers and politicians. But he acknowledges the *"massive uncertainties"* and suggests that *"coping with climate change is a worthy challenge for us all."* This is all a far cry from his suggestion a few years back when he suggested that climate change was of no consequence to the US as they had air conditioning and shopping malls. Later he suggested that spotted-owl-equivalents would do just as well as money for the numeraire in the global cost benefit analyses of climate change. It was the one moment of mirth in the period of "stressful learning"; - no one at the UN could understand how 15 dead Chinamen equalled one dead Englishman if a spotted owl equalled a spotted owl. Economics is sometimes more daft than dismal.

So what do social scientists – and most particularly the neoclassical economists - now suggest is the solution to the international distributional struggle? How do we establish the pattern of the ownership of the entitlements to consume a future global carbon budget that is finite and contracting by around 60% so as to be consistent with the objective of the UNFCCC? Is the role of politicians simply to relay the wisdom of social scientists to the negotiations at the UN and so deliver the climate treaty? It certainly has not worked so far. And critically the US is seeking the general acceptance of the international tradability of pollution permits and the assigning of property rights in the global commons is essential to the exercise. Simply trading margins off the existing trends of pollution in the globally inequitable status quo will not bring about the reductions to which the developed countries are now legally committed. Nor particularly will it encourage the involvement of the developing countries on whose participation the process and its success so obviously depends and whose participation in an unavoidable way depends on the issue of distributional fairness globally. As the end-game of Kyoto clearly demonstrates, China, India and the Africa Group of nations are making *"equitable allocations"* and the acceptance of linking Contraction to Convergence a precondition of their acceptance of emissions trading and their involvement in any global solution. This is not an invitation to have another century of neoclassical economics. It is about limits. Its more than a worthy challenge, it is wholly unprecedented in human affairs. Just as capitalism surrounded and contained communism, now the massive uncertainties of climate change engendered by them both, surround and contain them both. Discovering the way forward is a challenge more rigorous than merely worthy.

CLOSING FALSE DICHOTOMIES CAN OPEN THE WAY TO COMMON SENSE

Given disputes over early efforts, (the "descriptive" acceptance of CO_2 property rights proportional to GDP, mortality evaluation being made proportional to income and so on), the allegedly left/right relationship of what is presented as "prescriptive" as against "descriptive" should be re-evaluated in a common sense way. The free-market model is described as being free choice in action and largely unfettered by prescription. People vote with their dollars every time they make a purchase in this market. And this market described as GDP, also currently describes the human causation of climate change. Here is a descriptive example of the distribution of dollars globally in the global free-market in 1990. One third of global population responsible for 90% of fossil fuel emissions had 94% of the dollar-equivalent purchasing power, while the other two thirds responsible for 10% of fossil fuel emissions had the remaining 6% of the dollar-equivalent purchasing power. Notwithstanding, the SAR economists from the wealthy group describing this market with cost-benefit analysis revealed that it was cheaper to adapt to climate change than not. One consequence was that the dollar vote of Bangladesh for example was not big enough to weight the cost/benefit ratio towards prevention and away from adaptation.

The analysis claimed to be merely descriptive and free of prescription, but for the Bangladeshis it was a prescription about learning to adapt to rising sea level. The distinction between what is descriptive and prescriptive is not as clear as the convenience of cost/benefit analysis requires. In the now altered circumstances of human-induced climate change, it is a false dichotomy. Being in any way rational and particularly having recourse to measurements of any kind whatsoever, is being both by definition. Heisenberg clarified that.

This does not mean we should all try and seek immediate nescience. And nor does invite a continued tendency to tolerate the unfortunate free-market requirement for the liquidation of people who have not got any money. What common sense requires is a re-appraisal of our collective prospects freed from some of the absurdities above. It is not wrong to openly contemplate our prospects in terms of a requirement for negotiated prescriptivity in global solutions to global problems.

This re-appraisal may have already begun. At the meeting of IPCC WG3 in Oslo in August 1997, the Energy Modeling Forum (EMF) at Stanford University introduced specifications for modelers that for the first time included the device of a "prescriptive" requirement on the future distribution of global emissions entitlements. It was inter alia that the future costs and benefits of climate change control measures be evaluated in the context of carbon budgets which had been internationally distributed on the basis of a deliberate convergence to equal per capita shares globally by various predetermined dates. One of the core group members, Richard Richels, made the sanguine point that no economist can come up with sensible numbers for the international distribution of the costs of climate change policy *"until the economists had been given the rules of distribution"*. In the absence of agreement yet at the political level at the UNFCCC, the EMF had discontinued the pattern of "descriptive-only" distributional assumptions, as continued in the SAR, and admitted the expedient of at least theoretically prescribing a variety of formal convergence-based examples of distribution. Elsewhere on the theme of "Burden Sharing", in the IPIECA "Symposium on The Economics of Climate Change" (1997) he and his colleague Alan Manne, were even more specific. *"We begin with one widely discussed proposal: a transition to equal per capita emissions rights (globally) by 2030,"* again allowing the expedient of a "prescription" to "solve" what is otherwise insoluble.

It is worth emphasising that there are three key decisions here. One is that an assumption has been made that "prescriptivity" is unavoidably part of the process. Two is that the "prescriptivity" is the result of the application of a central organising principle. Three is that the choice of central organising principle (the convergence to equal per capita) is one which has been *"widely discussed"* which could be read as evidence of the reasonableness of the idea and that these economists share that judgement about that reasonableness.

GCI's CAMPAIGN FOR EQUITY AND SURVIVAL

We have actively advocated the linking of carbon contraction to per capita convergence of emissions entitlements globally for many years. We see this as the practical expression of recognising the global link between equity and survival. It is not equity just for its own sake but the equity of necessity. And we have also pleaded the wisdom or more probably just common sense of recognising the lack of any viable alternatives to this undoubted political novelty. Consequently GCI welcomes the positive attention being given to these ideas by these academics, and also now many bureaucrats, politicians, diplomats and other NGOs.

Our attempt to make a positive contribution to this debate has been not just to pose the need for *"Contraction and Convergence"*. We have also provided and promulgated a planning model with a central organising principle for distributional equity that can demonstrate many scenarios for the generation of integrated global accounts for emissions entitlements. In these, after the given initial distribution of CO_2 emissions entitlements, they are progressively distributed under any chosen (and even revisable) global cap so that the available entitlements become more proportional to population each year. This creates a pattern of international convergence to any chosen future date, from which point forward international emissions entitlements would contract pro rata on an equal per capita basis as determined by the global cap and any revision thereof. "*Contraction and Convergence*" is not a prescription per se, it is way of demonstrating how a global prescription could be negotiated and organised.

At the First Conference of the Parties (CoP1) in April 1995, the Indian Government, drawing on GCI's equity-focused contributions to the IPCC WG3 sections of the SAR, proposed a solution to future global carbon budgeting. It is significant that this happened at a time when the terms of the Berlin Mandate were being drawn up. While the mandate foresaw only strengthened commitments on greenhouse gas (GHG) abatement for the Industrial Countries of Annex One – that is with no GHG abatement commitments being contemplated for the developing countries of Non-Annex One – it is significant that the Indian Environment Minister declared:

> *We face the actuality of scarce resources and the increasing potential for conflict. Protecting the world's environment requires that development be sustainable. It also implies the implementation of a programme for convergence at equitable and sustainable par values for the use of environmental space on a per capita basis globally. In our view equal rights to carbon usage is fundamental to the convention [this is "Contraction and Convergence"]. Policy instruments such as tradable emissions, carbon taxes, and joint implementation may well serve to make matters worse unless they are properly referenced to targets and timetables to be observed by those responsible for the damage to the atmosphere and biosphere. The social, financial and ecological inter-relationships of equity should guide the route to global ecological recovery.*

Since CoP1 in March 1995:

1) GCI has generated the model *"Contraction and Convergence"* Options [CCOptions] to facilitate negotiation in these terms of *"Contraction and Convergence"* approach (see Annex C) and
2) Propagated the international uptake of this approach and the use of the model, through extensive outreach and international travel and attendance at international negotiations under the UNFCCC and related IPCC workshops.
3) The European Parliament adopted a resolution for the United Nations General Assembly Special Session (UNGASS) in June 1997 explicitly based on the approach.
4) The US government tabled a draft protocol 1997 in at the AGBM in April 1997. It enables *"Contraction and Convergence"* to be agreed so for this reason we supported the US draft.
5) The Chinese, Indian, and US Governments all have the CCOptions model with technical support as a result of their requests to GCI to provide it.
6) The Byrd-Hagel resolution was passed through the US Senate also enabling *"Contraction and Convergence"* with much debate recognising the need for controlled growth of emissions entitlements in developing countries in the medium term alongside the need for real reductions in developed countries.
7) The Africa Group of Nations adopted *"Contraction and Convergence"* as its position for the negotiations at CoP3 in Kyoto (see Appendix C) and re-iterated this call during the final plenary session.
8) IPCC WG3 adopted modeling specifications that included per capita convergence paths under different contraction scenarios acknowledging the normative character of this and its novelty within IPCC.
9) The Chinese State Councillor with responsibility for Climate Change and Population policy (Dr Song Jian) has publicly affirmed (October 5th 1997) his government's position in per capita terms which are precisely and explicitly consistent with *"Contraction and Convergence"* (see Appendix D).
10) GLOBE International adopted principles explicitly based on this approach. GLOBE also made the following proposal to a plenary session of the Third Conference of the Parties (CoP3) in Kyoto. *"Let the Conference of the Parties resolve to agree to negotiate a legally binding "Equity Protocol" establishing the principle that the apportionment of global emissions entitlements be deliberately converged to a point of equal per capita shares."* (Full speech is Appendix E)
11) Representatives of the AFLCIO, UNCTAD and the Chicago Board of Trade have publicly expressed support for *"Contraction and Convergence" as* a viable basis on which to negotiate the allocations necessary for emissions trading.
12) The prestigious science journal "Nature" produced a special edition for Kyoto endorsing the approach.

"Contraction and Convergence" is ever more widely seen as a way of overcoming the negotiating impasse between the United States of America, the other Annex One Countries and the Non Annex countries. Potentially it resolves USA's insistence on emissions trading and "globality", where "all" or "key" developing countries must be pre-figured for abatement commitments if the general Kyoto settlement, is to achieve the necessary ratification, whilst meeting the developing countries requirement for "equitable allocations".

The test of whether global emissions trading is relevant or irrelevant is not merely "avoided emissions at least cost." Stated thus, it is not indexed to the objective of the UNFCCC. The test is stabilisation of atmospheric greenhouse gas concentrations at levels that avoid disruption of the global climate system at least cost. Here "cost" means both damage cost (regardless of whether these costs are or can be monetised or not) and emissions abatement cost required for this. However, effecting the "relevant" trade plan is contingent on establishing globally inclusive QELROs and GCI asserts is not possible without *"Contraction and Convergence"*. Thus pursuing "relevant" emissions trade commands by definition convergence as well as reducing the costs of contraction. This way gives the triple win. The first and second parties, the trading partners, win but because of the equitable distribution so do the third parties. In other words all parties and the planet win because through *"Contraction and Convergence"* with trade climate change is avoided at least cost globally.

GLOBAL EQUITY DAWNS AT KYOTO's DARKEST HOUR

At the end of the Kyoto negotiations, the entire debate came to centre on the issues of trade and the assigning of property rights in the future carbon budget. By definition, emissions trading cannot occur until the principle of property rights has been agreed and the entitlements have been assigned and ratified.

At 3.00am when the negotiations were already into injury time, the paragraph in the draft Kyoto Protocol relating to trade came up for acceptance. The US re-iterated their insistence on the acceptance of emissions trading. The governments of China and India, contrary to people's expectations, did not rebut the idea. Instead they responded by saying that acceptance of trade depended on the issue of "equitable allocations" of emissions entitlements. The Africa group of countries intervened, re-iterating that this was why they wanted *"Contraction and Convergence"*. The US replied by saying that they were attentive to the call for *"Contraction and Convergence"* but felt it was too soon. This underlined the remarks made at the GLOBE International workshop in Bonn in October 1997 by US Ambassador Mark Hambley to an international gathering of Parliamentarians, that the idea for such a programme was being taken seriously in Washington.

The meeting went into recess for half and hour. When it was called to order again, Chairman Estrada read out a prepared text (now known as article 16 bis). In effect the COP issues instructions to SBI and SBSTA to elucidate during 1998 the rules, principles, modalities etc relating to trade, in time for CoP4 in Buenos Aires in November 1998. (A simple GCI module of global emissions trading consistent with the objective of the UNFCCC is suggested in Appendix F).

In order to make decisive progress and avoid another effective deferment at CoP4, the most urgent task for 1998 is to have these principles and rules developed in terms of the logic of *"Contraction and Convergence"* and to promote the widest uptake, acceptance and application of this approach. (The GCI Draft Protocol for *"Contraction and Convergence"* is attached in Appendix G).

(The *"Contraction and Convergence"* model is described in Appendix H. The model can be down loaded from GCI's website at http://www.gn.apc.org/gci with a detailed technical description of all its features).

CHARTS 5 & 6 below show gross & per capita *"Contraction and Convergence"* for the world in 6 regions.

The contraction profile is from 2000 for an outcome of atmospheric CO_2 concentrations at 450 ppmv by 2100.

The convergence pattern is from given levels at 2000 to equal per capita shares globally by 2030. Here it is achieved exponentially with population growth included until that date.

The model will compute (1) any budget with (2) any integral with (3) any take-off slope and (4) any landing slope for (5) any number of countries or (6) groupings thereof with (7) linear convergence or (8) any rate of exponential convergence to (9) any chosen target date with (10) the option to "freeze" population growth beyond a chosen date. Moreover (11) each budget can be revised "in progress".

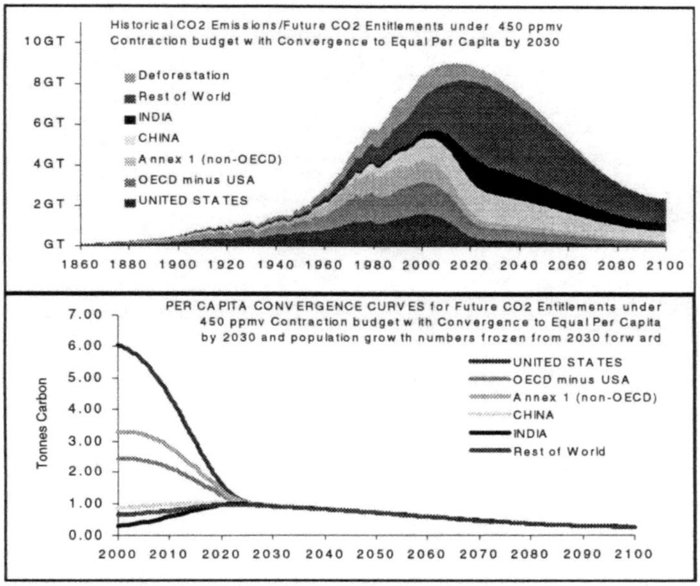

Chart 5: An example of gross and per capita "*Contraction and Convergence*" for 6 world regions

APPENDIX A

Consideration of IPCC CO_2 Scenarios for Carbon Emissions

IPCC have published scenarios for different atmospheric GHG concentration levels resulting from different CO_2 emissions scenarios, as in the graph below. The scenarios run through years 1990 to 2500 and are expressed in gigatonnes carbon (GTC) from CO_2 annually on the left-hand axis. The accumulated emissions (integrals), between 1990 and 2100 only, are summarised in the top right hand corner of the graphic. These are also expressed in GTC. The atmospheric CO_2 concentration curves are not shown, but in each case stabilisation occurs after the respective emissions contraction path of each has completed.

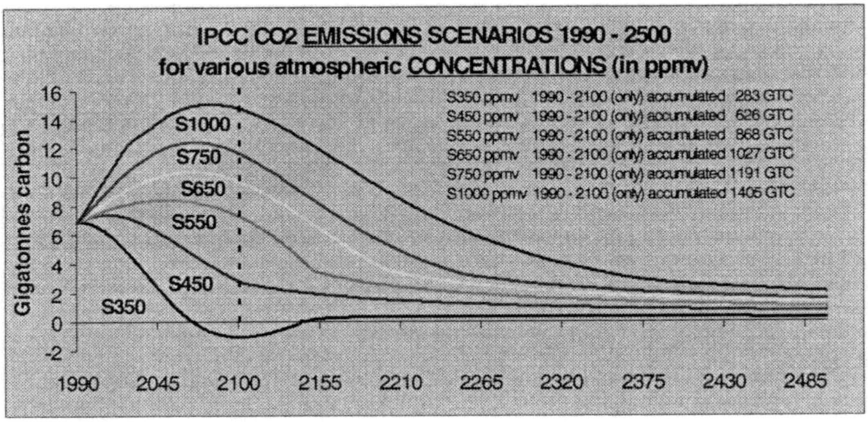

Chart 6: IPCC emissions scenarios 1990-2500 for various atmospheric concentrations

Integrals quoted in the right-hand corner of the graphic are obtained from the data supplied by IPCC. They are different from the integrals published by IPCC in their table, which follows. IPCC's published ranges for accumulated emissions against atmospheric concentration curves are in the table below.

Table 1: IPCC's ranges for accumulated Co_2 emissions

Atmospheric concentration of CO_2 expressed in parts per million by volume (ppmv)	Ranges of accumulated CO_2 emissions expressed in gigatonnes carbon (GTC)
350	300 to 430
450	630 to 650
550	870 to 890
650	1030 to 1190
750	1200 to 1300

APPENDIX B

POLICY MAKERS' SUMMARY SOCIAL COSTS OF CLIMATE CHANGE

Chapter 6 Summary for Policy Makers IPCC WG3 SAR

The literature on the subject in this section is controversial and mainly based on research done on developed countries, often extrapolated to developing countries. There is no consensus about how to value statistical lives or how to aggregate statistical lives across countries. Monetary valuation should not obscure the human consequences of anthropogenic climate change damages, because the value of life has meaning beyond monetary value. It should be noted that the Rio Declaration and Agenda 21 call for human beings to remain at the centre of sustainable development. The approach taken to this valuation might affect the scale of damage reduction strategies. It may be noted that in virtually all of the literature discussed in this section 1) the developing country statistical lives have not been valued equally at the developed country value 2) other damages in developing countries are also not equally valued at the developing country value. Because national circumstances, including opportunity costs, differ, economists sometimes evaluate certain kinds of impacts differently amongst countries.

The benefits of limiting greenhouse gas emissions and enhancing sinks are (a) the climate change damages avoided and (b) the secondary benefits associated with the relevant policies. Secondary benefits include reductions in other pollutants jointly produced with greenhouse gases and biological diversity conserved. Net climate change damages include both market and non-market impacts as far as they can be quantified at the present and, in some cases, adaptation costs. Damages are expressed in net terms to account for the fact that there are some beneficial impacts of global warming as well which are however, dominated by the damage costs. Non-market impacts, such as human health, risk of human mortality and damage to ecosystems, form an important component of available estimates of the social costs of climate change. The literature on monetary valuation of such non-market effects reflects a number of divergent views and approaches. The estimates of non-market damages, however, are highly speculative and not comprehensive.

Non-market damage estimates are a source of major uncertainty in assessing the implications of global climate change for human welfare. While some regard monetary valuation of such impacts as essential to sound decision making, others reject monetary valuation of some impacts, such as risk of human mortality, on ethical grounds. Additionally, there is a danger that entire unique cultures may be obliterated. This is not something that can be considered in monetary terms, but becomes a question of loss of human diversity for which we have no indicators to measure economic value.

The assessed literature contains only a few estimates of the monetised damages associated with doubled CO_2 equivalent concentration scenarios. The estimates are aggregated to a global scale and illustrate the potential impacts of climate change under selected scenarios. Aggregating individual monetised damages to obtain total social welfare impacts implies difficult decisions about equity amongst countries. Global estimates are based upon an aggregation of monetary damages across countries (damages which are themselves implicit aggregations across individuals) that reflect inter-country differences in wealth and income - this fundamentally affects the monetary valuation of damages. Taking income differences as given implies that an equivalent impact in two countries (such as an equal increase in human mortality) would receive very different weights in the calculations of global damages.

To enable choices between different ways of promoting human welfare to be made on a consistent basis, economists have for many years sought to express a wide range of human and environmental impacts in terms of monetary equivalents, using various techniques. The most commonly used of those techniques is an approach based on the observed willingness to pay for various non-market benefits. This is the approach that has been taken in most of the assessed literature.

Human life is an element outside the market and societies may want to preserve it in an equal way. An approach which includes equal valuation of impacts on human life wherever they occur may yield different aggregate global estimates than those reported below. For example, equalising the value of a statistical life at a global average could leave total global damage unchanged but would increase markedly the share of these damages borne by the developing world. Equalising the value at the level typical in the developing countries would increase the monetised damages several times, and would further increase the share of the developing countries in the total damage estimate.

Other aggregation methods can be used to adjust for differences in the wealth or incomes of countries in the calculations of monetary damages. Because the estimates of monetary damage tend to be a higher percentage of national GDP in low-income countries than for high-income countries, aggregation schemes, which adjust for wealth or income effects, are expected to yield higher estimates of global damage than those estimated in this report.

The assessed literature quantifying total damages from 2 to 3 degrees Celsius warming provide a wide range of point estimates for damages given the presumed change in atmospheric greenhouse gas concentrations. The aggregate estimates tend to be a few percent of world GDP, with in general, considerably higher estimates of damage to developing countries as a share of their GDP. The aggregate estimates are subject to considerable uncertainty, but the range of uncertainty cannot be gauged from the literature. The range of estimates cannot be interpreted as a confidence interval given the widely different assumptions and methodologies in the studies. As noted above, aggregation is likely to mask even greater uncertainties about damage components.

Regional or sectoral approaches to estimating the consequences of climate change include a much wider range of estimates of the net economic effects. For some areas, damages are estimated to be significantly greater and could negatively affect economic development. For others, climate change is estimated to increase economic production and present opportunities for economic development. For countries having a diversified industrial economy and an educated and flexible labour force, the limited set of published estimates of damages are of the order of one to a few percent of GDP. For countries having a specialised and natural resources based economy (e.g., heavily emphasising agriculture or forestry), and a poorly developed land-tied labour force, the limited set of published estimates of damages are several times larger. Small islands and low-lying coastal areas are particularly vulnerable. Damages for possible large-scale catastrophes, such as major changes in ocean circulation, are not reflected in these estimates. There is little agreement across studies about the exact magnitude of each category of damages or relative ranking of the damage categories. Climate changes of this magnitude are not expected to be realised for several decades and damages in the interim could be smaller. Damages over a longer period of time might be greater.

IPCC does not endorse any particular range of values for the marginal damage of CO_2 emissions, but published estimates range between \$5 and \$125 (1990 US) per tonne of carbon emitted now. This range of estimates does not represent the full range of uncertainty. The estimates are also based on models that remain simplistic and are limited representations of the actual climate processes in being and are based on earlier IPCC scientific reports. The wide range of damage estimates reflects variations in model scenarios, discount rates and other assumptions. It must be emphasised that the social costs estimates have a wide range of uncertainty because of limited knowledge of impacts, uncertain future technological and socio-economic developments and the possibility of catastrophic events or surprises.

1) The value of a statistical life is defined as the value people assign to a change in the risk of death among a population.
2) The concept of willingness to pay is indicative, based on expressed desires, available resources and information of a human being at a certain point in time. The values may change over time. Also other concepts, such as willingness to accept compensation for damage, have been advanced, but not yet widely applied in the literature and the interpretation and application of willingness to pay and other concepts to the climate problem may evolve.
3) Due to time lags between findings in the natural sciences, their use in determination of potential physical and biological impacts, and subsequent incorporation into economic analysis of climate change, the estimates of climate change damages are based mainly on the scientific results from the 1990 and 1992 IPCC reports.
4) See the Assessment Reports of Working Groups One and Two.

APPENDIX C

TEXT OF THE AFRICA GROUP OF NATIONS STATEMENT TO THE AUGUST 1997 AGBM

Mr. Chairman - Let me begin by adding the Africa Group support for the statement made by the Chairman of the Group of G-77 speaking on behalf of the G-77 and China. Speaking on behalf of the Africa Group, I wish to commend you on the manner in which you have presided over the negotiations in the AGBM process. This has been an extremely difficult session of meetings. However, what is crucial is to try to evaluate whether of not the Parties have made any real headway in trying to strengthen the commitments under Article 4.2 (a) and (b) and advance the implementation of Article 4.1 as was mandated to us by the Berlin Mandate.

We shall pack our bags and return home with a sense of concern about the pace of progress that has been made. Unfavourable climatic conditions will continue to plague our economies, our crops will continue to fail, national external debts will remain a problem to us and our basic social infrastructure will continue to suffer as a result of the impacts of climate change. Yet the Annex One Parties - in particular those parties that have chosen to refrain from giving us their numbers - will go home smiling, celebrating their success in holding back the negotiation process.

We are grateful to those Parties who have given us their proposals and we look forward to evaluating these proposals in order to assess the impacts they will have on our socio-economic infrastructures. Some of our countries are already in the process of implementing activities to address the problem of greenhouse gas emissions (GHG) emissions. We wait in anticipation for Annex One countries to show the necessary commitment. As we negotiate the reduction of GHG, the countries of Africa believe that there should be certain principles that need to be clearly defined.

First: There must be limits on all GHGs if the danger to our climate is to be averted. The IPCC scientific assessment report provides us with the basis for global consensus on such limits. The contrary view therefore does not enjoy much emotional, political, or indeed scientific support.

Second: A globally agreed ceiling of GHG emissions can only be achieved by adopting the principle of per capita emissions rights that fully take into account the reality of population growth and the principle of differentiation.

Third: Achievement of a safe limit to global GHG emissions can be achieved by reducing the emissions of Annex One while at the same time ensuring that there is controlled growth of future emissions from Non-Annex One countries, reflecting our legitimate right to sustainable economic growth. We strongly believe that this will take us along a path to responsible climate management that allows us to reach our goal of defining a mutually agreed point of convergence and sustainable development. Such a convergence, Mr. Chairman, must ensure that we maintain a global ceiling on emissions to prevent dangerous interference with the climate system.

Fourth: When we look at time frames, we believe that insufficient commitment by Annex One countries will only result in delaying our influence on the climate system. If this course is maintained, then we will all suffer and the burden will be even greater for humanity in general. The burden for any future mitigation efforts on those who have not been historically and currently responsible for creating the problem will be greater.

Mr. Chairman we recognise that per capita emissions rights, as a form of differentiation is not an easy goal. It calls for deliberate actions to attain reduction targets over time by Annex One Parties and sustainable growth in the Non-Annex One Parties. To do this Africa would need predictable financial resources, technology transfer, education, training, and public awareness, systematic observation and research. We look forward to renewed co-operation with other Parties in implementing our commitments under Article 4.1.

Mr. Chairman, we must focus our attention on the most appropriate, reasonable, and acceptable time frame for action. There is an over-riding pre-requisite. The time frame can not be too far away into the future if we are to avoid at all costs the dangers that global climate change poses. The current scientific evidence indicates that Africa faces decline in water resources, agricultural production, and economic performance. It is therefore for this reason that we wish to register the seriousness with which we view the effective implementation of the Convention and future agreements emanating from it.

Finally, Mr. Chairman, we would request that the Secretariat take note of the views expressed in this statement on behalf of the African Group of Nations and Parties to the Convention. We look forward to meaningful targets and timeframes for consideration at the next session of the AGBM. - I thank you. (Mrs. Karamanzira - Zimbabwe).

APPENDIX D

EXTRACT FROM SPEECH BY DR SONG JIAN (STATE COUNCILLOR WITH RESPONSIBILITY FOR CLIMATE CHANGE AND POPULATION) AT THE CLOSING CEREMONY OF THE CHINA COUNCIL FOR INTERNATIONAL CO-OPERATION ON ENVIRONMENT AND DEVELOPMENT (OCT 5TH 1997)

We fully understand the worldwide campaign to battle the climate change spearheaded by the European Union and Nordic Countries. The voice of small island states also brooks no ignorance. According to the United Nations Framework

Convention on Climate Change (UNFCCC) and the Berlin Mandate, China bears no responsibility for reducing greenhouse gas emissions. When we ask the opinions of people from all circles, many people, in particular the scientists think that the emissions control standard should be formulated on a per capita basis. According to the UN Charter, everybody is born equal, and has inalienable rights to enjoy modern technological civilisation. Today the per capita consumption is just one tenth of that of the developed countries, one eighth of that of medium developed countries. It is estimated 30-40 years would be needed for China to catch up with the level of medium developed countries. No one is entitled to prohibit families from using refrigerators or those who live or work in such a high temperature of 40 degrees Celsius from enjoying air-conditioning. However the Chinese people and government do have realised their due responsibilities for the global climate change and committed to make efforts to lower down the increase rate of greenhouse gas emissions in the following way:

1) To control the growth of population and stop China's population from growing within thirty years. This is the decisive guarantee of protecting climate and the environment.
2) To put the exploitation of hydroelectric power resources on the top priority, and to explore other renewable energy sources, develop clean coal technologies and reduce the emissions of unit energy consumption.
3) To continue the campaign of afforestations, plant more trees, attend pasture lands, develop ecological agriculture, protect ecological environment, and preserve and strengthen the ability of sinks for absorbing carbon dioxide from the atmosphere. The forest coverage is planned to be doubled to exceed thirty percent of the total territory of the country in the 30-40 years to come.
4) To raise extensively the efficiency of energy utilisation, develop new and less energy-consuming industries and reduce the consumption for per unit of GDP.

In all these fields, the scientific and industrial communities of China shall try the best to make their own contributions, and at the same time, we need international co-operation. I hope that all the members and the experts of the working groups could put forward opinions and creative suggestions.

APPENDIX E

SPEECH BY TOM SPENCER, MEP TO THE PLENARY SESSION OF THE UNFCCC. Kyoto, 9 December 1997

I had the honour to address CoP1 in Berlin. I had the impertinence to suggest that future meetings be held on the great Antarctic iceberg as a way of dramatising the urgency of our task. Kyoto is much more pleasant than an iceberg, but the urgency has intensified.

The nature of the challenge of climate change is now clear. It is not scientific. It is not technological - we are an adaptive species. As John Prescott said yesterday, the challenge is political, and I would add, it is intellectual, institutional, and ethical. After a week of debate about important details, I suggest that the time is right for a statement of even more important principle, After years of debate, about "efficiency" I suggest that your results cannot be brought to success without a statement about equity. The "sense of the Senate" resolution made the fair political point that they could not ratify a treaty, that was not seen to be fair by the American people. A treaty would have to include meaningful participation by developing countries. I would say to my friends in Congress that you cannot ask for that involvement on the basis of efficiency alone. You must specify that the nature of that involvement will be equitable. I am not a Government. I speak this morning only for an organisation of environmentally- committed parliamentarians from 47 countries. With all the humility appropriate to a non-negotiator, may I suggest a text to this great gathering of negotiators.

Many of you know the *"Contraction and Convergence"* analysis. It offers a framework for an answer. It offers an envelope of equity within which we can trade and barter our way to collective sanity in the coming decades.

Let us make a start in this direction. Let the Conference of Parties resolve " to agree to negotiate a legally binding "Equity Protocol" establishing the principle that the apportionment of global emission entitlements be deliberately converged to a point of equal per capita shares at a date to be agreed."

I invite the Government of Japan to propose such a text, which is in line with their policy statement at AGBM7. I invite the European Union, whose efforts entitle it to take a leadership role, to propose such a text. I invite the United States, which rightly takes its stand on the basis of a "global solution to a global problem" to draw the logical conclusion of its own approach, and to propose such a text.

And in response: May I invite the Africa Group whose statement in August led the way to respond to such a text. May I invite the AOSIS states whose very survival depends on our collective success to back such a proposal. May I invite the Governments of China and India to seize this opportunity of an equity protocol that would entrench in our process the principle they have correctly and courageously fought for.

In life the right thing to do is the right thing to do. It is occasionally true in diplomacy that an ethically just answer is also the only available way out of an impasse.

APPENDIX F

The Ideas behind *"Contraction and Convergence"*

Fundamental Assumptions

1/ Survival and IPCC Scenarios

In accordance with the FCCC and best scientific evidence as reported by the IPCC and elsewhere, we assume that total anthropogenic emissions of CO_2 over the 110-year period 1990-2100 must be in line with those set out in IPCC working group 1's 1994[1] and 1995[2] reports under the 'S350' or 'S450' scenarios.

We regard S350 as a necessary target to give a reasonable degree of belief that large-scale damage to the world economy, human lives, and natural ecosystems can be averted. We also used the S450 scenario as an upper limit for consideration; under which there is a chance that damage, though serious, will be containable.

The S550 and higher scenarios we ignore, as it is clear that going for those presents a high probability that positive feedback loops, admittedly underrepresented in the modeling underlying the IPCC scenario calculations, will lead to catastrophic failure of ecosystems and of human societies. Additionally, when a chart of an S550 scenario is examined for the period of the 21st century, rather than for the much longer period used in the IPCC reports, it is clear that, even ignoring these feedbacks, S550 is virtually equivalent to business-as-usual for a large chunk of that century. In other words, adopting it is little different from a do-nothing approach.

Finally we note that, if an aim of 550ppm were agreed, and later it was desired on the basis of new evidence to change policy and aim for 350ppm instead, it would be almost impossible to do so after about 2005. Cutting back from a 450ppm target to 350, on the other hand, could be done up to about 2015.

The CCOptions workbook enables the validity of these assertions to be checked.

2/ Contraction

To implement the above, and based on data reported in [1] and [2], we select a target value for CO_2 emissions to stabilise at and a target year to stabilise in. We select the target value as the highest value, up to a maximum of 40% of the start-year value, which yields a curve that does not dip below the target value. We select the target year to be as far into the future as looks workable. For an S350 scenario we suggest 2050; for an S450 scenario 2070. To compute CO_2 allocations one needs also to make assumptions about future population growth. In (7) below we set out precisely how we have dealt with this issue. Our algorithm enables a schedule of total global emissions to be laid out which adds up to a desired total over the 110-year period and is normally chosen to be in line with tables presented in [1] and [2]. Given the assertion in [1] that the total climate impact of a pattern of CO_2 emissions depends to a first approximation only on the total emissions, and not on the temporal pattern of those emissions, we have a reasonable basis for our assertion that our scenarios can legitimately be represented as S350 and S450 scenarios.

Scenarios where stabilisation occurs after 2100 can be represented by entering a global emissions decline rate for the year 2100. CCOptions does not attempt to look forward beyond 2100.

Note that there is no necessity to stick to 350/450/550 as targets. 400ppm might be a practically negotiable target; from which a change of plan to 350ppm looks to be possible until about 2025.

3/ Convergence

To enable international negotiations to have some chance of success, these emissions need to be allocated by countries in a way that is both achievable and is seen by all to be fair. To that end we specify that the per capita emissions converge from their present diverse values to a standard world value, to be the same for each country, in a 'convergence year' set as 2045. 2045 is chosen partly because using it gave a good balance between what is achievable and what can be seen to be fair. Equally importantly, 2045 is chosen because it is the United Nations' centenary year; and it seems a fruitful idea to invest that year with being the target date for achieving this limited but significant measure of international equity.

Note that there is no algorithmic need for the convergence year to be earlier or later than the contraction year. It might make for ease of negotiation to set them to be the same; but at this stage that seems an unnecessary restriction to us.

4/ Consistency with Rio Commitments

We have assumed that the 'Annex1' countries as defined in the 1992 FCCC will meet their commitment to return emissions to their 1990 levels by 2000. Even if they do not meet them, future allocations should be on that basis, otherwise the IPCC and the UN will be undermined. Modification of the model to be consistent with the Kyoto commitments is in hand, but will make only a minor impact on the process overall.

5/ Algorithms

The approach is a three-stage process, with an initial stage that extrapolates from the most recent year for which actual CO_2 emissions data is available up to 2000, the scope of the pre-Kyoto FCCC commitments. *"Contraction and Convergence"* proper are then started in 2000; a global contraction profile being determined first, and then a separate convergence criterion applied.

5.1/ The Initial Stage

As laid out in (4) above, we assume that Annex1 countries' emissions return to their 1990 values by 2000. For definiteness we assume that emissions from 1995 to 2000 retrace backwards the actual trajectories from 1990 to 1995. We have also had to estimate emissions for 1995, as data is only available from the UN as far as 1994 (as of May 1997). For the other countries for the remainder of the 90s decade we linearly extrapolate the trend of growth from the latest available figures.

5.2/ The Contraction Stage

For this we fit a curve of the form
$z = k+ly+my^2+ny^3+py^4$
where z is the total global industrial CO_2 emissions; y is the year and the parameters k,l,m,n, and p are jointly determined by the following five conditions:

i: z is set for the start year of 2000 as explained in (5.1) above.

ii: z is set at the contraction target value for the contraction target year as explained in (2) above.

iii: dz/dy was set to a target value for this target year (i.e., we assume that the emissions decline rate reaches a target -- commonly zero -- at that time).

iv: dz/dy is set to 1.5% in 2000, reflecting the actual global increase in that year implied by the method described in (5.1). In the workbook, this is an adjustable parameter.

v: The area under the curve, calculated by integration of the above formula, corresponding to the total global emissions over the 110-year period, was set as explained in (2) above.

These conditions yield a set of 5 simultaneous linear equations which are solved to compute the actual values of k,l,m,n, and p.

An allowance of 50GT in total is subtracted from the 110-year target taken from [1] or [2] to take account of deforestation emissions. As even the present annual emissions rate appears to be uncertain to within at least 50% or so; and the total amount of carbon in global forests is orders of magnitude less than that in unburnt fossil fuels, we think it both justifiable and politically helpful to make this very crude approximation. The actual figure of 50GT is towards the optimistic end of the range (30GT to 90GT) used in the six IS92a-f scenarios which were set out in IPCC WG1's 1992 report [3]. We are in the process of adding in also an allowance for 'bunker fuels' – fuel used in civil air and sea transport, which are not allocated to countries and have been excluded from most calculations up to now. We plan to treat these, like deforestation, as a world overhead, and do not suggest allocating them to countries. We suggest that reductions be projected to proceed at the same rate as total global industrial emissions. For similar reasons we have not taken account at all of other greenhouse gases. These shortfalls could easily be addressed in the future when or if accurate and acceptable data becomes available.

An option is also available to enter a preset profile of global emissions from 2000 to 2100 as a set of numbers. CCOptions then applies its convergence algorithm.

5.3/ The Convergence Stage

This, the process to allocate %-shares of global emissions to all the worlds' countries, is a little more complex. The ideas behind our algorithm are:

i/ to start out with 'actual' shares in 2000, as derived by the methods described above;

ii/ to converge all the shares on to actual proportions of global population in the convergence year, which, as we have explained we strongly urge be set to 2045. But the population figures used are subjected to a cap as set out in (7) below.

iii/ the actual degree of convergence allocated in each year to depend on the (potentially capped) population only for that and earlier years. This means that if these procedures were in use, the actual allocations for any given year would only depend on data then available, not on forecasts of population in some year that is then in the future.

The formula used is $s_{y+1}=s_y-(s_y - p_{y+1})\exp(-a(1-t))$

where s_y is the emissions share of a country in year y, p_y is its share of the global population (subject to the cap) in year y, t is the fractional time elapsed between 2000 and the target year (t=0 for 2000 and t=1 at the convergence year of, say, 2045), and a is an arbitrary parameter that determines the rate of convergence.

In the workbook a is adjustable. The higher the value the more the convergence happens towards the end of the convergence period, and vice-versa. Choosing a=4 gives an even balance.

This is intended to be the simplest formula that achieves the aims i-iii above.

An option has been added to converge linearly instead of using the above formula.

6/ Allocations

The actual industrial CO_2 emissions allocations are then made by multiplying the global total value derived from the contraction process by each country's shares derived from the convergence process.

7/ Population Assumptions

We have used UN median figures for forecast population growth by country.

We have then used a cap on population growth for the purposes of allocation of emissions rights. This was done by notionally freezing population numbers (only) for years beyond a 'population cutoff year' at the values for that year. Note there is no assumption being made about what populations will or should be beyond the cutoff year; merely that population growth after that year should not accrue additional emissions rights. In the workbook the year is adjustable. We hold it necessary to adopt some such cap criterion, as otherwise the system will be construed as giving national governments a positive incentive to encourage their populations to grow to obtain an increasing share of emissions rights.

8/ The CCOptions Workbook[2]

CCOptions models CO_2 emissions allocations from the present up to 2100, married with CDIAC data for historical CO_2 emissions.

The user is presented with a panel of parameters which are adjustable within limits; and whenever new values are typed in, graphs of countries' emissions are regenerated for the time-span from 1860 to 2100. Some of these graphs have been used to support well-received presentations at various climate-change conferences from CoP2 in Geneva in July 1996 on.

As well as the base C&C computation, the workbook also includes a feature to investigate what could happen if policy were to switch from a 400ppm initial target, say, to a 350ppm target at some future date. This is done by having a 'Phase two' for contraction, with user-settable start year, target 110-year CO_2 emissions, and target annual emissions. The convergence process is unaffected by this.

For a brief explanation of how to use the workbook see separate document, available from the web site.

---

2 The above ideas on *"contraction and convergence"* have been actualised in a Microsoft Excel workbook, CCOptions; which is available free from the GCI web site. Excel version 5 or later is required.

References

[1]CLIMATE CHANGE The IPCC 1990 and 1992 Assessments. WMO/UNEP June 1992.
[2]CLIMATE CHANGE 1994 Radiative Forcing of Climate Change *and* An Evaluation of the IPCC IS92 Emission Scenarios. Cambridge University Press for the IPCC. 1995
[3]CLIMATE CHANGE 1995 The Science of Climate Change, Summary for Policymakers *and* Technical Summary of the Working Group I report. IPCC. 1996.

APPENDIX G

1. Contraction of Global CO_2 Emissions (e.g., by 2100 as shown for 450 ppmv)

Stabilisation of atmospheric CO_2 concentrations – according to the ultimate objective of the UNFCCC – requires cuts in human emissions of CO_2 in the order of 60%. This level of reductions requires a long-term budget framework with a sequence of short-term controls under constant scientific and political review.

By definition, the solution to the global problem of human-induced climate change must be a *global* solution. On a differentiated basis, all countries must be involved if the solution is to be achievable.

2. Convergence to Equal Per Capita Entitlements by a Negotiated Date (e.g., shown 2030)

Adherence to a globally contracting CO_2 budget over several generations requires unprecedented international co-operation. The international distribution of future emissions entitlements inevitably needs to be determined on the basis of equity between the peoples of all nations. This suggests a process, which corrects the historically and currently polarised patterns of consumption globally.

We know of no method for doing this that has any chance of achieving this co-operation other than a negotiated convergence to equal per capita emissions entitlements globally.

3. International Emissions Allocations

The combination of the global emissions budget and global convergence of international emissions paths gives allocations to all countries. Countries move from their current levels of emissions along trajectories determined by the combination of the *"Contraction and Convergence"* (C&C) curves.

The example shown here (450 ppmv by 2100 with Convergence by 2030) is consistent with the short-term goals of the Berlin Mandate. Annex One contracts by 15% by 2010 while Non-Annex One countries grow collectively at an average of 3% per annum until 2030.

4. Trade of Emissions Permits

National allocations are for emissions permits that can be traded. Contraction rates in developed countries can be reduced through the purchase of permits from developing countries. The resultant cash-flow could be directed to non-fossil energy based development (leap-frogging) in developing countries, while also giving a general incentive to CO_2 emissions avoidance because of the value of permits.

All this provides that global emissions abatement can be achieved at least cost, thus meeting the requirements of global efficiency in a globally equitable and sustainable manner.

APPENDIX H

CONTRACTION AND CONVERGENCE

Draft Proposals for a Climate Change Protocol

Contents

Overview

Part I: Core tasks to achieve "*Contraction and Convergence*":

1 Recognise the prevention of dangerous climate change as an essential global security interest
2 Agree Danger Indicators
3 Decide CO_2 concentration target and timetable
4 Set annual global CO_2 emission budgets according to a contraction formula
5 Calculate relative national shares of the global budget according to a convergence formula
6 Allocate national CO_2 emissions quotas
7 Establish criteria and mechanisms for quota management

Part II: Proposals to achieve *"Contraction and Convergence"*

A. Draft proposals for control of greenhouse gas emissions

1 Set up a system for emissions trading
2 Require International Airline and Shipping companies to purchase CO_2 emission quotas
3 Consider national targets for anthropogenic emissions of other greenhouse gases
4 Devise potential sanctions, penalties, and compensation

B. Draft proposals on climate damages

1 Monitor climate damages
2 Plan for emergencies
3 Consider options for damage compensation and historic debt

C. Draft proposals on Policies and Measures to aid Implementation

1 Establish financial mechanisms to aid implementation
2 Establish mechanism for development and transfer of sustainable technologies
3 Phase out fossil fuel subsidies
4 Require consistency in international policy making
5 Establish a forum for local governments

D. Proposals on research and education

1 Enhance education, training, and awareness
2 Strengthen climate research particularly into feedback processes
3 Examine responsibilities of transnational corporations and finance

Documents available separately on request
Appendix A: contraction formula
Appendix B: convergence formula
Appendix C: cumulative credit/debit formula
Definition of terms

Overview

The Prevention of dangerous climate change is now an essential global security interest. Recognising this interest, this GCI draft document sets out some key tasks necessary for a protocol to stabilise greenhouse gas concentrations at a level that would prevent dangerous anthropogenic interference with the climate system.

Pre-eminently, this requires a rapid "*Contraction*" of all human CO_2 emissions globally. Moreover, unprecedented international co-operation will be required to achieve this. Such a "*comprehensive approach*" is only likely to be adopted by most if not all nations, if it is linked to the simultaneous task of "*Convergence*" towards an equitable and sustainable level of emissions on a per capita basis globally.

This interlinked configuration of "*Contraction and Convergence*" is fundamental to GCI's view of the entire climate change dilemma and its solution. GCI has crafted such an approach, which provides the basis for this document. Part One of the document presents the core tasks of this approach. The first proposal develops the well-established concept of national security interest to include dangerous climate change as a *global* security interest. This emphasises the urgency of this issue and has practical implications for the decision making process and technology development and transfer. The next six proposals spell out practical steps from agreement of danger indicators through to a mechanism for the management of national quota allocations according to a scientific assessment, a contraction formula, and a convergence formula.

Additionally, the document presents further proposals for a protocol. These have been drafted in response to events at the Second Conference of Parties (CoP2) to the Framework Convention on Climate Change in Geneva in July 1996. While we received remarkably wide-ranging support for our proposals for *"Contraction and Convergence"*, many questions were raised about the detailed implications of such an approach.

These further proposals are an attempt to map out some of these implications and are set out in Part Two. The first set concern emissions trading, air/shipping bunker fuels, other greenhouse gases and measures to deter non-compliance. The next set outlines the assessment of and compensation for climate damages, past, present and future. The next five points outline measures to aid implementation, including funding and technology transfer. The final three are concerned with education and research.

At CoP2, the US Government stated that all protocols currently on the table were *"unrealistic and unachievable"*. This effectively sank the AOSIS Protocol proposed by the island-states most vulnerable to climate change. However, many states considered that our *"Global Commons Initiative"* for *"Contraction and Convergence"* was the most plausible basis for a comprehensive long-term protocol.

I Core tasks for achieving "*Contraction and Convergence*"

1 Recognise the prevention of dangerous climate change as an essential global security interest

The reduction of greenhouse gas emissions shall be regarded as an essential global security interest for humanity.

RATIONALE AND POLICY IMPLICATIONS

This proposal generalises the accepted concept of "essential security interest" to the whole of humanity. The gravity of global climate change is greater than the essential security interests of any nation and fundamental to the maintenance of international peace and security during the coming century. This justifies the urgent development of measures proposed in this protocol and places a duty on governments to regard climate change as an international emergency. This also means that action may be taken to secure the use of sustainable technology for the benefit of humanity in much the same way as governments may protect the use of military technology under the Security Exceptions for intellectual property rights as set out in Article 73 of Annex 1C of the Final Act of the Uruguay Round of Multilateral Trade Negotiations as well as national security legislation.

2 Agree danger indicators

Agree a list of specific, quantifiable indicators to define "dangerous anthropogenic interference to the climate system" as stated in Article 2 of the Framework Convention. The indicators shall be listed in the protocol and their quantitative values shall be reviewed annually by SBSTA with advice from IPCC. The values should reflect the precautionary principle and take account of the time lag between emissions and climatic changes. The indicators shall be applicable on any geographical scale to include local damages induced by global climate change.

CoP shall commission detailed recommendations for suitable indicators and their appropriate values. The following suggestions are put forward as a starting point:

a) relative sea level rise to a threatening level at any location;

b) increased coastal erosion that forces evacuation of inhabited land or loss of wetlands;

c) global mean surface temperature rise or significant regional temperature changes over a given time period;

d) an annual increase in the number of tornadoes or tropical cyclones in any region beyond current natural variability;

e) a significant reduction in permafrost area, resulting in release of natural methane;

f) significant changes in regional or sub-regional precipitation resulting in prolonged droughts or frequent flooding;

g) an increase in bush and forest fires above natural variability;

h) loss of marine and terrestrial ecosystems and species;

i) substantial prolonged reduction in marine primary production (plankton, algae);

j) a significant contraction of either polar ice cap and / or glaciers;

k) a major prolonged change in ocean circulation, such as the north Atlantic current or El Nino / Southern Oscillation;

l) a spread attributable to climate change of any human, animal, or plant disease;

m) direct impact of climate change on human health;

n) economic damages attributable to climate change.

These danger indicators should be reviewed on a regular basis by IPCC as new scientific evidence becomes available and revised by the CoP when prudent to do so.

RATIONALE AND POLICY IMPLICATIONS

This Task puts key scientific indicators for dangerous climate change at the centre of the protocol. The "definition of adverse climate change" (FCCC Article. 1, 1) and "dangerous anthropogenic interference with the climate system" (Article. 2) need to be set out in the form of specific danger indicators based on best scientific advice and the precautionary principle (Article 3).

Each indicator should reflect a distinct impact resulting from climate change and should be defined in quantifiable terms on a global, regional and subregional scale. Where possible they should indicate both the danger threshold and potentially dangerous rates of change. The choice of indicators should take into account the considerable time lag between greenhouse gas emissions and subsequent climatic response.

The prospect of breaching any one of the indicators should be sufficient to require preventive action, on the grounds of the precautionary principle, even though it may only affect one area of the world directly. On the basis of equity, damages must be avoided on a local scale since many of the most vulnerable countries are neither responsible for global warming nor in a position to adapt to the harmful effects. Local climate-change-induced damages may not be traded off against the pursuit of global economic growth, because it is impossible to sensibly create a consensus around the quantification of such damages in terms of rising risks of ecological and political instability (see also II.A.1).

3 Decide CO_2 concentration target and timetable

A stabilisation target and timetable shall be agreed for atmospheric CO_2 concentrations. The target should be set by applying the precautionary principle to avoid the danger indicators agreed under Section I.1 (above). This target shall initially be 350 ppmv to be achieved by the year 2100 . The target shall be reviewed every five years by IPCC as new scientific evidence becomes available and revised by the CoP if the danger indicators clearly show that it is prudent to do so.

RATIONALE AND POLICY IMPLICATIONS

Dangerous changes in the climate system are a result of specific greenhouse gas concentrations in the atmosphere, so it is necessary to set a specific ceiling on the amount of CO_2 in the atmosphere. In absence of better scientific understanding of climatic feedback processes (listed in section II.D.2), the initial target of atmospheric CO_2 stabilisation at 350ppmv by the year 2100 is chosen such that the system remains close to the bounds of our present knowledge. This can be achieved by following a future emissions scenario in which the cumulative CO_2 emissions are similar to those of IPCC S350.

When reviewing this target, IPCC should take into account predicted changes in the concentration of other greenhouse gases (considered further in section II.A.3).

4 Set annual global CO_2 emissions budgets according to a "contraction" formula

The Conference of the Parties shall set a net global anthropogenic carbon emissions budget for each year throughout the period of contraction to meet the stabilisation objective as defined in I.3 above. This shall be calculated according to a mathematical formula which defines an emissions scenario that leads to stabilisation of emissions at around 2045. The formula may be reviewed annually five years in advance to take account of revisions to the stabilisation target set under proposal 3 as well as changes in natural sinks and sources based on scientific advice of the IPCC.

RATIONALE AND POLICY IMPLICATIONS

Stabilisation of CO_2 concentrations requires a global ceiling on emissions which contracts over time until they reach the target of 350 ppmv. The "contraction" formula would define a realistic emissions scenario which avoids both unachievable annual emissions reductions and temporary net negative emissions in the future. The formula should refer to all anthropogenic sources and sinks of CO_2, i.e. changes in emissions resulting directly from human activity, including anthropogenic changes of natural sinks and sources. Natural "equilibrium" sources and sinks should be considered in the science behind the emissions scenario, but should not be included in this global anthropogenic emissions budget.

The formula will need to be reviewed annually to take account of changes both to the concentration target and of actual sinks and sources of CO_2. However, in order to reduce uncertainty for both planning and trading purposes each annual review will adjust the budgets five years in advance.

Implementation will require setting up a Scientific Panel drawn from the SBSTA and approved by the CoP to agree the formula and recommend annual CO_2 budgets. The target date of 2045 is suggested because it lies within the window for realistic contraction and coincides with the centenary of the United Nations. For the purpose of illustration only, a sample "contraction" formula is available from GCI.

5 Calculate relative national shares of the global budget according to a "convergence" formula

Each country shall be allocated an annual, relative share of the global emissions budget (set according to section I.4 above) using a consistent formula to calculate the proportion for each country for each year. The allocation shall be set such that national shares move gradually from present emissions levels to equal per capita emissions levels by a fixed "convergence" year (e.g., 2045). Provision shall be made for bunker fuels for shipping and air transport (see II.A.2).

RATIONALE AND POLICY IMPLICATIONS

This task gives effect to the principle of equity set out in Article 3.1 of the FCCC while recognising that equal per capita allocations would be neither acceptable nor feasible for Annex I countries if implemented immediately. The formula therefore provides a predictable and viable method of achieving a convergence to equity. National shares would be based initially on current emissions levels, or for Annex 1 countries, those levels already specified by commitments under the UNFCCC, and would then converge to the same per capita level by the target date of 2045. After this relative shares would remain constant.

Calculation with the convergence formula will use the UN median population estimates. These population statistics may be reviewed if necessary at the request of a majority of the CoP.

However, after a fixed year (e.g., the convergence date) population figures could be frozen.

The relative national shares are independent of the annual scientific reviews, although the actual allocations of emissions (allocated in section I.6) will of course vary according to the global budget agreed. Annual shares would be calculated for all countries, whether or not they are Parties to the Protocol, and shall be set out in an Annex to the protocol as tables.

For the purpose of illustration only, a sample "convergence" formula is available from GCI.

The establishment of fixed national emissions budgets will encourage investment in appropriate technology and allow for the possibility of orderly emissions trading, but this depends crucially on having finite net CO_2 budgets calculated according to a formula that produces a predictable level of permissible emissions from the present to the stabilisation date, to achieve the concentration target set in Section I.3 above.

6 Allocate national CO_2 emissions quotas

National emissions quotas shall be calculated for each year by multiplying each country's relative share set by the convergence formula (agreed according to Section I.5) by the annual global emissions budget (set by the "contraction" formula agreed according to Section I.4). These quotas shall be measured in tonnes of carbon.

RATIONALE AND POLICY IMPLICATIONS

The national allocations are described as 'quotas' rather than entitlements or rights to emit CO_2 in order to emphasise that the atmosphere and climate system are a global commons which cannot be appropriated by any state or person but whose use must now be shared by common consent. The formula approach provides for the most effective way of establishing a consistent and equitable allocation of emissions quotas.

7 Establish criteria and mechanisms for quota management

Establish a mechanism for the international management of quota allocation, accounting and verification based on criteria consistent with these proposals. The quotas shall cover "net anthropogenic emissions" resulting from human activities, including reduction of natural sinks minus deliberate natural sinks enhancement.

RATIONALE AND POLICY IMPLICATIONS

The allocations would refer to "net anthropogenic emissions" of CO_2. This is defined as all emissions resulting from human activities, including reduction of natural sinks minus deliberate natural sinks enhancement. However, this should exclude changes in natural sources and sinks caused directly by global climate change. This definition encourages sustainable forestry, for example, but avoids crediting the existing natural resource endowment of each country. We recognise that some countries have deliberately retained such resources whilst others have already diminished them, therefore this will be accounted for in the task concerning historic debt (section II.B.3). Natural sink enhancement will only be credited within national boundaries, not within any global commons. For example, credit will not be given for enhancement of the CO_2 sink into the ocean.

The proposed mechanisms would also arbitrate in case of dispute over budgeting anthropogenic sources and sinks.

It is not necessary to allow for crediting of sink enhancement or emissions reduction within other countries, known as Joint Implementation, since this can effectively be achieved through the emissions quotas trading system (section II.A.1).

This proposal completes the procedure for applying *"Contraction and convergence"*.

II. Proposals to achieve *"Contraction and Convergence"*

A. Draft Proposals for Control of Greenhouse Gas Emissions

1 Set up a system for emissions trading

Establish mechanisms for real-time emissions trading between parties to the Protocol under strict conditions of "Contraction and Convergence". Trading shall be restricted to a proportion of the annual emissions quotas, defined as permits, and limited to one year in advance. The mechanism should be transparent and avoid financial feedbacks that would undermine the ultimate aim of the Convention and its protocols.

RATIONALE AND POLICY IMPLICATIONS

The criteria for a trading regime should set out the basis on which possible trading regimes will be assessed. These will inevitably need to balance the different interests and needs of parties to the Convention. In deciding on a suitable mechanism, parties should consider the work already done by UNCTAD in this area and continue to involve it in the development thereof.

A trading regime would initially be developed under the SBI which would continue to supervise the process for CoP. However, once agreed, emissions trading would require a set of robust institutions capable of acting on behalf and under the supervision of CoP. Trading must be open, transparent, efficient and well regulated. The regulatory bodies must also ensure that TEPs are not bought or sold under coercion of any kind.

Emissions trading would also address the issue of "carbon leakage" to countries with relatively lower emissions. This would particularly be the case if purchasing rights of permits were extended to TNCs. As the mechanism comes into effect, Joint Implementation will no longer become an issue as the trade will help to redress emission imbalances, while the incentive to invest in climate friendly technology remains by releasing quotas for trade. The mechanism must also ensure that trading is developed primarily as an efficient means of reducing emissions and must not compromise future generations on the principles of inter-generational equity. The implications of this international trade on intra-national equity will also need to be examined and addressed. The increased value of emission quotas from international trade must not detrimentally affect the disadvantaged in national societies. *"Contraction and Convergence"* applies within countries as well as between them.

Given the historic link between growth of monetary GDP and CO_2 emissions, there is a danger that trade in CO_2 quotas and any other increase in financial activity as a result of this Protocol will simply increase global purchasing power leading to an increase in CO_2 emissions. This would be contradictory to the purpose of the Convention. Another danger is that trade in Emissions Quotas increases international financial liquidity to produce inflation or other instability, as occurred following the OPEC oil price rises. These dangers might be addressed by the creation of a carbon-free "green currency." A study of the potential of carbon-free currency should be commissioned for consideration by CoP in future.

Moreover, existing purchasing power disparities between developing and developed countries can only be aggravated by creating a trade mechanism which continues to exploit the arbitrary advantage enjoyed in the international markets of economies based on hard currencies. This is especially relevant in view of the fact that it ignores the much higher efficiencies of soft currency based economies when national dollars-per-ton efficiencies are adjusted for purchasing power parity (PPP).

2 Require International Airline and Shipping companies to purchase CO_2 emission quotas

CO_2 emitted by all aircraft or shipping must be accounted for within the global emissions budget by requiring international transport companies to purchase emissions permits.

RATIONALE AND POLICY IMPLICATIONS

Currently, aircraft on international flights, or shipping in international waters, can purchase untaxed fossil fuel which is not included in any national carbon account. Air transport in particular is one of the fastest growing sectors of global CO_2 emissions, and must be constrained in a similar manner to other carbon intensive economic activities. Fiscal measures which might achieve this, such as an international tax on bunker fuels, would require a global authority to predict and control demand. On the other hand, if airline or shipping companies have to purchase emissions permits the market will ensure a "level playing field" with land-based transport.

This measure will also ensure that all emissions are constrained within the contraction / convergence global budget (sections I. 3,4,5), whilst the price of the quotas is passed on to the consumer of the transport, rather than becoming the responsibility of governments.

It may be possible to extend this option to purchase emissions quotas to other TransNational Corporations (TNCs). This would have the advantage of discouraging "leakage" or carbon-intensive production to countries where emissions are cheaper, since the TNCs could purchase emissions quotas from countries with a surplus without needing to relocate.

3 Consider national targets for anthropogenic emissions of other greenhouse gases

Draw up a timetable for agreeing constraints on concentrations of greenhouse gases other than CO_2, with specific targets for each gas, as scientific knowledge of their biogeochemical cycles becomes sufficiently reliable. The allocation of budgets should be based on the same task of equity as used above for CO_2, whilst giving special consideration to each country's need to exploit its natural resources and agriculture.

RATIONALE AND POLICY IMPLICATIONS

Concentrations of greenhouse gases other than CO_2 are rising fast and must be brought under the FCCC. Gases with similar sources and atmospheric lifetimes may be grouped together and some substitution of these may be possible within the national budgets. Some international trading of emissions quotas may also be considered. However, agreement on one greenhouse gas should not be delayed whilst awaiting better knowledge of the other greenhouse gases.

Greenhouse gases other than CO_2 fall into two main categories:

1) For wholly man-made gases such as most CFCs, HCFCs and SF6, an early agreement could be reached. Production of some of these gases is already constrained under the Montreal Protocol
2) For protection of stratospheric ozone. Their major sources and sinks are already sufficiently quantifiable. National budgets for these gases should be allocated using the same principles of "*Contraction and Convergence*", allowing for trading if necessary, as outlined for CO_2 in Sections I.2 through to I.5 above. Some of these gases have long lifetimes, and therefore their Global Warming Potential relative to each other is effectively independent of the time horizon used. These gases could be substituted within national budgets. However, on the basis of inter-generational equity and long-term sustainability, production of the very long-lived gases should be tightly constrained and phased out as soon as possible. If a time horizon were agreed, the global emissions budget for the shorter-lived gases could be linked to that for CO_2 by means of their Global Warming Potential (defined by IPCC).

Methane (CH_4) and Nitrous Oxide (N_2O) are significant greenhouse gases produced by a mixture of anthropogenic and natural sources. The sources and sinks are still poorly defined. Methane emissions are rapidly increasing both as a result of changing agricultural practice (cattle, irrigation) and leakage from natural gas installations. However, Methane has a short atmospheric lifetime and therefore it is not helpful to define a Global Warming Potential relative to CO_2. To account for the greater potency of Methane as a greenhouse gas, international standards of best practice should be agreed for industry and agriculture. Financial penalties collected when these standards are breached, could be used to fund the development and transfer of improved technology to reduce Methane emissions.

Nitrous Oxide has a longer lifetime than Methane but makes a smaller contribution to current global warming, and is less well understood. More research is urgently needed on the cycles of both of these gases.

4 Devise potential sanctions, penalties, and compensation

Request the Secretariat to draw up options for a system of proportional progressive sanctions and penalties for non-compliance with the protocol, taking account of experience of international, regional, and national legal instruments and the review of selected non-compliance, dispute resolution, and implementation review procedures (FCCC/CP/1995/Misc.2) prepared by the interim secretariat.

Income raised from penalties could contribute to measures to aid implementation and relieve damages, as listed in Section II.B.3 below.

RATIONALE AND POLICY IMPLICATIONS

There are at present no incentives to comply with the FCCC or the protocol. By contrast, the 1994 GATT agreement contains extensive procedures for dispute settlement, including conciliation, mediation, arbitration, establishment of panels, rights of third parties, remedies, and compensation.

Penalties should be in proportion to excess emissions of greenhouse gases, and considerably higher than the current purchase price of tradable emissions entitlements or investment benefits from the excessive CO_2 emissions. The virtue of a tough system of penalties is that it will encourage compliance and reduce the likelihood of it being used.

B. Draft Proposals on Climate Damages

1 Monitor climate damages

Require Parties to prepare an inventory of damages and damage trends, both past and present, directly attributable to climate change. These should include human health and mortality, economic impacts, loss of habitats, species and biodiversity, impact on agriculture, and coastal erosion.

RATIONALE AND POLICY IMPLICATIONS

Damage due to climate change has already begun and estimates of possible future damage, casualties and refugees have been made. The aim of this measure is to compile a comprehensive database of damages which would provide both a benchmark for the danger indicators proposed in proposal 2 above. Research on climate-related damages should be funded by an international programme whose emphasis should be on impacts to developing countries.

Data on damages should be presented in their original units rather than using monetised values. Such values based on the method of "willingness to pay" imply rights by income which is fundamentally inequitable. International aggregation of damage data for the purpose of a global cost-benefit analysis is not appropriate, since the majority of damages will be inflicted on developing countries whereas most of the CO_2 emissions, and hence mitigation costs, are currently the responsibility of the industrialised countries.

2 Plan for emergencies

Require all parties to draw up contingency provisions for future emergencies which may arise from climate change, such as flooding, drought, crop failure, or disease.

RATIONALE AND POLICY IMPLICATIONS

Damage due to climate change has already begun and estimates of possible future damage, casualties, and refugees have been made. These should be updated regularly as a basis for arranging emergency relief and compensation payments. Liability for compensation payments is considered in Section II.B.3.

Contingency plans should be also prepared for the potential relocation of entire populations from small island states and low-lying regions to the territory of Annex 1 countries.

3 Consider options for damage compensation and historic debt

Request that the Secretariat prepare a study of options for damage compensation due to climate change based on best practice in insurance and national compensation schemes, and for historic debt in relation to emissions by Annex I countries prior to 1990, for consideration at CoP4.

RATIONALE AND POLICY IMPLICATIONS

Insurance companies and governments are currently making large scale payments in respect of damages caused by asbestos, radiation, and other anthropogenic causes. European and US law also include provision for civil, statutory, and criminal liability for environmental damage. Countries and companies which emit CO_2 above the danger level should be under no illusion that they can avoid paying for damage caused by excess CO_2 emissions, thus increasing incentives to comply with the Convention and Protocol. Careful consideration needs to be given to historic emissions when foreseeability of damages could not reasonably have been expected and the precedents under tort thus become less relevant. The argument of historic debt, nevertheless, still holds as developing countries will, in future, not have the option of unrestrained emissions on which developed countries based their historic growth.

For these reasons, Annex 1 Parties should be required to make provision for climate change related damages in their national budgets and planning mechanism. CO_2 has a long lifetime in the atmosphere, and historic data shows that a constant fraction of emissions has remained airborne, although there is no guarantee for this fraction to remain constant in the future. Therefore, to a first approximation, a country's responsibility for global warming depends on its cumulative emissions integrated over time. Industrialised countries have thus accumulated an historic debt compared to developing countries. Applying the principle of per-capita equity to historic data, it is possible to create a formula for calculating cumulative debits or credits, which might be used for allocating damage liability.

An international panel should be set up by SBSTA to resolve disputes over damage claims; this should include advice from both climate scientists and insurance experts.

For the purpose of such calculations, estimates of cumulative emissions of CO_2 should include historic deforestation and other land use changes. Some countries have preserved much of their natural forest resources, whereas others have exploited them and consequently have more land on which to replant new forests. For consistency in accounting, it is necessary to include this form of historic debt if national emission entitlements are to be based on net anthropogenic emissions (i.e., including changes in natural sources and sinks).

C. Draft Proposals on Policies and Measures to Aid Implementation

1 Establish financial mechanisms to aid implementation

The SBI shall establish mechanisms by which money can be reliably collected and distributed for global programmes to implement the Convention. Decisions on spending shall be made through a fair and transparent mechanism accountable to CoP. Funding is required for the following:

Climate Research (see section II.D.2)
Education, training, and awareness (see section II.D.1).
Monitoring climate damage (see section II.B.1).
Technology Transfer (see section II.C.2).
Activities currently funded by the GEF
Administration of the CoP and the Secretariat
Administration of emissions trading (see section I.7 and II.A.1)
Emergency Relief and Damage Compensation (see section II.B.2 and II.B.3)

Funding sources may include:

A tax on trading of emissions entitlements (as in section II.A.1)
Penalties for non-compliance (as in section II.A.4)
According to cumulative historic debt (as in section II.B.3)

RATIONALE AND POLICY IMPLICATIONS

Activities which are critical to the success of implementing the Convention should be able to rely on secure funding. This is already required for the Financial Mechanism under Art. 11 of the FCCC, and the commitments set out in Art. 4 c, d and h. However, present arrangements are unsatisfactory as funds are reliant on the goodwill of a few Parties which then control their use. Binding mechanisms must be set up to enable money to flow directly from the cause of the climate change problem (i.e., greenhouse gas emissions) towards funding its solution. This would encourage a reduction in CO_2 emissions, although the main mechanism for achieving this should remain the allocation of emission entitlements according to Contraction (section I.3) and Convergence (section I.4).

Liability to pay compensation for damages should be linked directly to cumulative historic debt as outlined in section II.B.3.

2 Establish mechanism for development and transfer of sustainable technologies

The development, diffusion, and use of the most sustainable technologies, practices and processes which minimise greenhouse gas emissions shall be regarded as an essential global security interest as defined in section I.1. To this end, a mechanism should be established under the protocol to aid the development within and transfer to developing countries of sustainable technologies.

The transfer of outdated or second-hand, carbon-inefficient technology should be controlled and preferably prohibited.

RATIONALE AND POLICY IMPLICATIONS

Explicit measures and positive incentives are urgently needed to stimulate the development and transfer of the most climate-friendly sustainable technology. Such technology should:

(a) be carbon efficient or based on renewable energy sources,
(b) be an appropriate scale for the local community which it serves,
(c) be independent of expensive supplies or repairs from distant companies,
(d) not damage the local environment (as do, for example, large hydropower schemes),

(e) not entail unacceptable risk (as do nuclear power stations).

Measures to stimulate the development and diffusion of climate-friendly technology would include:

(a) establish a technology transfer fund (paid for according to section II.C.1),
(b) fund research, development and diffusion of sustainable technologies, particularly in developing countries,
(c) create an international inventory of climate-friendly technology,
(d) promote best practice and sharing expertise between countries,
(e) identify gaps and opportunities in national and multilateral technology programmes,
(f) limit the period for which patents on relevant technologies may be held without being exploited for the benefit of humanity.

A substantial transfer of outdated, carbon inefficient technology from developed to developing countries is currently widespread. This has the effect of prolonging the detrimental contribution of this machinery or technology on global carbon emissions.

3 Phase out fossil fuel subsidies

Agree mechanisms to phase out all subsidies for fossil-fuels. Transitional procedures and financial support shall be made available to developing countries in order to achieve a smooth transition and avoid penalising the poor.

RATIONALE AND POLICY IMPLICATIONS

Subsidies for the use of fossil fuels both increase global warming and distort the efficient allocation of resources through markets. Definitions of subsidies should be carefully established and the needs of the poor fully taken into consideration.

4 Require consistency in international policy-making

Set up an international task force to ensure that all international agencies, treaties, and agreements actively contribute to reducing CO_2 emissions as an essential global security interest. This should be backed by a resolution to the United Nations General Assembly to require all international agencies, treaties, and agreements take the climatic implications of their actions fully into account and to support the implementation of the UNFCCC and its protocols.

RATIONALE AND POLICY IMPLICATIONS

There is a danger that trade liberalisation under the 1994 GATT agreement, investment by the World Bank and economic measures by the IMF and central banking system as well as actions by other agencies increase CO_2 emissions or otherwise undermine the objectives of the FCCC. The aim of this task is to ensure that all international decision-making takes full account of the Climate Convention. A precedent for this task was established by the Ministerial Declaration on the Contribution of the World Trade Organisation to Achieving Greater Coherence in Global Economic Policymaking of 15 December 1993.

5 Establish a forum for local governments

Recognising the important contribution local governments make in implementing climate friendly policies and technologies at a local level, an international forum on climate change for local governments should be established with rights of representation as an observer to CoP under Art. 7 (6). The role of this forum would be to allow sharing of experiences and making relevant policy recommendations to national governments and CoP. This body could also discuss issues of intra-national equity arising from the restraint of carbon emissions.

RATIONALE AND POLICY IMPLICATIONS

A considerable proportion of sustainable policies and technologies are most effectively implemented at a local level and local governments can, therefore, play an important role in emissions reductions. The exchange of information on a local level would complement the transfer of technology on a national level as outlined in Section 2.9. The forum would also be the ideal organ to voice any problems of intra-national inequities arising from national and international implementation of the protocol and specifically the trade mechanism.

D. Tasks on Research and Education

1 Enhance education, training, and awareness

Require all Parties to increase their commitment to education, training, and public awareness of climate change under Article 6. This should encourage an holistic perspective emphasising the links between local and global processes, both natural and economic, and convey the full scale of the climate change problem as a survival issue. This must be underpinned by better understanding of the basic science of climate change.

Commission a feasibility study to establish options for an international system of greenhouse gas labelling to provide immediate public awareness of the extent to which a particular product or service contributes to emissions of CO_2 or other greenhouse gases and to create incentives to reduce carbon consumption.

Require parties to report annually to the CoP on the following matters:

(a) the extent and nature of education, training and public awareness undertaken;
(b) organisation, finance, and target audience thereof;
(c) independent evaluation of the effectiveness of such education and training;
(d) results of an independent poll of public awareness of climate change, its effects, and measures needed to reduce emissions to a sustainable level;
(e) progress on greenhouse gas labelling;
(f) actions to curb promotion of carbon consumption.

RATIONALE AND POLICY IMPLICATIONS

Most countries have done relatively little to fulfill their commitments under Article 6 of the UNFCCC. Raising awareness of the way in which fossil fuels and other products increase global warming has a significant role in motivating individuals, companies and countries to reduce such consumption. If all goods and services carried a conspicuous and unambiguous statement of the CO_2 or other greenhouse gases emitted by their production and disposal, this would enable people to make more informed choices. Public awareness and education on global warming is in constant competition with the advertising power of the air, car, and fossil fuel industries. In view of the gravity of climate change, serious consideration should also be given to measures which curb the promotion of activities responsible for carbon emissions, similar to those currently applied to alcohol, tobacco, pharmaceuticals, and other drugs. There is much public confusion over the science and impacts of climate change. This should be tackled both as a core part of school curricula, and as specifically tailored training programmes for employees, stressing measures for alleviating impacts of their field of work on global climate change, and vice versa.

2 Strengthen climate research, particularly into feedback processes

Increase research into physical, biogeochemical, social, and economic climatic feedback processes. Such research requires international consistency to ensure that the different processes can be combined in integrated models, whilst also encouraging researchers to investigate new processes. This should be co-ordinated principally through the International Geosphere Biosphere Programme

(IGBP) and World Climate Research Programme (WCRP). IPCC and SBSTA (when calculating the emissions ceiling according to section I.2) should attempt to take account of the cumulative risk from low-probability positive feedbacks. When IPCC presents predictions or scenarios of global climate changes, there should also be less emphasis on global average temperature trends, and more on regional differences, particularly regarding the sulphate aerosol effect.

RATIONALE AND POLICY IMPLICATIONS

There are particular dangers that the effects of climate change may trigger uncontrollable feedback loops that accelerate global climate change further than would be expected from current IPCC predictions, and exacerbate regional anomalies. Some known examples of such natural climatic feedback processes are listed below:

a) Clouds: Different levels of clouds can have opposite effects on climate, since they both trap terrestrial radiation and reflect solar radiation. Clouds and water vapour also transport much latent heat. Thus it is difficult to resolve the feedbacks. Clouds may also be seeded by sulphate aerosols, both natural and anthropogenic.
b) Ice sheets: Polar warming may increase ice melt, but also increase snowfall. This affects:
   * Global albedo (proportion of sunlight reflected to space).
   * Sea level.
   * Local ocean salinity and albedo and hence circulation and ecology.
c) Ocean circulation:
   * Arctic warming or increased freshwater input to the North Atlantic (ice melt, rainfall) could halt deep water formation, weaken the North Atlantic Current, and thus make NW Europe much colder.
   * Increased frequency of El Nino circulation in the Pacific affects climate world-wide.
d) Ocean ecology: Changes in phytoplankton ecology might be caused by:
   * increased stratification of the water column, due to surface warming. would reduce nutrient (Nitrate, Phosphate, Iron) supply from bottom waters.
   * increased nutrient runoff from rivers due to changing land-use.
   * increased UV-B flux due to stratospheric ozone loss.

   The effect of this might be to change:
   * the export of Carbon from surface water to deep water by the "biological pump".
   * the alkalinity of surface water due to calcifying algae. Calcification puts CO_2 back into the atmosphere.
   * the production of greenhouse gases (N_2O, CH_4, other hydrocarbons)

* the production of Dimethyl Sulphide which (as sulphate aerosols) seeds clouds over the ocean and thus significantly influences global albedo.
* ocean fish stocks.

e) Terrestrial ecology: warming and drying in high latitudes may cause:
  * release of greenhouse gases CH_4 and N_2O from melting permafrost.
  * release of CO_2 from peat bogs, if drying allows aeration.
  * increased forest fires and subsequent CO_2 release.
  * reduced snow cover, particularly if forest replaces tundra, with consequent change in albedo.

  The "CO_2 fertilisation effect" may cause:
  * increased CO_2 uptake by terrestrial vegetation.
  * changes in evapotranspiration and hence local rainfall and groundwater.

  Vegetation changes will affect albedo and also surface roughness which affects wind.
f) Flooding of coastal wetlands by rising sea level, particularly in the tropics, may cause substantial release of CH_4 and N_2O.

There is presently much confusion among policymakers about the cooling effect of sulphate aerosols produced by fossil fuel combustion. It should be stressed that the aerosol effect is short-lived and local, whereas the warming by CO_2 is long-lived and global. Such differences are obscured by an emphasis on global average temperature trends, which should be remedied within IPCC.

3 Examine responsibilities of transnational corporations and finance

Set up two working groups to examine the role of international finance and transnational corporations respectively in relation to CO_2 emissions, in order to report on:

(a) the extent to which their activities contribute to or abate global warming.
(b) differential responsibilities between corporate sectors.
(c) examples or models of good practice in regulation, incentives, statutory, or voluntary codes, reporting requirements or other measures for encouraging corporations and investment fund managers to cut CO_2 emissions.
(d) the contribution or otherwise of small and medium businesses to global warming.
(e) recommendations for specific measures to enhance the contribution of investment finance and transnational corporations to stabilising CO_2 emissions.

RATIONALE AND POLICY IMPLICATIONS

Representatives of transnational corporations are active participants in the climate change negotiations and major players in both sustainable and carbon energy industries. The majority of world trade and a substantial proportion of global production is conducted by transnational corporations, while international financial flows are the driving force in trade and industry throughout the world, often more powerful than governments. The aim of these two study groups would be to examine the specific role of these two major types of economic agents to identify measures to enhance their contribution to stabilising greenhouse gas emissions.

ANNEX

A1 UNITED NATIONS FRAMEWORK CONVENTION ON CLIMATE CHANGE

UNITED NATIONS

1992

The Parties to this Convention,

Acknowledging that change in the Earth's climate and its adverse effects are a common concern of humankind,

Concerned that human activities have been substantially increasing the atmospheric concentrations of greenhouse gases, that these increases enhance the natural greenhouse effect, and that this will result on average in an additional warming of the Earth's surface and atmosphere and may adversely affect natural ecosystems and humankind,

Noting that the largest share of historical and current global emissions of greenhouse gases has originated in developed countries, that per capita emissions in developing countries are still relatively low and that the share of global emissions originating in developing countries will grow to meet their social and development needs,

Aware of the role and importance in terrestrial and marine ecosystems of sinks and reservoirs of greenhouse gases,

Noting that there are many uncertainties in predictions of climate change, particularly with regard to the timing, magnitude and regional patterns thereof,

Acknowledging that the global nature of climate change calls for the widest possible cooperation by all countries and their participation in an effective and appropriate international response, in accordance with their common but differentiated responsibilities and respective capabilities and their social and economic conditions,

Recalling the pertinent provisions of the Declaration of the United Nations Conference on the Human Environment, adopted at Stockholm on 16 June 1972,

Recalling also that States have, in accordance with the Charter of the United Nations and the principles of international law, the sovereign right to exploit their own resources pursuant to their own environmental and developmental policies, and the responsibility to ensure that activities within their jurisdiction or control do not cause damage to the environment of other States or of areas beyond the limits of national jurisdiction,

Reaffirming the principle of sovereignty of States in international cooperation to address climate change,

Recognizing that States should enact effective environmental legislation, that environmental standards, management objectives and priorities should reflect the environmental and developmental context to which they apply, and that standards applied by some countries may be inappropriate and of unwarranted economic and social cost to other countries, in particular developing countries,

Recalling the provisions of General Assembly resolution 44/228 of 22 December 1989 on the United Nations Conference on Environment and Development, and resolutions 43/53 of 6 December 1988, 44/207 of 22 December 1989, 45/212 of 21 December 1990 and 46/169 of 19 December 1991 on protection of global climate for present and future generations of mankind,

Recalling also the provisions of General Assembly resolution 44/206 of 22 December 1989 on the possible adverse effects of sea-level rise on islands and coastal areas, particularly low-lying coastal areas and the pertinent provisions of General Assembly resolution 44/172 of 19 December 1989 on the implementation of the Plan of Action to Combat Desertification,

Recalling further the Vienna Convention for the Protection of the Ozone Layer, 1985, and the Montreal Protocol on Substances that Deplete the Ozone Layer, 1987, as adjusted and amended on 29 June 1990,

Noting the Ministerial Declaration of the Second World Climate Conference adopted on 7 November 1990,

Conscious of the valuable analytical work being conducted by many States on climate change and of the important contributions of the World Meteorological Organization, the United Nations Environment Programme and other organs, organizations and bodies of the United Nations system, as well as other international and intergovernmental bodies, to the exchange of results of scientific research and the coordination of research,

Recognizing that steps required to understand and address climate change will be environmentally, socially and economically most effective if they are based on relevant scientific, technical and economic considerations and continually re-evaluated in the light of new findings in these areas,

Recognizing that various actions to address climate change can be justified economically in their own right and can also help in solving other environmental problems,

Recognizing alsothe need for developed countries to take immediate action in a flexible manner on the basis of clear priorities, as a first step towards comprehensive response strategies at the global, national and, where agreed, regional levels that take into account all greenhouse gases, with due consideration of their relative contributions to the enhancement of the greenhouse effect,

Recognizing further that low-lying and other small island countries, countries with low-lying coastal, arid and semi-arid areas or areas liable to floods, drought and desertification, and developing countries with fragile mountainous ecosystems are particularly vulnerable to the adverse effects of climate change,

Recognizing the special difficulties of those countries, especially developing countries, whose economies are particularly dependent on fossil fuel production, use and exportation, as a consequence of action taken on limiting greenhouse gas emissions,

Affirming that responses to climate change should be coordinated with social and economic development in an integrated manner with a view to avoiding adverse impacts on the latter, taking into full account the legitimate priority needs of developing countries for the achievement of sustained economic growth and the eradication of poverty,

Recognizing that all countries, especially developing countries, need access to resources required to achieve sustainable social and economic development and that, in order for developing countries to progress towards that goal, their energy consumption will need to grow taking into account the possibilities for achieving greater energy efficiency and for controlling greenhouse gas emissions in general, including through the application of new technologies on terms which make such an application economically and socially beneficial,

Determined to protect the climate system for present and future generations,

Have agreed as follows:

ARTICLE 1

DEFINITIONS [1]

For the purposes of this Convention:

1. "Adverse effects of climate change" means changes in the physical environment or biota resulting from climate change which have significant deleterious effects on the composition, resilience or productivity of natural and managed ecosystems or on the operation of socio-economic systems or on human health and welfare.

2. "Climate change" means a change of climate which is attributed directly or indirectly to human activity that alters the composition of the global atmosphere and which is in addition to natural climate variability observed over comparable time periods.

3. "Climate system" means the totality of the atmosphere, hydrosphere, biosphere and geosphere and their interactions.

4. "Emissions" means the release of greenhouse gases and/or their precursors into the atmosphere over a specified area and period of time.

5. "Greenhouse gases" means those gaseous constituents of the atmosphere, both natural and anthropogenic, that absorb and re-emit infrared radiation.

6. "Regional economic integration organization" means an organization constituted by sovereign States of a given region which has competence in respect of matters governed by this Convention or its protocols and has been duly authorized, in accordance with its internal procedures, to sign, ratify, accept, approve or accede to the instruments concerned.

7. "Reservoir" means a component or components of the climate system where a greenhouse gas or a precursor of a greenhouse gas is stored.

8. "Sink" means any process, activity or mechanism which removes a greenhouse gas, an aerosol or a precursor of a greenhouse gas from the atmosphere.

9. "Source" means any process or activity which releases a greenhouse gas, an aerosol or a precursor of a greenhouse gas into the atmosphere.

---

[1] Titles of articles are included solely to assist the reader.

ARTICLE 2

OBJECTIVE

The ultimate objective of this Convention and any related legal instruments that the Conference of the Parties may adopt is to achieve, in accordance with the relevant provisions of the Convention, stabilization of greenhouse gas concentrations in the atmosphere at a level that would prevent dangerous anthropogenic interference with the climate system. Such a level should be achieved within a time-frame sufficient to allow ecosystems to adapt naturally to climate change, to ensure that food production is not threatened and to enable economic development to proceed in a sustainable manner.

ARTICLE 3

PRINCIPLES

In their actions to achieve the objective of the Convention and to implement its provisions, the Parties shall be guided, inter alia, by the following:

1. The Parties should protect the climate system for the benefit of present and future generations of humankind, on the basis of equity and in accordance with their common but differentiated responsibilities and respective capabilities. Accordingly, the developed country Parties should take the lead in combating climate change and the adverse effects thereof.

2. The specific needs and special circumstances of developing country Parties, especially those that are particularly vulnerable to the adverse effects of climate change, and of those Parties, especially developing country Parties, that would have to bear a disproportionate or abnormal burden under the Convention, should be given full consideration.

3. The Parties should take precautionary measures to anticipate, prevent or minimize the causes of climate change and mitigate its adverse effects. Where there are threats of serious or irreversible damage, lack of full scientific certainty should not be used as a reason for postponing such measures, taking into account that policies and measures to deal with climate change should be cost-effective so as to ensure global benefits at the lowest possible cost. To achieve this, such policies and measures should take into account different socio-economic contexts, be comprehensive, cover all relevant sources, sinks and reservoirs of greenhouse gases and adaptation, and comprise all economic sectors. Efforts to address climate change may be carried out cooperatively by interested Parties.

4. The Parties have a right to, and should, promote sustainable development. Policies and measures to protect the climate system against human-induced change should be appropriate for the specific conditions of each Party and should be

integrated with national development programmes, taking into account that economic development is essential for adopting measures to address climate change.

5. The Parties should cooperate to promote a supportive and open international economic system that would lead to sustainable economic growth and development in all Parties, particularly developing country Parties, thus enabling them better to address the problems of climate change. Measures taken to combat climate change, including unilateral ones, should not constitute a means of arbitrary or unjustifiable discrimination or a disguised restriction on international trade.

ARTICLE 4

COMMITMENTS

1. All Parties, taking into account their common but differentiated responsibilities and their specific national and regional development priorities, objectives and circumstances, shall:

(a) Develop, periodically update, publish and make available to the Conference of the Parties, in accordance with Article 12, national inventories of anthropogenic emissions by sources and removals by sinks of all greenhouse gases not controlled by the Montreal Protocol, using comparable methodologies to be agreed upon by the Conference of the Parties;

(b) Formulate, implement, publish and regularly update national and, where appropriate, regional programmes containing measures to mitigate climate change by addressing anthropogenic emissions by sources and removals by sinks of all greenhouse gases not controlled by the Montreal Protocol, and measures to facilitate adequate adaptation to climate change;

(c) Promote and cooperate in the development, application and diffusion, including transfer, of technologies, practices and processes that control, reduce or prevent anthropogenic emissions of greenhouse gases not controlled by the Montreal Protocol in all relevant sectors, including the energy, transport, industry, agriculture, forestry and waste management sectors;

(d) Promote sustainable management, and promote and cooperate in the conservation and enhancement, as appropriate, of sinks and reservoirs of all greenhouse gases not controlled by the Montreal Protocol, including biomass, forests and oceans as well as other terrestrial, coastal and marine ecosystems;

(e) Cooperate in preparing for adaptation to the impacts of climate change; develop and elaborate appropriate and integrated plans for coastal zone management, water resources and agriculture, and for the protection and rehabilitation of areas, particularly in Africa, affected by drought and desertification, as well as floods;

(f) Take climate change considerations into account, to the extent feasible, in their relevant social, economic and environmental policies and actions, and employ appropriate methods, for example impact assessments, formulated and determined nationally, with a view to minimizing adverse effects on the economy, on public health and on the quality of the environment, of projects or measures undertaken by them to mitigate or adapt to climate change;

(g) Promote and cooperate in scientific, technological, technical, socio-economic and other research, systematic observation and development of data archives related to the climate system and intended to further the understanding and to reduce or eliminate the remaining uncertainties regarding the causes, effects, magnitude and timing of climate change and the economic and social consequences of various response strategies;

(h) Promote and cooperate in the full, open and prompt exchange of relevant scientific, technological, technical, socio-economic and legal information related to the climate system and climate change, and to the economic and social consequences of various response strategies;

(i) Promote and cooperate in education, training and public awareness related to climate change and encourage the widest participation in this process, including that of non-governmental organizations; and

(j) Communicate to the Conference of the Parties information related to implementation, in accordance with Article 12.

2. The developed country Parties and other Parties included in Annex I commit themselves specifically as provided for in the following:

(a) Each of these Parties shall adopt national[2] policies and take corresponding measures on the mitigation of climate change, by

---

[2] This includes policies and measures adopted by regional economic integration organizations.

limiting its anthropogenic emissions of greenhouse gases and protecting and enhancing its greenhouse gas sinks and reservoirs. These policies and measures will demonstrate that developed countries are taking the lead in modifying longer-term trends in anthropogenic emissions consistent with the objective of the Convention, recognizing that the return by the end of the present decade to earlier levels of anthropogenic emissions of carbon dioxide and other greenhouse gases not controlled by the Montreal Protocol would contribute to such modification, and taking into account the differences in these Parties' starting points and approaches, economic structures and resource bases, the need to maintain strong and sustainable economic growth, available technologies and other individual circumstances, as well as the need for equitable and appropriate contributions by each of these Parties to the global effort regarding that objective. These Parties may implement such policies and measures jointly with other Parties and may assist other Parties in contributing to the achievement of the objective of the Convention and, in particular, that of this subparagraph;

(b) In order to promote progress to this end, each of these Parties shall communicate, within six months of the entry into force of the Convention for it and periodically thereafter, and in accordance with Article 12, detailed information on its policies and measures referred to in subparagraph (a) above, as well as on its resulting projected anthropogenic emissions by sources and removals by sinks of greenhouse gases not controlled by the Montreal Protocol for the period referred to in subparagraph (a), with the aim of returning individually or jointly to their 1990 levels these anthropogenic emissions of carbon dioxide and other greenhouse gases not controlled by the Montreal Protocol. This information will be reviewed by the Conference of the Parties, at its first session and periodically thereafter, in accordance with Article 7;

(c) Calculations of emissions by sources and removals by sinks of greenhouse gases for the purposes of subparagraph (b) above should take into account the best available scientific knowledge, including of the effective capacity of sinks and the respective contributions of such gases to climate change. The Conference of the Parties shall consider and agree on methodologies for these calculations at its first session and review them regularly thereafter;

(d) The Conference of the Parties shall, at its first session, review the adequacy of subparagraphs (a) and (b) above. Such review shall be carried out in the light of the best available scientific information and assessment on climate change and its impacts, as well as relevant technical, social and economic information. Based on this review, the Conference of the Parties shall take appropriate action, which may include the adoption of amendments to the commitments in subparagraphs (a) and (b) above. The Conference of the Parties, at its first session, shall also take decisions regarding criteria for joint implementation as indicated in subparagraph (a) above. A second review of subparagraphs (a) and (b) shall take place not later than 31 December 1998, and thereafter at regular intervals determined by the Conference of the Parties, until the objective of the Convention is met;

(e) Each of these Parties shall :

(i) Coordinate as appropriate with other such Parties, relevant economic and administrative instruments developed to achieve the objective of the Convention; and

(ii) Identify and periodically review its own policies and practices which encourage activities that lead to greater levels of anthropogenic emissions of greenhouse gases not controlled by the Montreal Protocol than would otherwise occur;

(f) The Conference of the Parties shall review, not later than 31 December 1998, available information with a view to taking decisions regarding such amendments to the lists in Annexes I and II as may be appropriate, with the approval of the Party concerned;

(g) Any Party not included in Annex I may, in its instrument of ratification, acceptance, approval or accession, or at any time thereafter, notify the Depositary that it intends to be bound by subparagraphs (a) and (b) above. The Depositary shall inform the other signatories and Parties of any such notification.

3. The developed country Parties and other developed Parties included in Annex II shall provide new and additional financial resources to meet the agreed full costs incurred by developing country Parties in complying with their obligations under Article 12, paragraph 1. They shall also provide such financial resources, including for the transfer of technology, needed by the developing country Parties to meet the agreed full incremental costs of implementing measures that are covered by paragraph 1 of this Article and that are agreed between a developing country Party and the international entity or entities referred to in Article 11, in accordance with that Article. The implementation of these

commitments shall take into account the need for adequacy and predictability in the flow of funds and the importance of appropriate burden sharing among the developed country Parties.

4. The developed country Parties and other developed Parties included in Annex II shall also assist the developing country Parties that are particularly vulnerable to the adverse effects of climate change in meeting costs of adaptation to those adverse effects.

5. The developed country Parties and other developed Parties included in Annex II shall take all practicable steps to promote, facilitate and finance, as appropriate, the transfer of, or access to, environmentally sound technologies and know-how to other Parties, particularly developing country Parties, to enable them to implement the provisions of the Convention. In this process, the developed country Parties shall support the development and enhancement of endogenous capacities and technologies of developing country Parties. Other Parties and organizations in a position to do so may also assist in facilitating the transfer of such technologies.

6. In the implementation of their commitments under paragraph 2 above, a certain degree of flexibility shall be allowed by the Conference of the Parties to the Parties included in Annex I undergoing the process of transition to a market economy, in order to enhance the ability of these Parties to address climate change, including with regard to the historical level of anthropogenic emissions of greenhouse gases not controlled by the Montreal Protocol chosen as a reference.

7. The extent to which developing country Parties will effectively implement their commitments under the Convention will depend on the effective implementation by developed country Parties of their commitments under the Convention related to financial resources and transfer of technology and will take fully into account that economic and social development and poverty eradication are the first and overriding priorities of the developing country Parties.

8. In the implementation of the commitments in this Article, the Parties shall give full consideration to what actions are necessary under the Convention, including actions related to funding, insurance and the transfer of technology, to meet the specific needs and concerns of developing country Parties arising from the adverse effects of climate change and/or the impact of the implementation of response measures, especially on:

(a) Small island countries;

(b) Countries with low-lying coastal areas;

(c) Countries with arid and semi-arid areas, forested areas and areas liable to forest decay;

(d) Countries with areas prone to natural disasters;

(e) Countries with areas liable to drought and desertification;

(f) Countries with areas of high urban atmospheric pollution;

(g) Countries with areas with fragile ecosystems, including mountainous ecosystems;

(h) Countries whose economies are highly dependent on income generated from the production, processing and export, and/or on consumption of fossil fuels and associated energy-intensive products; and

(i) Land-locked and transit countries.

Further, the Conference of the Parties may take actions, as appropriate, with respect to this paragraph.

9. The Parties shall take full account of the specific needs and special situations of the least developed countries in their actions with regard to funding and transfer of technology.

10. The Parties shall, in accordance with Article 10, take into consideration in the implementation of the commitments of the Convention the situation of Parties, particularly developing country Parties, with economies that are vulnerable to the adverse effects of the implementation of measures to respond to climate change. This applies notably to Parties with economies that are highly dependent on income generated from the production, processing and export, and/or consumption of fossil fuels and associated energy-intensive products and/or the use of fossil fuels for which such Parties have serious difficulties in switching to alternatives.

ARTICLE 5

RESEARCH AND SYSTEMATIC OBSERVATION

In carrying out their commitments under Article 4, paragraph 1(g), the Parties shall:

(a) Support and further develop, as appropriate, international and intergovernmental programmes and networks or organizations aimed at defining, conducting, assessing and financing research, data collection and systematic observation, taking into account the need to minimize duplication of effort;

(b) Support international and intergovernmental efforts to strengthen systematic observation and national scientific and technical research capacities and capabilities, particularly in developing countries, and to promote access to, and the exchange of, data and analyses thereof obtained from areas beyond national jurisdiction; and

(c) Take into account the particular concerns and needs of developing countries and cooperate in improving their endogenous capacities and capabilities to participate in the efforts referred to in subparagraphs (a) and (b) above.

ARTICLE 6

EDUCATION, TRAINING AND PUBLIC AWARENESS

In carrying out their commitments under Article 4, paragraph 1(i), the Parties shall:

(a) Promote and facilitate at the national and, as appropriate, subregional and regional levels, and in accordance with national laws and regulations, and within their respective capacities:

(i) The development and implementation of educational and public awareness programmes on climate change and its effects;

(ii) Public access to information on climate change and its effects;

(iii) Public participation in addressing climate change and its effects and developing adequate responses; and

(iv) Training of scientific, technical and managerial personnel.

(b) Cooperate in and promote, at the international level, and, where appropriate, using existing bodies:

(i) The development and exchange of educational and public awareness material on climate change and its effects; and

(ii) The development and implementation of education and training programmes, including the strengthening of national institutions and the exchange or secondment of personnel to train experts in this field, in particular for developing countries.

ARTICLE 7

CONFERENCE OF THE PARTIES

1. A Conference of the Parties is hereby established.

2. The Conference of the Parties, as the supreme body of this Convention, shall keep under regular review the implementation of the Convention and any related legal instruments that the Conference of the Parties may adopt, and shall make, within its mandate, the decisions necessary to promote the effective implementation of the Convention. To this end, it shall:

(a) Periodically examine the obligations of the Parties and the institutional arrangements under the Convention, in the light of the objective of the Convention, the experience gained in its implementation and the evolution of scientific and technological knowledge;

(b) Promote and facilitate the exchange of information on measures adopted by the Parties to address climate change and its effects, taking into account the differing circumstances, responsibilities and capabilities of the Parties and their respective commitments under the Convention;

(c) Facilitate, at the request of two or more Parties, the coordination of measures adopted by them to address climate change and its effects, taking into account the differing circumstances, responsibilities and capabilities of the Parties and their respective commitments under the Convention;

(d) Promote and guide, in accordance with the objective and provisions of the Convention, the development and periodic refinement of comparable methodologies, to be agreed on by the Conference of the Parties, inter alia, for preparing inventories of greenhouse gas emissions by sources and removals by sinks, and for evaluating the effectiveness of measures to limit the emissions and enhance the removals of these gases;

(e) Assess, on the basis of all information made available to it in accordance with the provisions of the Convention, the implementation of the Convention by the Parties, the overall effects of the measures taken pursuant to the Convention, in particular environmental, economic and social effects as well as their cumulative impacts and the extent to which progress towards the objective of the Convention is being achieved;

(f) Consider and adopt regular reports on the implementation of the Convention and ensure their publication;

(g) Make recommendations on any matters necessary for the implementation of the Convention;

(h) Seek to mobilize financial resources in accordance with Article 4, paragraphs 3, 4 and 5, and Article 11;

(i) Establish such subsidiary bodies as are deemed necessary for the implementation of the Convention;

(j) Review reports submitted by its subsidiary bodies and provide guidance to them;

(k) Agree upon and adopt, by consensus, rules of procedure and financial rules for itself and for any subsidiary bodies;

(l) Seek and utilize, where appropriate, the services and cooperation of, and information provided by, competent international organizations and intergovernmental and non-governmental bodies; and

(m) Exercise such other functions as are required for the achievement of the objective of the Convention as well as all other functions assigned to it under the Convention.

3. The Conference of the Parties shall, at its first session, adopt its own rules of procedure as well as those of the subsidiary bodies established by the Convention, which shall include decision-making procedures for matters not already covered by decision-making procedures stipulated in the Convention. Such procedures may include specified majorities required for the adoption of particular decisions.

4. The first session of the Conference of the Parties shall be convened by the interim secretariat referred to in Article 21 and shall take place not later than one year after the date of entry into force of the Convention. Thereafter, ordinary sessions of the Conference of the Parties shall be held every year unless otherwise decided by the Conference of the Parties.

5. Extraordinary sessions of the Conference of the Parties shall be held at such other times as may be deemed necessary by the Conference, or at the written request of any Party, provided that, within six months of the request being

communicated to the Parties by the secretariat, it is supported by at least one third of the Parties.

6. The United Nations, its specialized agencies and the International Atomic Energy Agency, as well as any State member thereof or observers thereto not Party to the Convention, may be represented at sessions of the Conference of the Parties as observers. Any body or agency, whether national or international, governmental or non-governmental, which is qualified in matters covered by the Convention, and which has informed the secretariat of its wish to be represented at a session of the Conference of the Parties as an observer, may be so admitted unless at least one third of the Parties present object. The admission and participation of observers shall be subject to the rules of procedure adopted by the Conference of the Parties.

ARTICLE 8

SECRETARIAT

1. A secretariat is hereby established.

2. The functions of the secretariat shall be:

(a) To make arrangements for sessions of the Conference of the Parties and its subsidiary bodies established under the Convention and to provide them with services as required;

(b) To compile and transmit reports submitted to it;

(c) To facilitate assistance to the Parties, particularly developing country Parties, on request, in the compilation and communication of information required in accordance with the provisions of the Convention;

(d) To prepare reports on its activities and present them to the Conference of the Parties;

(e) To ensure the necessary coordination with the secretariats of other relevant international bodies;

(f) To enter, under the overall guidance of the Conference of the Parties, into such administrative and contractual arrangements as may be required for the effective discharge of its functions; and

(g) To perform the other secretariat functions specified in the Convention and in any of its protocols and such other functions as may be determined by the Conference of the Parties.

3. The Conference of the Parties, at its first session, shall designate a permanent secretariat and make arrangements for its functioning.

ARTICLE 9

SUBSIDIARY BODY FOR SCIENTIFIC AND TECHNOLOGICAL ADVICE

1. A subsidiary body for scientific and technological advice is hereby established to provide the Conference of the Parties and, as appropriate, its other subsidiary bodies with timely information and advice on scientific and technological matters relating to the Convention. This body shall be open to participation by all Parties and shall be multidisciplinary. It shall comprise government representatives competent in the relevant field of expertise. It shall report regularly to the Conference of the Parties on all aspects of its work.

2. Under the guidance of the Conference of the Parties, and drawing upon existing competent international bodies, this body shall:

(a) Provide assessments of the state of scientific knowledge relating to climate change and its effects;

(b) Prepare scientific assessments on the effects of measures taken in the implementation of the Convention;

(c) Identify innovative, efficient and state-of-the-art technologies and know-how and advise on the ways and means of promoting development and/or transferring such technologies;

(d) Provide advice on scientific programmes, international cooperation in research and development related to climate change, as well as on ways and means of supporting endogenous capacity-building in developing countries; and

(e) Respond to scientific, technological and methodological questions that the Conference of the Parties and its subsidiary bodies may put to the body.

3. The functions and terms of reference of this body may be further elaborated by the Conference of the Parties.

ARTICLE 10

SUBSIDIARY BODY FOR IMPLEMENTATION

1. A subsidiary body for implementation is hereby established to assist the Conference of the Parties in the assessment and review of the effective implementation of the Convention. This body shall be open to participation by all Parties and comprise government representatives who are experts on matters related to climate change. It shall report regularly to the Conference of the Parties on all aspects of its work.

2. Under the guidance of the Conference of the Parties, this body shall:

(a) Consider the information communicated in accordance with Article 12, paragraph 1, to assess the overall aggregated effect of the steps taken by the Parties in the light of the latest scientific assessments concerning climate change;

(b) Consider the information communicated in accordance with Article 12, paragraph 2, in order to assist the Conference of the Parties in carrying out the reviews required by Article 4, paragraph 2(d); and

(c) Assist the Conference of the Parties, as appropriate, in the preparation and implementation of its decisions.

ARTICLE 11

FINANCIAL MECHANISM

1. A mechanism for the provision of financial resources on a grant or concessional basis, including for the transfer of technology, is hereby defined. It shall function under the guidance of and be accountable to the Conference of the Parties, which shall decide on its policies, programme priorities and eligibility criteria related to this Convention. Its operation shall be entrusted to one or more existing international entities.

2. The financial mechanism shall have an equitable and balanced representation of all Parties within a transparent system of governance.

3. The Conference of the Parties and the entity or entities entrusted with the operation of the financial mechanism shall agree upon arrangements to give effect to the above paragraphs, which shall include the following:

(a) Modalities to ensure that the funded projects to address climate change are in conformity with the policies, programme priorities and eligibility criteria established by the Conference of the Parties;

(b) Modalities by which a particular funding decision may be reconsidered in light of these policies, programme priorities and eligibility criteria;

(c) Provision by the entity or entities of regular reports to the Conference of the Parties on its funding operations, which is consistent with the requirement for accountability set out in paragraph 1 above; and

(d) Determination in a predictable and identifiable manner of the amount of funding necessary and available for the implementation of this Convention and the conditions under which that amount shall be periodically reviewed.

4. The Conference of the Parties shall make arrangements to implement the above-mentioned provisions at its first session, reviewing and taking into account the interim arrangements referred to in Article 21, paragraph 3, and shall decide whether these interim arrangements shall be maintained. Within four years thereafter, the Conference of the Parties shall review the financial mechanism and take appropriate measures.

5. The developed country Parties may also provide and developing country Parties avail themselves of, financial resources related to the implementation of the Convention through bilateral, regional and other multilateral channels.

ARTICLE 12

COMMUNICATION OF INFORMATION RELATED TO IMPLEMENTATION

1. In accordance with Article 4, paragraph 1, each Party shall communicate to the Conference of the Parties, through the secretariat, the following elements of information:

(a) A national inventory of anthropogenic emissions by sources and removals by sinks of all greenhouse gases not controlled by the Montreal Protocol, to the extent its capacities permit, using comparable methodologies to be promoted and agreed upon by the Conference of the Parties;

(b) A general description of steps taken or envisaged by the Party to implement the Convention; and

(c) Any other information that the Party considers relevant to the achievement of the objective of the Convention and suitable for inclusion in its communication, including, if feasible, material relevant for calculations of global emission trends.

2. Each developed country Party and each other Party included in Annex I shall incorporate in its communication the following elements of information:

(a) A detailed description of the policies and measures that it has adopted to implement its commitment under Article 4, paragraphs 2(a) and 2(b); and

(b) A specific estimate of the effects that the policies and measures referred to in subparagraph (a) immediately above will have on anthropogenic emissions by its sources and removals by its sinks of greenhouse gases during the period referred to in Article 4, paragraph 2(a).

3. In addition, each developed country Party and each other developed Party included in Annex II shall incorporate details of measures taken in accordance with Article 4, paragraphs 3, 4 and 5.

4. Developing country Parties may, on a voluntary basis, propose projects for financing, including specific technologies, materials, equipment, techniques or practices that would be needed to implement such projects, along with, if possible, an estimate of all incremental costs, of the reductions of emissions and increments of removals of greenhouse gases, as well as an estimate of the consequent benefits.

5. Each developed country Party and each other Party included in Annex I shall make its initial communication within six months of the entry into force of the Convention for that Party. Each Party not so listed shall make its initial communication within three years of the entry into force of the Convention for that Party, or of the availability of financial resources in accordance with Article 4, paragraph 3. Parties that are least developed countries may make their initial communication at their discretion. The frequency of subsequent communications by all Parties shall be determined by the Conference of the Parties, taking into account the differentiated timetable set by this paragraph.

6. Information communicated by Parties under this Article shall be transmitted by the secretariat as soon as possible to the Conference of the Parties and to any subsidiary bodies concerned. If necessary, the procedures for the communication of information may be further considered by the Conference of the Parties.

7. From its first session, the Conference of the Parties shall arrange for the provision to developing country Parties of technical and financial support, on request, in compiling and communicating information under this Article, as well as

in identifying the technical and financial needs associated with proposed projects and response measures under Article 4. Such support may be provided by other Parties, by competent international organizations and by the secretariat, as appropriate.

8. Any group of Parties may, subject to guidelines adopted by the Conference of the Parties, and to prior notification to the Conference of the Parties, make a joint communication in fulfilment of their obligations under this Article, provided that such a communication includes information on the fulfilment by each of these Parties of its individual obligations under the Convention.

9. Information received by the secretariat that is designated by a Party as confidential, in accordance with criteria to be established by the Conference of the Parties, shall be aggregated by the secretariat to protect its confidentiality before being made available to any of the bodies involved in the communication and review of information.

10. Subject to paragraph 9 above, and without prejudice to the ability of any Party to make public its communication at any time, the secretariat shall make communications by Parties under this Article publicly available at the time they are submitted to the Conference of the Parties.

ARTICLE 13

RESOLUTION OF QUESTIONS REGARDING IMPLEMENTATION

The Conference of the Parties shall, at its first session, consider the establishment of a multilateral consultative process, available to Parties on their request, for the resolution of questions regarding the implementation of the Convention.

ARTICLE 14

SETTLEMENT OF DISPUTES

1. In the event of a dispute between any two or more Parties concerning the interpretation or application of the Convention, the Parties concerned shall seek a settlement of the dispute through negotiation or any other peaceful means of their own choice.

2. When ratifying, accepting, approving or acceding to the Convention, or at any time thereafter, a Party which is not a regional economic integration organization may declare in a written instrument submitted to the Depositary that, in respect of any dispute concerning the interpretation or application of the

Convention, it recognizes as compulsory ipso facto and without special agreement, in relation to any Party accepting the same obligation:

(a) Submission of the dispute to the International Court of Justice, and/or

(b) Arbitration in accordance with procedures to be adopted by the Conference of the Parties as soon as practicable, in an annex on arbitration.

A Party which is a regional economic integration organization may make a declaration with like effect in relation to arbitration in accordance with the procedures referred to in subparagraph (b) above.

3. A declaration made under paragraph 2 above shall remain in force until it expires in accordance with its terms or until three months after written notice of its revocation has been deposited with the Depositary.

4. A new declaration, a notice of revocation or the expiry of a declaration shall not in any way affect proceedings pending before the International Court of Justice or the arbitral tribunal, unless the parties to the dispute otherwise agree.

5. Subject to the operation of paragraph 2 above, if after twelve months following notification by one Party to another that a dispute exists between them, the Parties concerned have not been able to settle their dispute through the means mentioned in paragraph 1 above, the dispute shall be submitted, at the request of any of the parties to the dispute, to conciliation.

6. A conciliation commission shall be created upon the request of one of the parties to the dispute. The commission shall be composed of an equal number of members appointed by each party concerned and a chairman chosen jointly by the members appointed by each party. The commission shall render a recommendatory award, which the parties shall consider in good faith.

7. Additional procedures relating to conciliation shall be adopted by the Conference of the Parties, as soon as practicable, in an annex on conciliation.

8. The provisions of this Article shall apply to any related legal instrument which the Conference of the Parties may adopt, unless the instrument provides otherwise.

ARTICLE 15

AMENDMENTS TO THE CONVENTION

1. Any Party may propose amendments to the Convention.

2. Amendments to the Convention shall be adopted at an ordinary session of the Conference of the Parties. The text of any proposed amendment to the Convention shall be communicated to the Parties by the secretariat at least six months before the meeting at which it is proposed for adoption. The secretariat shall also communicate proposed amendments to the signatories to the Convention and, for information, to the Depositary.

3. The Parties shall make every effort to reach agreement on any proposed amendment to the Convention by consensus. If all efforts at consensus have been exhausted, and no agreement reached, the amendment shall as a last resort be adopted by a three-fourths majority vote of the Parties present and voting at the meeting. The adopted amendment shall be communicated by the secretariat to the Depositary, who shall circulate it to all Parties for their acceptance.

4. Instruments of acceptance in respect of an amendment shall be deposited with the Depositary. An amendment adopted in accordance with paragraph 3 above shall enter into force for those Parties having accepted it on the ninetieth day after the date of receipt by the Depositary of an instrument of acceptance by at least three fourths of the Parties to the Convention.

5. The amendment shall enter into force for any other Party on the ninetieth day after the date on which that Party deposits with the Depositary its instrument of acceptance of the said amendment.

6. For the purposes of this Article, "Parties present and voting" means Parties present and casting an affirmative or negative vote.

ARTICLE 16

ADOPTION AND AMENDMENT OF ANNEXES TO THE CONVENTION

1. Annexes to the Convention shall form an integral part thereof and, unless otherwise expressly provided, a reference to the Convention constitutes at the same time a reference to any annexes thereto. Without prejudice to the provisions of Article 14, paragraphs 2(b) and 7, such annexes shall be restricted to lists, forms and any other material of a descriptive nature that is of a scientific, technical, procedural or administrative character.

2. Annexes to the Convention shall be proposed and adopted in accordance with the procedure set forth in Article 15, paragraphs 2, 3 and 4.

3. An annex that has been adopted in accordance with paragraph 2 above shall enter into force for all Parties to the Convention six months after the date of the communication by the Depositary to such Parties of the adoption of the annex, except for those Parties that have notified the Depositary, in writing, within that period of their non-acceptance of the annex. The annex shall enter into force for Parties which withdraw their notification of non-acceptance on the ninetieth day after the date on which withdrawal of such notification has been received by the Depositary.

4. The proposal, adoption and entry into force of amendments to annexes to the Convention shall be subject to the same procedure as that for the proposal, adoption and entry into force of annexes to the Convention in accordance with paragraphs 2 and 3 above.

5. If the adoption of an annex or an amendment to an annex involves an amendment to the Convention, that annex or amendment to an annex shall not enter into force until such time as the amendment to the Convention enters into force.

ARTICLE 17

PROTOCOLS

1. The Conference of the Parties may, at any ordinary session, adopt protocols to the Convention.

2. The text of any proposed protocol shall be communicated to the Parties by the secretariat at least six months before such a session.

3. The requirements for the entry into force of any protocol shall be established by that instrument.

4. Only Parties to the Convention may be Parties to a protocol.

5. Decisions under any protocol shall be taken only by the Parties to the protocol concerned.

ARTICLE 18

RIGHT TO VOTE

1. Each Party to the Convention shall have one vote, except as provided for in paragraph 2 below.

2. Regional economic integration organizations, in matters within their competence, shall exercise their right to vote with a number of votes equal to the number of their member States that are Parties to the Convention. Such an organization shall not exercise its right to vote if any of its member States exercises its right, and vice versa.

ARTICLE 19

DEPOSITARY

The Secretary-General of the United Nations shall be the Depositary of the Convention and of protocols adopted in accordance with Article 17.

ARTICLE 20

SIGNATURE

This Convention shall be open for signature by States Members of the United Nations or of any of its specialized agencies or that are Parties to the Statute of the International Court of Justice and by regional economic integration organizations at Rio de Janeiro, during the United Nations Conference on Environment and Development, and thereafter at United Nations Headquarters in New York from 20 June 1992 to 19 June 1993.

ARTICLE 21

INTERIM ARRANGEMENTS

1. The secretariat functions referred to in Article 8 will be carried out on an interim basis by the secretariat established by the General Assembly of the United Nations in its resolution 45/212 of 21 December 1990, until the completion of the first session of the Conference of the Parties.

2. The head of the interim secretariat referred to in paragraph 1 above will cooperate closely with the Intergovernmental Panel on Climate Change to ensure that the Panel can respond to the need for objective scientific and technical advice. Other relevant scientific bodies could also be consulted.

3. The Global Environment Facility of the United Nations Development Programme, the United Nations Environment Programme and the International Bank for Reconstruction and Development shall be the international entity entrusted with the operation of the financial mechanism referred to in Article 11 on an interim basis. In this connection, the Global Environment Facility should be appropriately restructured and its membership made universal to enable it to fulfil the requirements of Article 11.

ARTICLE 22

RATIFICATION, ACCEPTANCE, APPROVAL OR ACCESSION

1. The Convention shall be subject to ratification, acceptance, approval or accession by States and by regional economic integration organizations. It shall be open for accession from the day after the date on which the Convention is closed for signature. Instruments of ratification, acceptance, approval or accession shall be deposited with the Depositary.

2. Any regional economic integration organization which becomes a Party to the Convention without any of its member States being a Party shall be bound by all the obligations under the Convention. In the case of such organizations, one or more of whose member States is a Party to the Convention, the organization and its member States shall decide on their respective responsibilities for the performance of their obligations under the Convention. In such cases, the organization and the member States shall not be entitled to exercise rights under the Convention concurrently.

3. In their instruments of ratification, acceptance, approval or accession, regional economic integration organizations shall declare the extent of their competence with respect to the matters governed by the Convention. These organizations shall also inform the Depositary, who shall in turn inform the Parties, of any substantial modification in the extent of their competence.

ARTICLE 23

ENTRY INTO FORCE

1. The Convention shall enter into force on the ninetieth day after the date of deposit of the fiftieth instrument of ratification, acceptance, approval or accession.

2. For each State or regional economic integration organization that ratifies, accepts or approves the Convention or accedes thereto after the deposit of the fiftieth instrument of ratification, acceptance, approval or accession, the Convention shall enter into force on the ninetieth day after the date of deposit by

such State or regional economic integration organization of its instrument of ratification, acceptance, approval or accession.

3. For the purposes of paragraphs 1 and 2 above, any instrument deposited by a regional economic integration organization shall not be counted as additional to those deposited by States members of the organization.

ARTICLE 24

RESERVATIONS

No reservations may be made to the Convention.

ARTICLE 25

WITHDRAWAL

1. At any time after three years from the date on which the Convention has entered into force for a Party, that Party may withdraw from the Convention by giving written notification to the Depositary.

2. Any such withdrawal shall take effect upon expiry of one year from the date of receipt by the Depositary of the notification of withdrawal, or on such later date as may be specified in the notification of withdrawal.

3. Any Party that withdraws from the Convention shall be considered as also having withdrawn from any protocol to which it is a Party.

ARTICLE 26

AUTHENTIC TEXTS

The original of this Convention, of which the Arabic, Chinese, English, French, Russian and Spanish texts are equally authentic, shall be deposited with the Secretary-General of the United Nations.

IN WITNESS WHEREOF the undersigned, being duly authorized to that effect, have signed this Convention.

DONE at New York this ninth day of May one thousand nine hundred and ninety-two.

Annex I

Australia
Austria
Belarus[a/]
Belgium
Bulgaria[a/]
Canada
Czechoslovakia[a/]
Denmark
European Economic Community
Estonia[a/]
Finland
France
Germany
Greece
Hungary[a/]
Iceland
Ireland
Italy
Japan
Latvia[a/]
Lithuania[a/]
Luxembourg
Netherlands
New Zealand
Norway
Poland[a/]
Portugal
Romania[a/]
Russian Federation[a/]
Spain
Sweden
Switzerland
Turkey
Ukraine[a/]
United Kingdom of Great Britain and Northern Ireland
United States of America

[a/] Countries that are undergoing the process of transition to a market economy.

Annex II

Australia
Austria
Belgium
Canada
Denmark
European Economic Community
Finland
France
Germany
Greece
Iceland
Ireland
Italy
Japan
Luxembourg
Netherlands
New Zealand
Norway
Portugal
Spain
Sweden
Switzerland
Turkey
United Kingdom of Great
Britain and Northern Ireland
United States of America

A2 KYOTO PROTOCOL TO THE UNITED NATIONS FRAMEWORK CONVENTION ON CLIMATE CHANGE

The Parties to this Protocol,

Being Parties to the United Nations Framework Convention on Climate Change, hereinafter referred to as the Convention,

In pursuit of the ultimate objective of the Convention as stated in its Article 2,

Recalling the provisions of the Convention,

Being guided by Article 3 of the Convention,

Pursuant to the Berlin Mandate adopted by decision 1/CP.1 of the Conference of the Parties to the Convention at its first session,

Have agreed as follows:

Article 1

For the purposes of this Protocol, the definitions contained in Article 1 of the Convention shall apply. In addition:

1. Conference of the Parties means the Conference of the Parties to the Convention.

2. Convention means the United Nations Framework Convention on Climate Change, adopted in New York on 9 May 1992.

3. Intergovernmental Panel on Climate Change means the Intergovernmental Panel on Climate Change established in 1988 jointly by the World Meteorological Organization and the United Nations Environment Programme.

4. Montreal Protocol means the Montreal Protocol on Substances that Deplete the Ozone Layer, adopted in Montreal on 16 September 1987 and as subsequently adjusted and amended.

5. Parties present and voting means Parties present and casting an affirmative or negative vote.

6. Party means, unless the context otherwise indicates, a Party to this Protocol.

7. Party included in Annex I means a Party included in Annex I to the Convention, as may be amended, or a Party which has made a notification under Article 4, paragraph 2(g), of the Convention.

Article 2

1. Each Party included in Annex I, in achieving its quantified emission limitation and reduction commitments under Article 3, in order to promote sustainable development, shall:

(a) Implement and/or further elaborate policies and measures in accordance with its national circumstances, such as:

(i) Enhancement of energy efficiency in relevant sectors of the national economy;

(ii) Protection and enhancement of sinks and reservoirs of greenhouse gases not controlled by the Montreal Protocol, taking into account its commitments under relevant international environmental agreements; promotion of sustainable forest management practices, afforestation and reforestation;

(iii) Promotion of sustainable forms of agriculture in light of climate change considerations;

(iv) Research on, and promotion, development and increased use of, new and renewable forms of energy, of carbon dioxide sequestration technologies and of advanced and innovative environmentally sound technologies;

(v) Progressive reduction or phasing out of market imperfections, fiscal incentives, tax and duty exemptions and subsidies in all greenhouse gas emitting sectors that run counter to the objective of the Convention and application of market instruments;

(vi) Encouragement of appropriate reforms in relevant sectors aimed at promoting policies and measures which limit or reduce emissions of greenhouse gases not controlled by the Montreal Protocol;

(vii) Measures to limit and/or reduce emissions of greenhouse gases not controlled by the Montreal

Protocol in the transport sector;

(viii) Limitation and/or reduction of methane emissions through recovery and use in waste management, as well as in the production, transport and distribution of energy;

(b) Cooperate with other such Parties to enhance the individual and combined effectiveness of their policies and measures adopted under this Article, pursuant to Article 4, paragraph 2(e)(i), of the Convention. To this end, these Parties shall take steps to share their experience and exchange information on such policies and measures, including developing ways of improving their comparability, transparency and effectiveness. The Conference of the Parties serving as the meeting of the Parties to this Protocol shall, at its first session or as soon as practicable thereafter, consider ways to facilitate such cooperation, taking into account all relevant information.

2. The Parties included in Annex I shall pursue limitation or reduction of emissions of greenhouse gases not controlled by the Montreal Protocol from aviation and marine bunker fuels, working through the International Civil Aviation Organization and the International Maritime Organization, respectively.

3. The Parties included in Annex I shall strive to implement policies and measures under this Article in such a way as to minimize adverse effects, including the adverse effects of climate change, effects on international trade, and social, environmental and economic impacts on other Parties, especially developing country Parties and in particular those identified in Article 4, paragraphs 8 and 9, of the Convention, taking into account Article 3 of the Convention. The Conference of the Parties serving as the meeting of the Parties to this Protocol may take further action, as appropriate, to promote the implementation of the provisions of this paragraph.

4. The Conference of the Parties serving as the meeting of the Parties to this Protocol, if it decides that it would be beneficial to coordinate any of the policies and measures in paragraph 1(a) above, taking into account different national circumstances and potential effects, shall consider ways and means to elaborate the coordination of such policies and measures.

Article 3

1. The Parties included in Annex I shall, individually or jointly, ensure that their aggregate anthropogenic carbon dioxide equivalent emissions of the greenhouse gases listed in Annex A do not exceed their assigned amounts, calculated pursuant to their quantified emission limitation and reduction commitments inscribed in Annex B and in accordance with the provisions of this Article, with a view to reducing their overall emissions of such gases by at least 5

per cent below 1990 levels in the commitment period 2008 to 2012.

2. Each Party included in Annex I shall, by 2005, have made demonstrable progress in achieving its commitments under this Protocol.

3. The net changes in greenhouse gas emissions by sources and removals by sinks resulting from direct human-induced land-use change and forestry activities, limited to afforestation, reforestation and deforestation since 1990, measured as verifiable changes in carbon stocks in each commitment period, shall be used to meet the commitments under this Article of each Party included in Annex I. The greenhouse gas emissions by sources and removals by sinks associated with those activities shall be reported in a transparent and verifiable manner and reviewed in accordance with Articles 7 and 8.

4. Prior to the first session of the Conference of the Parties serving as the meeting of the Parties to this Protocol, each Party included in Annex I shall provide, for consideration by the Subsidiary Body for Scientific and Technological Advice, data to establish its level of carbon stocks in 1990 and to enable an estimate to be made of its changes in carbon stocks in subsequent years. The Conference of the Parties serving as the meeting of the Parties to this Protocol shall, at its first session or as soon as practicable thereafter, decide upon modalities, rules and guidelines as to how, and which, additional human-induced activities related to changes in greenhouse gas emissions by sources and removals by sinks in the agricultural soils and the land-use change and forestry categories shall be added to, or subtracted from, the assigned amounts for Parties included in Annex I, taking into account uncertainties, transparency in reporting, verifiability, the methodological work of the Intergovernmental Panel on Climate Change, the advice provided by the Subsidiary Body for Scientific and Technological Advice in accordance with Article 5 and the decisions of the Conference of the Parties. Such a decision shall apply in the second and subsequent commitment periods. A Party may choose to apply such a decision on these additional human-induced activities for its first commitment period, provided that these activities have taken place since 1990.

5. The Parties included in Annex I undergoing the process of transition to a market economy whose base year or period was established pursuant to decision 9/CP.2 of the Conference of the Parties at its second session shall use that base year or period for the implementation of their commitments under this Article. Any other Party included in Annex I undergoing the process of transition to a market economy which has not yet submitted its first national communication under Article 12 of the Convention may also notify the Conference of the Parties serving as the meeting of the Parties to this Protocol that it intends to use an historical base year or period other than 1990 for the implementation of its commitments under this Article. The Conference of the Parties serving as the meeting of the Parties to this Protocol shall decide on the acceptance of such notification.

6. Taking into account Article 4, paragraph 6, of the Convention, in the implementation of their commitments under this Protocol other than those under this Article, a certain degree of flexibility shall be allowed by the Conference of the Parties serving as the meeting of the Parties to this Protocol to the Parties included in Annex I undergoing the process of transition to a market economy.

7. In the first quantified emission limitation and reduction commitment period, from 2008 to 2012, the assigned amount for each Party included in Annex I shall be equal to the percentage inscribed for it in Annex B of its aggregate anthropogenic carbon dioxide equivalent emissions of the greenhouse gases listed in Annex A in 1990, or the base year or period determined in accordance with paragraph 5 above, multiplied by five. Those Parties included in Annex I for whom land-use change and forestry constituted a net source of greenhouse gas emissions in 1990 shall include in their 1990 emissions base year or period the aggregate anthropogenic carbon dioxide equivalent emissions by sources minus removals by sinks in 1990 from land-use change for the purposes of calculating their assigned amount.

8. Any Party included in Annex I may use 1995 as its base year for hydrofluorocarbons, perfluorocarbons and sulphur hexafluoride, for the purposes of the calculation referred to in paragraph 7 above.

9. Commitments for subsequent periods for Parties included in Annex I shall be established in amendments to Annex B to this Protocol, which shall be adopted in accordance with the provisions of Article 21, paragraph 7. The Conference of the Parties serving as the meeting of the Parties to this Protocol shall initiate the consideration of such commitments at least seven years before the end of the first commitment period referred to in paragraph 1 above.

10. Any emission reduction units, or any part of an assigned amount, which a Party acquires from another Party in accordance with the provisions of Article 6 or of Article 17 shall be added to the assigned amount for the acquiring Party.

11. Any emission reduction units, or any part of an assigned amount, which a Party transfers to another Party in accordance with the provisions of Article 6 or of Article 17 shall be subtracted from the assigned amount for the transferring Party.

12. Any certified emission reductions which a Party acquires from another Party in accordance with the provisions of Article 12 shall be added to the assigned amount for the acquiring Party.

13. If the emissions of a Party included in Annex I in a commitment period are less than its assigned amount under this Article, this difference shall, on request of that Party, be added to the assigned amount for that Party for subsequent commitment periods.

14. Each Party included in Annex I shall strive to implement the commitments mentioned in paragraph 1 above in such a way as to minimize adverse social, environmental and economic impacts on developing country Parties, particularly those identified in Article 4, paragraphs 8 and 9, of the Convention. In line with relevant decisions of the Conference of the Parties on the implementation of those paragraphs, the Conference of the Parties serving as the meeting of the Parties to this Protocol shall, at its first session, consider what actions are necessary to minimize the adverse effects of climate change and/or the impacts of response measures on Parties referred to in those paragraphs. Among the issues to be considered shall be the establishment of funding, insurance and transfer of technology.

Article 4

1. Any Parties included in Annex I that have reached an agreement to fulfil their commitments under Article 3 jointly, shall be deemed to have met those commitments provided that their total combined aggregate anthropogenic carbon dioxide equivalent emissions of the greenhouse gases listed in Annex A do not exceed their assigned amounts calculated pursuant to their quantified emission limitation and reduction commitments inscribed in Annex B and in accordance with the provisions of Article 3. The respective emission level allocated to each of the Parties to the agreement shall be set out in that agreement.

2. The Parties to any such agreement shall notify the secretariat of the terms of the agreement on the date of deposit of their instruments of ratification, acceptance or approval of this Protocol, or accession thereto. The secretariat shall in turn inform the Parties and signatories to the Convention of the terms of the agreement.

3. Any such agreement shall remain in operation for the duration of the commitment period specified in Article 3, paragraph 7.

4. If Parties acting jointly do so in the framework of, and together with, a regional economic integration organization, any alteration in the composition of the organization after adoption of this Protocol shall not affect existing commitments under this Protocol. Any alteration in the composition of the organization shall only apply for the purposes of those commitments under Article 3 that are adopted subsequent to that alteration.

5. In the event of failure by the Parties to such an agreement to achieve their total combined level of emission reductions, each Party to that agreement shall be responsible for its own level of emissions set out in the agreement.

6. If Parties acting jointly do so in the framework of, and together with, a regional economic integration organization which is itself a Party to this Protocol, each member State of that regional economic integration organization individually,

and together with the regional economic integration organization acting in accordance with Article 24, shall, in the event of failure to achieve the total combined level of emission reductions, be responsible for its level of emissions as notified in accordance with this Article.

Article 5

1. Each Party included in Annex I shall have in place, no later than one year prior to the start of the first commitment period, a national system for the estimation of anthropogenic emissions by sources and removals by sinks of all greenhouse gases not controlled by the Montreal Protocol. Guidelines for such national systems, which shall incorporate the methodologies specified in paragraph 2 below, shall be decided upon by the Conference of the Parties serving as the meeting of the Parties to this Protocol at its first session.

2. Methodologies for estimating anthropogenic emissions by sources and removals by sinks of all greenhouse gases not controlled by the Montreal Protocol shall be those accepted by the Intergovernmental Panel on Climate Change and agreed upon by the Conference of the Parties at its third session. Where such methodologies are not used, appropriate adjustments shall be applied according to methodologies agreed upon by the Conference of the Parties serving as the meeting of the Parties to this Protocol at its first session. Based on the work of, *inter alia*, the Intergovernmental Panel on Climate Change and advice provided by the Subsidiary Body for Scientific and Technological Advice, the Conference of the Parties serving as the meeting of the Parties to this Protocol shall regularly review and, as appropriate, revise such methodologies and adjustments, taking fully into account any relevant decisions by the Conference of the Parties. Any revision to methodologies or adjustments shall be used only for the purposes of ascertaining compliance with commitments under Article 3 in respect of any commitment period adopted subsequent to that revision.

3. The global warming potentials used to calculate the carbon dioxide equivalence of anthropogenic emissions by sources and removals by sinks of greenhouse gases listed in Annex A shall be those accepted by the Intergovernmental Panel on Climate Change and agreed upon by the Conference of the Parties at its third session. Based on the work of, *inter alia*, the Intergovernmental Panel on Climate Change and advice provided by the Subsidiary Body for Scientific and Technological Advice, the Conference of the Parties serving as the meeting of the Parties to this Protocol shall regularly review and, as appropriate, revise the global warming potential of each such greenhouse gas, taking fully into account any relevant decisions by the Conference of the Parties. Any revision to a global warming potential shall apply only to commitments under Article 3 in respect of any commitment period adopted subsequent to that revision.

Article 6

1. For the purpose of meeting its commitments under Article 3, any Party included in Annex I may transfer to, or acquire from, any other such Party emission reduction units resulting from projects aimed at reducing anthropogenic emissions by sources or enhancing anthropogenic removals by sinks of greenhouse gases in any sector of the economy, provided that:

(a) Any such project has the approval of the Parties involved;

(b) Any such project provides a reduction in emissions by sources, or an enhancement of removals by sinks, that is additional to any that would otherwise occur;

(c) It does not acquire any emission reduction units if it is not in compliance with its obligations under Articles 5 and 7; and

(d) The acquisition of emission reduction units shall be supplemental to domestic actions for the purposes of meeting commitments under Article 3.

2. The Conference of the Parties serving as the meeting of the Parties to this Protocol may, at its first session or as soon as practicable thereafter, further elaborate guidelines for the implementation of this Article, including for verification and reporting.

3. A Party included in Annex I may authorize legal entities to participate, under its responsibility, in actions leading to the generation, transfer or acquisition under this Article of emission reduction units.

4. If a question of implementation by a Party included in Annex I of the requirements referred to in this Article is identified in accordance with the relevant provisions of Article 8, transfers and acquisitions of emission reduction units may continue to be made after the question has been identified, provided that any such units may not be used by a Party to meet its commitments under Article 3 until any issue of compliance is resolved.

Article 7

1. Each Party included in Annex I shall incorporate in its annual inventory of anthropogenic emissions by sources and removals by sinks of greenhouse gases not controlled by the Montreal Protocol, submitted in accordance with the relevant decisions of the Conference of the Parties, the necessary supplementary information for the purposes of ensuring compliance with Article 3, to be determined in accordance with paragraph 4 below.

2. Each Party included in Annex I shall incorporate in its national communication, submitted under Article 12 of the Convention, the supplementary information necessary to demonstrate compliance with its commitments under this Protocol, to be determined in accordance with paragraph 4 below.

3. Each Party included in Annex I shall submit the information required under paragraph 1 above annually, beginning with the first inventory due under the Convention for the first year of the commitment period after this Protocol has entered into force for that Party. Each such Party shall submit the information required under paragraph 2 above as part of the first national communication due under the Convention after this Protocol has entered into force for it and after the adoption of guidelines as provided for in paragraph 4 below. The frequency of subsequent submission of information required under this Article shall be determined by the Conference of the Parties serving as the meeting of the Parties to this Protocol, taking into account any timetable for the submission of national communications decided upon by the Conference of the Parties.

4. The Conference of the Parties serving as the meeting of the Parties to this Protocol shall adopt at its first session, and review periodically thereafter, guidelines for the preparation of the information required under this Article, taking into account guidelines for the preparation of national communications by Parties included in Annex I adopted by the Conference of the Parties. The Conference of the Parties serving as the meeting of the Parties to this Protocol shall also, prior to the first commitment period, decide upon modalities for the accounting of assigned amounts.

Article 8

1. The information submitted under Article 7 by each Party included in Annex I shall be reviewed by expert review teams pursuant to the relevant decisions of the Conference of the Parties and in accordance with guidelines adopted for this purpose by the Conference of the Parties serving as the meeting of the Parties to this Protocol under paragraph 4 below. The information submitted under Article 7, paragraph 1, by each Party included in Annex I shall be reviewed as part of the annual compilation and accounting of emissions inventories and assigned amounts. Additionally, the information submitted under Article 7, paragraph 2, by each Party included in Annex I shall be reviewed as part of the review of communications.

2. Expert review teams shall be coordinated by the secretariat and shall be composed of experts selected from those nominated by Parties to the Convention and, as appropriate, by intergovernmental organizations, in accordance with guidance provided for this purpose by the Conference of the Parties.

3. The review process shall provide a thorough and comprehensive technical assessment of all aspects of the implementation by a Party of this Protocol. The

expert review teams shall prepare a report to the Conference of the Parties serving as the meeting of the Parties to this Protocol, assessing the implementation of the commitments of the Party and identifying any potential problems in, and factors influencing, the fulfilment of commitments. Such reports shall be circulated by the secretariat to all Parties to the Convention. The secretariat shall list those questions of implementation indicated in such reports for further consideration by the Conference of the Parties serving as the meeting of the Parties to this Protocol.

4. The Conference of the Parties serving as the meeting of the Parties to this Protocol shall adopt at its first session, and review periodically thereafter, guidelines for the review of implementation of this Protocol by expert review teams taking into account the relevant decisions of the Conference of the Parties.

5. The Conference of the Parties serving as the meeting of the Parties to this Protocol shall, with the assistance of the Subsidiary Body for Implementation and, as appropriate, the Subsidiary Body for Scientific and Technological Advice, consider:

(a) The information submitted by Parties under Article 7 and the reports of the expert reviews thereon conducted under this Article; and

(b) Those questions of implementation listed by the secretariat under paragraph 3 above, as well as any questions raised by Parties.

6. Pursuant to its consideration of the information referred to in paragraph 5 above, the Conference of the Parties serving as the meeting of the Parties to this Protocol shall take decisions on any matter required for the implementation of this Protocol.

Article 9

1. The Conference of the Parties serving as the meeting of the Parties to this Protocol shall periodically review this Protocol in the light of the best available scientific information and assessments on climate change and its impacts, as well as relevant technical, social and economic information. Such reviews shall be coordinated with pertinent reviews under the Convention, in particular those required by Article 4, paragraph 2(d), and Article 7, paragraph 2(a), of the Convention. Based on these reviews, the Conference of the Parties serving as the meeting of the Parties to this Protocol shall take appropriate action.

2. The first review shall take place at the second session of the Conference of the Parties serving as the meeting of the Parties to this Protocol. Further reviews shall take place at regular intervals and in a timely manner.

Article 10

All Parties, taking into account their common but differentiated responsibilities and their specific national and regional development priorities, objectives and circumstances, without introducing any new commitments for Parties not included in Annex I, but reaffirming existing commitments under Article 4, paragraph 1, of the Convention, and continuing to advance the implementation of these commitments in order to achieve sustainable development, taking into account Article 4, paragraphs 3, 5 and 7, of the Convention, shall:

(a) Formulate, where relevant and to the extent possible, cost-effective national and, where appropriate, regional programmes to improve the quality of local emission factors, activity data and/or models which reflect the socio-economic conditions of each Party for the preparation and periodic updating of national inventories of anthropogenic emissions by sources and removals by sinks of all greenhouse gases not controlled by the Montreal Protocol, using comparable methodologies to be agreed upon by the Conference of the Parties, and consistent with the guidelines for the preparation of national communications adopted by the Conference of the Parties;

(b) Formulate, implement, publish and regularly update national and, where appropriate, regional programmes containing measures to mitigate climate change and measures to facilitate adequate adaptation to climate change:

(i) Such programmes would, *inter alia*, concern the energy, transport and industry sectors as well as agriculture, forestry and waste management. Furthermore, adaptation technologies and methods for improving spatial planning would improve adaptation to climate change; and

(ii) Parties included in Annex I shall submit information on action under this Protocol, including national programmes, in accordance with Article 7; and other Parties shall seek to include in their national communications, as appropriate, information on programmes which contain measures that the Party believes contribute to addressing climate change and its adverse impacts, including the abatement of increases in greenhouse gas emissions, and enhancement of and removals by sinks, capacity building and adaptation measures;

(c) Cooperate in the promotion of effective modalities for the development, application and diffusion of, and take all practicable steps to promote, facilitate and finance, as appropriate, the transfer of, or access to,

environmentally sound technologies, know-how, practices and processes pertinent to climate change, in particular to developing countries, including the formulation of policies and programmes for the effective transfer of
environmentally sound technologies that are publicly owned or in the public domain and the creation of an enabling environment for the private sector, to promote and enhance the transfer of, and access to, environmentally sound technologies;

(d) Cooperate in scientific and technical research and promote the maintenance and the development of systematic observation systems and development of data archives to reduce uncertainties related to the climate system, the adverse impacts of climate change and the economic and social consequences of various response strategies, and promote the development and strengthening of endogenous capacities and capabilities to participate in international and intergovernmental efforts, programmes and networks on research and systematic observation, taking into account Article 5 of the Convention;

(e) Cooperate in and promote at the international level, and, where appropriate, using existing bodies, the development and implementation of education and training programmes, including the strengthening of national capacity building, in particular human and institutional capacities and the exchange or secondment of personnel to train experts in this field, in particular for developing countries, and facilitate at the national level public awareness of, and public access to information on, climate change. Suitable modalities should be developed to implement these activities through the relevant bodies of the Convention, taking into account Article 6 of the Convention;

(f) Include in their national communications information on programmes and activities undertaken pursuant to this Article in accordance with relevant decisions of the Conference of the Parties; and

(g) Give full consideration, in implementing the commitments under this Article, to Article 4, paragraph 8, of the Convention.

Article 11

1. In the implementation of Article 10, Parties shall take into account the provisions of Article 4, paragraphs 4, 5, 7, 8 and 9, of the Convention.

2. In the context of the implementation of Article 4, paragraph 1, of the Convention, in accordance with the provisions of Article 4, paragraph 3, and Article 11 of the Convention, and through the entity or entities entrusted with the operation of the financial mechanism of the Convention, the developed country Parties and other developed Parties included in Annex II to the Convention shall:

(a) Provide new and additional financial resources to meet the agreed full costs incurred by developing country Parties in advancing the implementation of existing commitments under Article 4, paragraph 1(a), of the Convention that are covered in Article 10, subparagraph (a); and

(b) Also provide such financial resources, including for the transfer of technology, needed by the developing country Parties to meet the agreed full incremental costs of advancing the implementation of existing commitments under Article 4, paragraph 1, of the Convention that are covered by Article 10 and that are agreed between a developing country Party and the international entity or entities referred to in Article 11 of the Convention, in accordance with that Article.

The implementation of these existing commitments shall take into account the need for adequacy and predictability in the flow of funds and the importance of appropriate burden sharing among developed country Parties. The guidance to the entity or entities entrusted with the operation of the financial mechanism of the Convention in relevant decisions of the Conference of the Parties, including those agreed before the adoption of this Protocol, shall apply *mutatis mutandis* to the provisions of this paragraph.

3. The developed country Parties and other developed Parties in Annex II to the Convention may also provide, and developing country Parties avail themselves of, financial resources for the implementation of Article 10, through bilateral, regional and other multilateral channels.

Article 12

1. A clean development mechanism is hereby defined.

2. The purpose of the clean development mechanism shall be to assist Parties not included in Annex I in achieving sustainable development and in contributing to the ultimate objective of the Convention, and to assist Parties included in Annex I in achieving compliance with their quantified emission limitation and reduction commitments under Article 3.

3. Under the clean development mechanism:

(a) Parties not included in Annex I will benefit from project activities resulting in certified emission reductions; and

(b) Parties included in Annex I may use the certified emission reductions accruing from such project activities to contribute to compliance with part of their quantified emission limitation and reduction commitments under Article 3, as determined by the Conference of the Parties serving as the meeting of the Parties to this Protocol.

4. The clean development mechanism shall be subject to the authority and

guidance of the Conference of the Parties serving as the meeting of the Parties to this Protocol and be supervised by an executive board of the clean development mechanism.

5. Emission reductions resulting from each project activity shall be certified by operational entities to be designated by the Conference of the Parties serving as the meeting of the Parties to this Protocol, on the basis of:

(a) Voluntary participation approved by each Party involved;

(b) Real, measurable, and long-term benefits related to the mitigation of climate change; and

(c) Reductions in emissions that are additional to any that would occur in the absence of the certified project activity.

6. The clean development mechanism shall assist in arranging funding of certified project activities as necessary.

7. The Conference of the Parties serving as the meeting of the Parties to this Protocol shall, at its first session, elaborate modalities and procedures with the objective of ensuring transparency, efficiency and accountability through independent auditing and verification of project activities.

8. The Conference of the Parties serving as the meeting of the Parties to this Protocol shall ensure that a share of the proceeds from certified project activities is used to cover administrative expenses as well as to assist developing country Parties that are particularly vulnerable to the adverse effects of climate change to meet the costs of adaptation.

9. Participation under the clean development mechanism, including in activities mentioned in paragraph 3(a) above and in the acquisition of certified emission reductions, may involve private and/or public entities, and is to be subject to whatever guidance may be provided by the executive board of the clean development mechanism.

10. Certified emission reductions obtained during the period from the year 2000 up to the beginning of the first commitment period can be used to assist in achieving compliance in the first commitment period.

Article 13

1. The Conference of the Parties, the supreme body of the Convention, shall serve as the meeting of the Parties to this Protocol.

2. Parties to the Convention that are not Parties to this Protocol may participate as observers in the proceedings of any session of the Conference of the Parties serving as the meeting of the Parties to this Protocol. When the Conference of the Parties serves as the meeting of the Parties to this Protocol, decisions under this Protocol shall be taken only by those that are Parties to this Protocol.

3. When the Conference of the Parties serves as the meeting of the Parties to this Protocol, any member of the Bureau of the Conference of the Parties representing a Party to the Convention but, at that time, not a Party to this Protocol, shall be replaced by an additional member to be elected by and from amongst the Parties to this Protocol.

4. The Conference of the Parties serving as the meeting of the Parties to this Protocol shall keep under regular review the implementation of this Protocol and shall make, within its mandate, the decisions necessary to promote its effective implementation. It shall perform the functions assigned to it by this Protocol and shall:

(a) Assess, on the basis of all information made available to it in accordance with the provisions of this Protocol, the implementation of this Protocol by the Parties, the overall effects of the measures taken pursuant to this Protocol, in particular environmental, economic and social effects as well as their cumulative impacts and the extent to which progress towards the objective of the Convention is being achieved;

(b) Periodically examine the obligations of the Parties under this Protocol, giving due consideration to any reviews required by Article 4, paragraph 2(d), and Article 7, paragraph 2, of the Convention, in the light of the objective of the Convention, the experience gained in its implementation and the evolution of scientific and technological knowledge, and in this respect consider and adopt regular reports on the implementation of this Protocol;

(c) Promote and facilitate the exchange of information on measures adopted by the Parties to address climate change and its effects, taking into account the differing circumstances, responsibilities and capabilities of the Parties and their respective commitments under this Protocol;

(d) Facilitate, at the request of two or more Parties, the coordination of measures adopted by them to address climate change and its effects, taking into account the differing circumstances, responsibilities and capabilities of the Parties and their respective commitments under this Protocol;

(e) Promote and guide, in accordance with the objective of the Convention and the provisions of this Protocol, and taking fully into account the relevant decisions by the Conference of the Parties, the development and periodic refinement of comparable methodologies for the effective implementation of this Protocol, to be agreed on by the Conference of the Parties serving as the meeting of the Parties to this Protocol;

(f) Make recommendations on any matters necessary for the implementation of this Protocol;

(g) Seek to mobilize additional financial resources in accordance with Article 11, paragraph 2;

(h) Establish such subsidiary bodies as are deemed necessary for the implementation of this Protocol;

(i) Seek and utilize, where appropriate, the services and cooperation of, and information provided by, competent international organizations and intergovernmental and non-governmental bodies; and

(j) Exercise such other functions as may be required for the implementation of this Protocol, and consider any assignment resulting from a decision by the Conference of the Parties.

5. The rules of procedure of the Conference of the Parties and financial procedures applied under the Convention shall be applied *mutatis mutandis* under this Protocol, except as may be otherwise decided by consensus by the Conference of the Parties serving as the meeting of the Parties to this Protocol.

6. The first session of the Conference of the Parties serving as the meeting of the Parties to this Protocol shall be convened by the secretariat in conjunction with the first session of the Conference of the Parties that is scheduled after the date of the entry into force of this Protocol. Subsequent ordinary sessions of the Conference of the Parties serving as the meeting of the Parties to this Protocol shall be held every year and in conjunction with ordinary sessions of the Conference of the Parties, unless otherwise decided by the Conference of the Parties serving as the meeting of the Parties to this Protocol.

7. Extraordinary sessions of the Conference of the Parties serving as the meeting of the Parties to this Protocol shall be held at such other times as may be deemed necessary by the Conference of the Parties serving as the meeting of the Parties to this Protocol, or at the written request of any Party, provided that, within six months of the request being communicated to the Parties by the secretariat, it is supported by at least one third of the Parties.

8. The United Nations, its specialized agencies and the International Atomic Energy Agency, as well as any State member thereof or observers thereto not party to the Convention, may be represented at sessions of the Conference of the Parties

serving as the meeting of the Parties to this Protocol as observers. Any body or agency, whether national or international, governmental or non-governmental, which is qualified in matters covered by this Protocol and which has informed the secretariat of its wish to be represented at a session of the Conference of the Parties serving as the meeting of the Parties to this Protocol as an observer, may be so admitted unless at least one third of the Parties present object. The admission and participation of observers shall be subject to the rules of procedure, as referred to in paragraph 5 above.

Article 14

1. The secretariat established by Article 8 of the Convention shall serve as the secretariat of this Protocol.

2. Article 8, paragraph 2, of the Convention on the functions of the secretariat, and Article 8, paragraph 3, of the Convention on arrangements made for the functioning of the secretariat, shall apply *mutatis mutandis* to this Protocol. The secretariat shall, in addition, exercise the functions assigned to it under this Protocol.

Article 15

1. The Subsidiary Body for Scientific and Technological Advice and the Subsidiary Body for Implementation established by Articles 9 and 10 of the Convention shall serve as, respectively, the Subsidiary Body for Scientific and Technological Advice and the Subsidiary Body for Implementation of this Protocol. The provisions relating to the functioning of these two bodies under the Convention shall apply *mutatis mutandis* to this Protocol. Sessions of the meetings of the Subsidiary Body for Scientific and Technological Advice and the Subsidiary Body for Implementation of this Protocol shall be held in conjunction with the meetings of, respectively, the Subsidiary Body for Scientific and Technological Advice and the Subsidiary Body for Implementation of the Convention.

2. Parties to the Convention that are not Parties to this Protocol may participate as observers in the proceedings of any session of the subsidiary bodies. When the subsidiary bodies serve as the subsidiary bodies of this Protocol, decisions under this Protocol shall be taken only by those that are Parties to this Protocol.

3. When the subsidiary bodies established by Articles 9 and 10 of the Convention exercise their functions with regard to matters concerning this Protocol, any member of the Bureaux of those subsidiary bodies representing a Party to the Convention but, at that time, not a party to this Protocol, shall be replaced by an additional member to be elected by and from amongst the Parties to

this Protocol.

Article 16

The Conference of the Parties serving as the meeting of the Parties to this Protocol shall, as soon as practicable, consider the application to this Protocol of, and modify as appropriate, the multilateral consultative process referred to in Article 13 of the Convention, in the light of any relevant decisions that may be taken by the Conference of the Parties. Any multilateral consultative process that may be applied to this Protocol shall operate without prejudice to the procedures and mechanisms established in accordance with Article 18.

Article 17

The Conference of the Parties shall define the relevant principles, modalities, rules and guidelines, in particular for verification, reporting and accountability for emissions trading. The Parties included in Annex B may participate in emissions trading for the purposes of fulfilling their commitments under Article 3. Any such trading shall be supplemental to domestic actions for the purpose of meeting quantified emission limitation and reduction commitments under that Article.

Article 18

The Conference of the Parties serving as the meeting of the Parties to this Protocol shall, at its first session, approve appropriate and effective procedures and mechanisms to determine and to address cases of non-compliance with the provisions of this Protocol, including through the development of an indicative list of consequences, taking into account the cause, type, degree and frequency of non-compliance. Any procedures and mechanisms under this Article entailing binding consequences shall be adopted by means of an amendment to this Protocol.

Article 19

The provisions of Article 14 of the Convention on settlement of disputes shall apply *mutatis mutandis* to this Protocol.

Article 20

1. Any Party may propose amendments to this Protocol.

2. Amendments to this Protocol shall be adopted at an ordinary session of the Conference of the Parties serving as the meeting of the Parties to this Protocol.

The text of any proposed amendment to this Protocol shall be communicated to the Parties by the secretariat at least six months before the meeting at which it is proposed for adoption. The secretariat shall also communicate the text of any proposed amendments to the Parties and signatories to the Convention and, for information, to the Depositary.

3. The Parties shall make every effort to reach agreement on any proposed amendment to this Protocol by consensus. If all efforts at consensus have been exhausted, and no agreement reached, the amendment shall as a last resort be adopted by a three-fourths majority vote of the Parties present and voting at the meeting. The adopted amendment shall be communicated by the secretariat to the Depositary, who shall circulate it to all Parties for their acceptance.

4. Instruments of acceptance in respect of an amendment shall be deposited with the Depositary. An amendment adopted in accordance with paragraph 3 above shall enter into force for those Parties having accepted it on the ninetieth day after the date of receipt by the Depositary of an instrument of acceptance by at least three fourths of the Parties to this Protocol.

5. The amendment shall enter into force for any other Party on the ninetieth day after the date on which that Party deposits with the Depositary its instrument of acceptance of the said amendment.

Article 21

1. Annexes to this Protocol shall form an integral part thereof and, unless otherwise expressly provided, a reference to this Protocol constitutes at the same time a reference to any annexes thereto. Any annexes adopted after the entry into force of this Protocol shall be restricted to lists, forms and any other material of a descriptive nature that is of a scientific, technical, procedural or administrative character.

2. Any Party may make proposals for an annex to this Protocol and may propose amendments to annexes to this Protocol.

3. Annexes to this Protocol and amendments to annexes to this Protocol shall be adopted at an ordinary session of the Conference of the Parties serving as the meeting of the Parties to this Protocol. The text of any proposed annex or amendment to an annex shall be communicated to the Parties by the secretariat at least six months before the meeting at which it is proposed for adoption. The secretariat shall also communicate the text of any proposed annex or amendment to an annex to the Parties and signatories to the Convention and, for information, to the Depositary.

4. The Parties shall make every effort to reach agreement on any proposed annex or amendment to an annex by consensus. If all efforts at consensus have been exhausted, and no agreement reached, the annex or amendment to an annex shall as a last resort be adopted by a three-fourths majority vote of the Parties present and voting at the meeting. The adopted annex or amendment to an annex shall be communicated by the secretariat to the Depositary, who shall circulate it to all Parties for their acceptance.

5. An annex, or amendment to an annex other than Annex A or B, that has been adopted in accordance with paragraphs 3 and 4 above shall enter into force for all Parties to this Protocol six months after the date of the communication by the Depositary to such Parties of the adoption of the annex or adoption of the amendment to the annex, except for those Parties that have notified the Depositary, in writing, within that period of their non-acceptance of the annex or amendment to the annex. The annex or amendment to an annex shall enter into force for Parties which withdraw their notification of non-acceptance on the ninetieth day after the date on which withdrawal of such notification has been received by the Depositary.

6. If the adoption of an annex or an amendment to an annex involves an amendment to this Protocol, that annex or amendment to an annex shall not enter into force until such time as the amendment to this Protocol enters into force.

7. Amendments to Annexes A and B to this Protocol shall be adopted and enter into force in accordance with the procedure set out in Article 20, provided that any amendment to Annex B shall be adopted only with the written consent of the Party concerned.

Article 22

1. Each Party shall have one vote, except as provided for in paragraph 2 below.

2. Regional economic integration organizations, in matters within their competence, shall exercise their right to vote with a number of votes equal to the number of their member States that are Parties to this Protocol. Such an organization shall not exercise its right to vote if any of its member States exercises its right, and vice versa.

Article 23

The Secretary-General of the United Nations shall be the Depositary of this Protocol.

Article 24

1. This Protocol shall be open for signature and subject to ratification, acceptance or approval by States and regional economic integration organizations which are Parties to the Convention. It shall be open for signature at United Nations Headquarters in New York from 16 March 1998 to 15 March 1999. This Protocol shall be open for accession from the day after the date on which it is closed for signature. Instruments of ratification, acceptance, approval or accession shall be deposited with the Depositary.

2. Any regional economic integration organization which becomes a Party to this Protocol without any of its member States being a Party shall be bound by all the obligations under this Protocol. In the case of such organizations, one or more of whose member States is a Party to this Protocol, the organization and its member States shall decide on their respective responsibilities for the performance of their obligations under this Protocol. In such cases, the organization and the member States shall not be entitled to exercise rights under this Protocol concurrently.

3. In their instruments of ratification, acceptance, approval or accession, regional economic integration organizations shall declare the extent of their competence with respect to the matters governed by this Protocol. These organizations shall also inform the Depositary, who shall in turn inform the Parties, of any substantial modification in the extent of their competence.

Article 25

1. This Protocol shall enter into force on the ninetieth day after the date on which not less than 55 Parties to the Convention, incorporating Parties included in Annex I which accounted in total for at least 55 per cent of the total carbon dioxide emissions for 1990 of the Parties included in Annex I, have deposited their instruments of ratification, acceptance, approval or accession.

2. For the purposes of this Article, Athe total carbon dioxide emissions for 1990 of the Parties included in Annex I means the amount communicated on or before the date of adoption of this Protocol by the Parties included in Annex I in their first national communications submitted in accordance with Article 12 of the Convention.

3. For each State or regional economic integration organization that ratifies, accepts or approves this Protocol or accedes thereto after the conditions set out in paragraph 1 above for entry into force have been fulfilled, this Protocol shall enter into force on the ninetieth day following the date of deposit of its instrument of ratification, acceptance, approval or accession.

4. For the purposes of this Article, any instrument deposited by a regional economic integration organization shall not be counted as additional to those deposited by States members of the organization.

Article 26

No reservations may be made to this Protocol.

Article 27

1. At any time after three years from the date on which this Protocol has entered into force for a Party, that Party may withdraw from this Protocol by giving written notification to the Depositary.

2. Any such withdrawal shall take effect upon expiry of one year from the date of receipt by the Depositary of the notification of withdrawal, or on such later date as may be specified in the notification of withdrawal.

3. Any Party that withdraws from the Convention shall be considered as also having withdrawn from this Protocol.

Article 28

The original of this Protocol, of which the Arabic, Chinese, English, French, Russian and Spanish texts are equally authentic, shall be deposited with the Secretary-General of the United Nations.

DONE at Kyoto this eleventh day of December one thousand nine hundred and ninety-seven.

IN WITNESS WHEREOF the undersigned, being duly authorized to that effect, have affixed their signatures to this Protocol on the dates indicated.

Annex A

Greenhouse gases

Carbon dioxide (CO_2)

Methane (CH_4)
Nitrous oxide (N_2O)
Hydrofluorocarbons (HFCs)
Perfluorocarbons (PFCs)
Sulphur hexafluoride (SF_6)

Sectors/source categories

Energy
- Fuel combustion
  - Energy industries
  - Manufacturing industries and construction
  - Transport
  - Other sectors
  - Other
- Fugitive emissions from fuels
  - Solid fuels
  - Oil and natural gas
  - Other

Industrial processes
- Mineral products
- Chemical industry
- Metal production
- Other production
- Production of halocarbons and sulphur hexafluoride
- Consumption of halocarbons and sulphur hexafluoride
- Other

Solvent and other product use

Agriculture
- Enteric fermentation
- Manure management
- Rice cultivation

Agricultural soils
Prescribed burning of savannas
Field burning of agricultural residues
Other

Waste

Solid waste disposal on land
Wastewater handling
Waste incineration
Other

Annex B

Party	**Quantified emission limitation or reduction commitment** (percentage of base year or period)
Australia	108
Austria	92
Belgium	92
Bulgaria*	92
Canada	94
Croatia*	95
Czech Republic*	92
Denmark	92
Estonia*	92
European Community	92
Finland	92
France	92
Germany	92
Greece	92
Hungary*	94
Iceland	110
Ireland	92
Italy	92
Japan	94
Latvia*	92
Liechtenstein	92
Lithuania*	92
Luxembourg	92
Monaco	92
Netherlands	92
New Zealand	100
Norway	101
Poland*	94
Portugal	92
Romania*	92
Russian Federation*	100
Slovakia*	92
Slovenia*	92
Spain	92
Sweden	92
Switzerland	92
Ukraine*	100
United Kingdom of Great Britain and Northern Ireland	92
United States of America	93

* Countries that are undergoing the process of transition to a market economy.